Macromolecular Symposia 215

Proceedings of the 2003 International Symposium on Ionic Polymerization and Related Processes

Boston, USA
June 30–July 4, 2003

Symposium Editors:
J. W. Mays, R. F. Storey, Knoxville, USA

pp. 1–393 · August 2004
ISBN 3-527-31048-7

Macromolecular Symposia publishes lectures given at international symposia and is issued irregularly, with normally 14 volumes published per year. For each symposium volume, an Editor is appointed. The articles are peer-reviewed. The journal is produced by photo-offset lithography directly from the authors' typescripts.
Further information for authors can be found at http://www.ms-journal.de
Suggestions or proposals for conferences or symposia to be covered in this series should also be sent to the Editorial office (E-mail: macro-symp@wiley-vch.de).

Macromolecular Symposia:
Annual subscription rates 2005
Macromolecular Full Package: including Macromolecular Chemistry & Physics (24 issues), Macromolecular Rapid Communications (24), Macromolecular Bioscience (12), Macromolecular Theory & Simulations (9), Macromolecular Materials and Engineering (12), Macromolecular Symposia (14):

Europe	Euro	7.088 / 7.797
Switzerland	Sfr	12.448 / 13.693
All other areas	US$	8.898 / 9.788

print only **or** electronic only / print **and** electronic

Postage and handling charges included. All Wiley-VCH prices are exclusive of VAT. Prices are subject to change.

Single issues and back copies are available. Please ask for details at: service@wiley-vch.de

Orders may be placed through your bookseller or directly at the publishers:
WILEY-VCH Verlag GmbH & Co. KGaA, P. O. Box 10 11 61, 69451 Weinheim, Germany, Tel. +49 (0) 62 01/6 06-400, Fax +49 (0) 62 01/60 61 84, E-mail: service@wiley-vch.de

For USA and Canada: Macromolecular Symposia (ISSN 1022-1360) is published with 14 volumes per year by WILEY-VCH Verlag GmbH & Co. KGaA, Boschstr. 12, 69451 Weinheim, Germany. Air freight and mailing in the USA by Publications Expediting Inc., 200 Meacham Ave., Elmont, NY 11003, USA. Application to mail at Periodicals Postage rate is pending at Jamaica, NY 11431, USA. POSTMASTER please send address changes to: Macromolecular Symposia, c/o Wiley-VCH, III River Street, Hoboken, NJ 07030, USA.

Macromolecular Symposia

Articles published on the web will appear several weeks before the print edition. They are available through:

www.ms-journal.de

www.interscience.wiley.com

International Symposium on Ionic Polymerization Boston (USA), 2003

Preface
J. W. Mays, R. F. Storey

Anionic Polymerization

Author Index

Preface

This volume contains key papers presented at the International Symposium on Ionic Polymerization, held in Boston, Massachusetts on June 30–July 4, 2003 under the auspices of the International Union of Pure and Applied Chemistry (IUPAC). This meeting, held at the Tremont Hotel, featured a series of invited and contributed papers and posters, covering the areas of anionic, cationic, and related polymerization processes. The papers dealt with these polymerization systems in the broadest sense, with contributions ranging from new syntheses – to mechanistic studies – to applications. The meeting brought together most of the leading experts in this field, from academia and industry, as well as a significant number of younger scientists and students.

This meeting would not have been possible without the hard work of the chairmen and principal organizers, Professors Rudolf Faust and Roderick P. Quirk. We would like to thank them for organizing an exceptional conference in the very pleasant surroundings of Boston in the middle of summer. They were aided in this task by the Organizing Committee: Drs. J. Crivello, T. Long, J. Mays, J. Puskas, T. Shaffer, and R. Storey. This meeting could also not have taken place without the generous support of various sponsors: IUPAC; American Chemical Society – Polymer Division; The University of Akron; University of Massachusetts, Lowell; National Science Foundation DMR; American Chemical Society – Petroleum Research Fund; Asahi Kasei Corp.; BASF; Bayer Corp.; Boston Scientific Corp.; Bridgestone/Firestone; Chemetall Foote Corp.; Chevron Phillips Chem. Co.; General Electric; Kaneka Corp.; Korea Kumho Petrochemical Co., Ltd.; Kraton Polymers; Sartomer; Synthomer; John Wiley & Sons, Inc.; Wyatt Technology Corp. We are indebted to these organizations for their support.

J. W. Mays

R. F. Storey

A New View of the Anionic Diene Polymerization Mechanism

A. Z. Niu,[1] *J. Stellbrink,*[1] *J. Allgaier,*[*1] *L. Willner,*[1] *D. Richter,*[1] *B. W. Koenig,*[2] *M. Gondorf,*[3] *S. Willbold,*[3] *L. J. Fetters,*[4] *R. P. May*[5]

[1] IFF, FZ Jülich GmbH, 52425 Jülich, Germany
[2] IBI 2, FZ Jülich GmbH, 52425 Jülich and Institut für Physikalische Biologie, Heinrich-Heine-Univ., 40225 Düsseldorf, Germany
[3] ZCH, FZ Jülich GmbH, 52425 Jülich, Germany
[4] School of Chemical and Biomolecular Engineering, Cornell University, Ithaca, NY 14583-5201, USA
[5] Institute Laue-Langevin, 38042 Grenoble, Cedex 9, France

Summary: We investigated the anionic polymerization of butadiene in d-heptane solvent using *tert*-butyl lithium as initiator. Two complementary techniques were used to follow the polymerization processes: ^{1}H NMR and small angle neutron scattering (SANS). The time resolved ^{1}H NMR measurements allowed us to evaluate quantitatively the kinetics of the processes involved. The initiation event commences slowly and then progressively accelerates. This indicates an autocatalytic mechanism. The microstructure of the first monomer units attached is to a high extent 1,2. The disappearance of initiator---at about 10% monomer conversion---signals the onset of the normal ~6% vinyl content of the chain. Small angle neutron scattering was used to study the aggregation behavior of the carbon lithium head groups. It is well known that the polar head groups aggregate and form micellar structures. For dienes in non-polar solvents the textbook mechanism assumes the formation of only tetramers during the propagation reaction. By combining ^{1}H NMR and SANS results we were able to determine quantitatively the aggregation number during all stages of the polymerization. Our measurements show the existence of large-scale structures during the initiation period. The initial degree of aggregation of more than 100 living polymer chains diminished as the polymerization progressed. In addition, even larger, giant structures with $N_{agg} \gg 1000$ and $R_g \approx 1000$Å were found.

Keywords: anionic polymerization; kinetics; neutron scattering; NMR; polybutadiene

Introduction

Organolithium compounds self-associate in organic solvents. The aggregation state depends on the solvent and the structure of the compound. It influences significantly the reactivity of organolithium reagents. In hydrocarbon solvents *sec*- and *tert*-butyl lithium form tetramers whereas *n*-butyl lithium aggregates as hexamers.[1,2] This behavior strongly influences the kinetics of the initiation reaction. The methods used in the past to follow the initiation kinetics were mainly UV spectroscopy[3-5] and to some extent gas chromatic analysis of the hydrolyzed samples taken at different times.[6] It was found that the initiation kinetics of the butyl lithium isomers with styrene and diene monomers are complex. It is first order on monomer concentration and of fractional order in initiator concentration:

$$-d[I]/dt = k_I[BuLi]^{1/n}[M] \tag{1}$$

Depending on initiator, *n* was found to vary between 4 and 6 in benzene and was correlated with the aggregation state of the butyl lithium initiator. It was assumed that there is an equilibrium between the aggregates and the unassociated molecules, the latter ones being the only species initiating the polymerization.[3] Most likely the initiation process is more complex and cross-associated species formed by the reaction of monomer with butyl lithium aggregates could be directly involved.[7,8] In analogy to the short chain initiators, the living polymer chains associate due to aggregation of the carbon lithium active centers was a point of controversy. In earlier studies, the association state of diene based head-groups were a point of controversy.[7-12] Dimers and tetramers were proposed as the association states for completely polymerized systems. The living polymer systems were investigated by concentrated solution viscosity measurements and light scattering.[7-12] Recent publications where small angle neutron scattering (SANS) was used reveal that the aggregation behavior is more complex. All told, the diene based systems can show (depending on the conditions) association degrees ranging from 2 to >100.[13-16]

In this work we analyze the polymerization of butadiene initiated with *tert*-butyl lithium in d-heptane using a combined *in situ* technique of ^{1}H NMR and SANS. ^{1}H NMR allows the simultaneous analysis of the initiation and chain propagation processes. Together, with the SANS results, it was possible to obtain a more detailed picture of the structures involved in the different processes.

Experimental

Solvent and monomer purification techniques have been described.[1,13-16] The polymerization reactions were carried out at 8°C in sealed ^{1}H NMR tubes and SANS quartz cells (Fig. 1). The apparatus equipped with Young® stopcocks was flamed under high vacuum conditions and transferred into the glove box (MBraun, Unilab®, <0.1 ppm O_2 and <0.1 ppm H_2O). There the initiator solution was filled into the small reactors with a microliter syringe. The Young® stopcocks allowed transfer of the reactors from the glove-box to the vacuum line without contamination with air. Butadiene and d-heptane were distilled from *n*-butyl lithium solutions into small flasks equipped with Young® stopcocks. This allowed the weights to be precisely measured with an analytical balance. Monomer and solvent were then distilled into the reactors quantitatively and the reactors then flame sealed at liquid nitrogen temperature. This was done for the SANS cells by first filling the ingredients into the container below the stopcock (see Fig. 1). After sealing from the vacuum line the contents were warmed to dry ice temperature, poured into the SANS cell and the cell sealed off at -78°C. This procedure was necessary to prevent the cells from cracking during the warm up procedure, of the frozen content, from liquid nitrogen temperature. These procedures allowed precise assay of monomer (10-20 mg) and initiator quantities in the sub μmol range. The monomer concentration was 0.73 mol/L for the ^1H -NMR samples and 0.69 mol/L for the SANS samples, the initiator concentration was 8.5×10^{-4} mol/L in all samples.

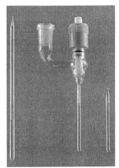

Figure 1. SANS and NMR glassware for the in situ polymerization reactions.

Polymerization reactions were carried out by warming up the samples from liquid nitrogen temperature or dry ice temperature to 8°C within 1 to 3 minutes and keeping the temperature at 8°C±0.7°C for at least two weeks monitoring the polymerization reaction by ^1H NMR and SANS. Then the reactors were opened in the glove-box and terminated with degassed methanol. The solutions were filtered, the solvents evaporated and the polymers dried. The samples were characterized via on-line GPC/light scattering in THF at 30°C. For ^{1}H NMR measurements the reaction mixtures were flame sealed in 4 mm outer diameter ^{1}H NMR tubes (Wilmad, Buena Vista, USA). For measurements in a 5 mm probe the sealed samples were placed into Wilmad 5 mm outer diameter ^{1}H NMR tubes.

Proton NMR spectra were recorded at a Larmor-Frequency of 600.14 MHz on a Bruker DMX600 using a 5 mm TXI probe optimized for proton observation. The free induction decay was recorded immediately after a non-selective 90°-pulse (7 μs) with a dwell time of 111 μs and 32k data points in the time domain. Four transients were added for each spectrum using the CYCLOPS phase cycle to avoid artifacts from quadrature detection. A long pre-scan delay of 300 sec was found to be necessary to ensure complete T_1-relaxation of all signals (strict requirement for quantitative interpretation of integral peak intensity). An exponential window function was applied prior to Fourier transformation and integration of the frequency domain signals.

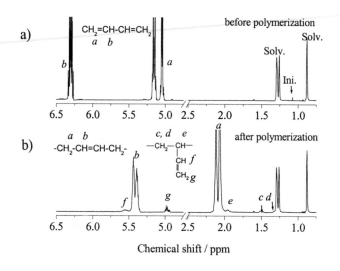

Figure 2. NMR traces of before and after polymerization.

Figures 2a and 2b show typical ^{1}H NMR spectra recorded before and after polymerization, respectively. Signal intensities at different reaction times were compared quantitatively via comparison with the residual proton signal of the solvent n-heptane signal at 0.88 ppm. The molarity of the solvent protons does not change during the polymerization in a sealed sample tube. The fraction of reacted *tert*-butyl lithium at time t was calculated from the signal intensities at 1.07 ppm of the initial spectrum ($I_{1.07ppm,t=0}$) and the spectrum at time t ($I_{1.07ppm,t}$). Conversion as a function of time, Con.(t), was calculated from the intensities of the monomer signal at ~6.3 ppm ($I_{6.3ppm,t}$) and the polymer signals at ~2.1 ppm ($I_{2.1ppm,t}$, 1,4 structure) and ~1.5 ppm ($I_{1.5ppm,t}$, 1,2 structure) at time t, e.g.:

$$\text{Con.(t)} = [P(t)]/([M(t)]+[P(t)]) = (I_{2.1ppm,t}/4+I_{1.5ppm,t})/(I_{6.3ppm,t}/2+I_{2.1ppm,t}/4+I_{1.5ppm,t}). \quad (2)$$

Here, [P(t)] is the polymer repeat unit concentration at time t and [M(t)] is monomer concentration at time t. The polymerization degree $D_p(t)$ was calculated from the signal intensities of the polymer signals at ~2.1 ppm and ~1.5 ppm and the initiator signals at 1.07 ppm as

$$D_p(t) = [P(t)]/([Ini(t = 0)]-[Ini(t)]) = (I_{2.1ppm}/4+I_{1.5ppm})/((I_{1.07ppm, t=0} - I_{1.07ppm,t})/9) \quad (3)$$

Here, [Ini(t)] is the initiator concentration at time t and [Ini(t= 0)] is initial initiator concentration. Microstructures were calculated from the signal intensities of the polymer signals at ~2.1 ppm and ~1.5 ppm, In all cases the number of protons correlated with the different signals was taken into account.

SANS measurements were performed initially on the KWS2 at FZ Jülich (Germany). These were followed by a series of measurements on D22 at ILL, Grenoble (France). In general, the scattering cross section $\dfrac{d\Sigma}{d\Omega}$ observed in a SANS experiment from polymers in dilute solution is given by:

$$\frac{d\Sigma}{d\Omega} = \frac{\Delta\rho^2}{N_a} \frac{\phi(1-\phi)}{\left[\dfrac{1}{V_w P(Q)} + 2A_2\Phi\right]} \quad (4)$$

Here, polymer concentration is given in terms of ϕ which denotes the polymer volume fraction, $P(Q)$ is the form factor of the polymer or the polymer aggregates, V_w the corresponding weight average molecular volume, A_2 the second virial coefficient, and N_a the Avogadro number. $\Delta\rho$ is the scattering contrast defined by:

$$\Delta\rho = \left[\frac{\Sigma b_s}{v_s} - \frac{\Sigma b_{mon}}{v_{mon}}\right] \quad (5)$$

The ratio $\Sigma b_s/v_s$ is the scattering length density of the solvent with b_s the scattering lengths of the atoms forming the solvent molecule and v_s the corresponding volume. $\Sigma b_{mon}/v_{mon}$ is the corresponding quantity for the repeat unit. To achieve maximum contrast and minimum incoherent background resulting from protonated material, we investigated h-butadiene in d-heptane ($\rho_{h\text{-butadiene}}$=4.12x10^{-9}cm^{-2}, $\rho_{d\text{-heptane}}$ = 6.26x10^{-10}cm^{-2}, $\Delta\rho$ = 5.85x10^{-10}cm^{-2}). Because all b_i are well defined and tabulated properties of the corresponding nucleus solely, SANS data obtained on an absolute scale can be interpreted quantitatively without any ambiguity. There is no need to determine a contrast parameter experimentally as a function of M_w (such as the *(dn/dc)* used in light scattering experiments).

In general all azimuthally averaged data were corrected for empty cell scattering and then normalized to absolute scattering cross section using a water standard. Contributions due to incoherent background and solvent scattering were subtracted from all data sets before analysis. A useful presentation of the theory and practice of SANS is available from Higgins and Benoit.[18] All experiments were performed at 8°C in order to slow the propagation event. This enables us to obtain time resolved data with good statistics including those at low scattering vectors encountered in the initiation period. This is not trivial since we are starting at $t = 0$ with a monomer solution where scattering intensity is only slightly above that of the solvent.

Combined in situ ¹H NMR and SANS Technique

In our study on anionic polymerization, we choose a combination of ¹H NMR and small angle neutron scattering (SANS). Using time resolved ¹H NMR we obtain quantitative information concerning initiator, monomer, and polymer concentration and polymerisation degree of the single chains as a function of time and conversion ([Ini(t)], [M(t)], [P(t)] and $D_p(t)$), but also structural information on a microscopic level, e.g. the 1,2-to-1,4 ratio. Time resolved SANS on the other hand provides us structural information on a mesoscopic <u>and</u> microscopic length scale, e.g. $R_g(t)$[19] of intermediate aggregates, and quantitative information concerning their M_w. This enables us to cross check kinetics as well as structural data, which makes this combination of experimental techniques especially appealing.

The crucial parameter in a SANS experiment, which determines the spatial resolution, is the scattering vector Q, given by $4\pi\lambda^{-1} \sin(\Theta/2)$, with θ the scattering angle and λ the neutron wavelength. Q has the dimensions of a reciprocal length and can therefore be regarded as an "inverse yard stick". For SANS experiments typically performed with different settings, i.e. different sample-to-detector distances and collimation lengths, a Q-range of nearly 2.5 orders of magnitude can be achieved, $1 \times 10^{-3} \leq Q \leq 0.2 \text{Å}^{-1}$. This corresponds to a spatial resolution $5\text{Å} \leq D = 1/Q \leq 1000$ Å. Assuming that the growing chains form only star-like aggregates with a mean functionality N_{agg}, ($N_{agg} = 4$ is predicted by the textbook reaction mechanism), we can derive a quantitative relation for the scattering intensity $I(Q,t)$ observed in a SANS experiment. The scattering intensity as a function of reaction time t and scattering vector Q is then given by:

$$I(Q,t) = \left(\frac{d\Sigma}{d\Omega} \right) \bigg/ \left(\frac{\Delta\rho^2}{N_a} \right) = \phi_p(t) \cdot N_{agg}(t) \cdot V_w(t) \cdot P_{star}(Q,t) \tag{6}$$

Here $P_{star}(Q,t)$ is the form factor of a Gaussian star polymer as given by Benoit[20], which is the only Q-dependent variable in Eq. (6) For simplicity we have neglected the concentration dependence (the second virial coefficient) of $I(Q,t)$. The expected evolution of the scattering intensity calculated $I(Q,t)$ for different conversion between 1% and 100% is illustrated in Fig. 3. One clearly recognises how well suited the chosen SANS setup with respect to Q-resolution is to cover the full polymerization event.

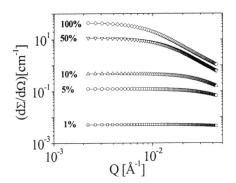

Figure 3. Expected evolution of the SANS scattering pattern with conversion assuming a time independent aggregation state of N_{agg} = 4.

In a previous publication,[16] we have shown that all time dependent quantities in Eq. (6) can be reduced to only one parameter, i.e. the number density of reacted monomers $N_p(t)$, which can therefore be calculated from the observed forward scattering $I(Q = 0,t)$. However, this is only valid under two conditions: (1). the number of living chains is time independent and equal to the initiator concentration, i.e. the initiation period has already passed, and (2). all reacted monomers participate in the same type of aggregates with time independent aggregation number. In this case

SANS data alone already give direct access to the reaction kinetics. If we want to extend our studies to the initiation period or if the investigated monomer forms different types of aggregates, we need additional quantitative information to unambiguously interpret our SANS data. The needed information can be independently obtained from time resolved ^{1}H NMR experiments, which gives us directly [Ini(t)], [M(t)], and [P(t)], from which we further can calculate $D_p(t)$ and $M_n(t)$. If we now use these data to analyse our SANS data, we can directly access the aggregation number as a function of time for all phases of the polymerization.

^{1}H NMR Measurements

Careful optimization of experimental conditions was crucial for quantitative interpretation of the ^{1}H NMR spectra. The design of the NMR sample cell, the filling factor of the sealed NMR tube, and the length of the relaxation period required between subsequent scans were found to be critical. ^{1}H NMR spectra were recorded during a first preliminary polymerization attempt with a pre-scan delay of 4 seconds, which turned out to be much too short for complete T_1 relaxation. As a result of this incorrect delay, the overall intensity of the polymer peaks after completion of the polymerization was three times higher than the overall intensity of the monomer peaks of the same sample at the beginning of the reaction. The T_1 relaxation times of the monomer peaks were determined to be (40 ± 5) sec at 8 °C. Spin-lattice relaxation times are significantly shorter for the polymer peaks and decreased with the length of the polymer chain as expected for rapidly tumbling molecules in solution. The most likely explanation for the rather long T_1 values of the solute is solvent deuteration. Nuclear dipole-dipole interaction is a very important relaxation mechanism for protons and the interaction with other nearby protons is particularly efficient due to the extraordinary high gyromagnetic ratio, γ, of protons. Replacing solvent protons with low γ deuterons will increase T_1 of the solute. Paramagnetic impurities, e.g. molecular oxygen, provide another efficient source of T_1 relaxation. However, during the experiments reported here trace amounts of oxygen were carefully removed from the samples to avoid unwanted side reactions. For the quantitative NMR experiments reported in this manuscript a pre-scan delay of 300 sec was used. Further increase of the delay to 1,000 sec did not change the integral intensity of any relevant peak in the spectrum by more than 1%.

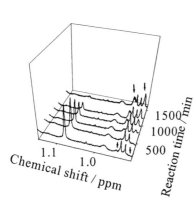

Figure 4. NMR signals of t-butyl lithium and the allyl-lithium head groups.

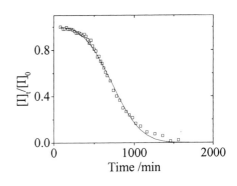

Figure 5. Concentration of reacted initiator normalized by initial initiator concentration $[I]_t/[I]_0$ as a function of time.

In the next polymerization reaction the overall polymer peak intensity was again 20% larger than the initial monomer peak intensity despite the correct pre-scan delay. This time the error could be attributed to the large void volume in the sample tube, which amounted to almost 70% of the tube volume. The void volume contains a substantial fraction of the volatile monomer. For this reason the design of the NMR tube was changed by putting a capillary on top of a 4 mm tube (see Fig. 1). The capillary allows the nearly quantitative filling of the tube while retaining enough distance between the hot glass and the cold liquid during the sealing-off procedure. The short 4 mm sample tube was inserted in a standard length 5 mm outer ^{1}H NMR tube for optimal positioning of the sample in the RF-coil of the probe. This design was used for preparation of all subsequent NMR samples and the overall intensities of the monomer and polymer peaks prior to and after the polymerization, respectively, were found to agree within experimental error. Fig. 4 shows a series of ^{1}H NMR traces during the initiation period at 8°C. With the disappearing initiator signal at 1.07 ppm signals between 0.91 and 0.96 ppm appear, which correspond to the initiated *tert*-butyl lithium. These signals are difficult to analyze quantitatively because of spurious signals appearing in the same ppm range. In Fig. 5 the fraction of reacted *tert*-butyl lithium is plotted against time. The result shows that the initiation reaction starts very slowly and then gradually is accelerated with proceeding initiator consumption. The change of initiation rate with time indicates a kind of autocatalytic process being involved in the initiation process. Similar results were obtained in the past for the system *sec*-butyl lithium-styrene-benzene using the increase in absorption (334nm) of poly(styryl lithium) to monitor the initiation reaction.[3]

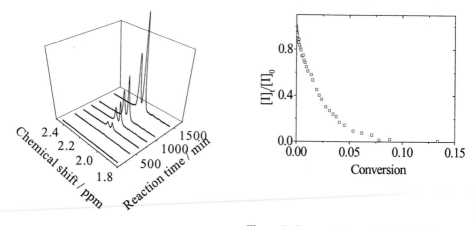

Figure 6. NMR signals of polymer peaks at the beginning of the polymerization reaction.

Figure 7. Concentration of residual initiator normalized by initial initiator concentration $[I]_t/[I]_0$ as a function of conversion rate.

Fig. 6 documents the beginning of the propagation reaction. It shows the signals of the 4 aliphatic cis and trans 1,4 protons at approximately ~2.1 ppm. The signals of the protons related to the 1,2 structures at ~1.5 ppm and ~1.9 ppm are hardly visible due to the small concentration. As expected at early reaction times no polymer signal is visible. First after approximately 300 min a hint for the appearance of the signal at ~2.1 ppm is detectable. From the comparison of Fig. 4 and Fig. 6 it can be concluded that before the initiation event is complete the chain propagation has advanced considerably. This is presented in more detail in Fig. 7 where the fraction of reacted *tert*-butyl lithium is plotted against the conversion. Although *tert*-butyl lithium is considered to be a good initiator for diene polymerization in non-polar solvents the initiation step is slower than chain propagation. After consumption of 50 % of the initiator the number average polymerization degree (D_p) of the initiated chains is 15 and after cessation of the initiation step it increases to 75

- 80 which translates to M_n = 4,000 - 4,300. The molecular weight distribution of the final product is not affected because of the 10 times higher overall molecular weight. Furthermore it should be mentioned that the monomer concentration dependence of the propagation rate is first order as it is expected for anionic diene polymerization.[17] From comparison of the polymer signal intensities after completion of the reaction with the initial initiator intensity it was possible to calculate M_n of the polymer. The value found was M_n = 46,400. This is in good agreement with the GPC analysis where M_n was 50,600 and M_w/M_n = 1.04.

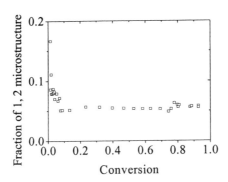

Figure 8. Fraction of 1, 2-microstructure as a function of conversion rate.

^{1}H NMR measurements are useful to investigate the microstructure of polydienes. It is known that in non-polar solvents polymers with mainly 1,4-enchainments are produced. The results obtained during the polymerization process of butadiene are presented in Fig. 8. At very low conversions, the fraction of 1,2-microstructure is between 15 % and 20 %. It then decreases and from 10 % conversion the fraction of 1,2-microstructure is 5-6 % and constant. Similar values as obtained for higher conversion rates of the living process are found for terminated polymers of respective molecular weights. The strong change in microstructure at early polymerization stages indicates that the first monomer units are attached to a high extent in the 1,2-sense. The results shown in Fig. 8 even underestimate the fraction of 1,2-enchainments because in parallel to the initiation event the much quicker chain propagation takes place. This means that smaller chains with a higher 1,2-content exist together with longer chains having a lower 1,2-content and the NMR measurement yields the average value.

SANS Measurements

Fig. 9 shows the temporal evolution of the SANS scattering pattern within the first 60 hours of the polymerisation for butadiene in d-heptane at 8°C. The start of the scattering measurements was synchronized with the beginning of the polymerization event. We collected data initially in 5 minute slices but found that under these experimental conditions the polymerization is quite slow. Within the first 3-5 hours no changes in the scattering pattern were observed. This coincides with ^{1}H NMR experiments, where an induction period of $\approx$5 hours was observed.

Therefore we relaxed our temporal resolution and averaged at each case 4 of these five minutes runs to improve statistics and to align our SANS experiments with the *in situ* ^{1}H NMR; see discussion in ^{1}H NMR section. Furthermore we used the data collected during the first 20 minutes of the polymerisation to correct all further data with this " background run". That is, the scattering intensities shown in Fig. 9 unambiguously result from the polymerization event alone and not from background scattering. The data show exactly the same fingerprint as observed in all previous experiments on butadienyl-lithium. The hallmark is the presence of two characteristic length scales present during all stages of the polymerisation.

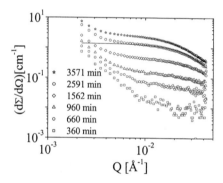
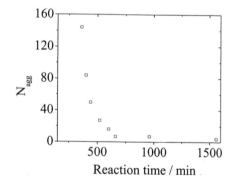

Figure 9. Measured time dependence of the SANS intensity in absolute units vs. scattering vector Q. The solid line represents a fit using Eq. (6) and ^{1}H NMR data.

Figure 10. The time dependence of the aggregation number N_{agg} of the smaller aggregates during the early stages of the polymerization.

In the initiation period the scattering pattern is dominated by a pronounced forward scattering at low $Q \leq 10^{-2}$Å^{-1}, which shows a power law decay as $I(Q) \sim Q^{-3}$. This high fractal order is attributed to very dense, large scale aggregates with $R_g \geq 1000$Å and $N_{agg} >> 1000$, probably consisting of mixtures of initiator molecules and initiated chains. As the Q range accessible by SANS is limited, it was not possible to analyse the size of the aggregates more quantitatively. With continuing polymerization, the scattering from these giant structures gets less prominent, but is still visible at the end of the initiation period at about 1500 min. Here D_p is ≈ 80.

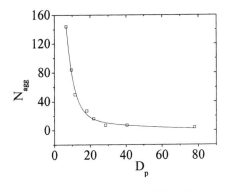

Figure 11. Aggregation number N_{agg} as a function of the polymerization degree D_p. The data fitting result (using the function of the second order exponential decay) yields $N_{agg} = 549.3\exp(-D_p/4.4) + 21.4\exp(-D_p/36.2)$.

The scattering intensity in the higher Q-range increases continuously during the initiation and propagation reaction. It is caused by the smaller aggregates. To analyse them more quantitatively, we applied the following approach. Using D_p of the single chains as a function of time obtained from ^{1}H NMR, we calculated the polymer volume fraction $\Phi(t)$ and an average $V_w(t)$. These results were used for fitting the SANS data to Eq. (6) using a Benoit form factor[20]. This gives direct access to N_{agg} of the aggregates as a function of time, the only adjustable parameter, which determines the scattering intensity. In Fig. 10, aggregation numbers are shown as a function of time. Starting from an average aggregation number of ≈ 140 in the beginning N_{agg} decays strongly during the initiation period. Fig. 11 shows that the aggregates with higher N_{agg} are composed of oligomeric butadienyllithium. With increasing chain length, N_{agg} gets smaller and levels off at ~4. This value is kept constant during the propagation stage. For the lower aggregation numbers the structure of the aggregates is star-like and is in agreement with the textbook mechanism which assumes tetramers. Experimentally, the limiting value for the dienyl head-group aggregation state is the dimer.[20-22] This is based upon oligomeric poly(butadienyllithium) in the melt[21] and high molecular weight concentrated solution viscosity measurements (carried out above the crossover concentration, c*, where chain overlap occurs and excluded volume effects are screened out).[22,23] Watanabe and co-workers[24] have concluded that the viscometric method will yield the correct aggregation state for concentrated solutions of high molecular weight dienes. These trends for the variation in aggregation states are in keeping with conclusions from the quantum chemical based calculations of Frischknecht and Milner.[25]

Conclusions

Time resolved ^{1}H NMR measurements allowed us to follow quantitatively the kinetics of anionic polymerization. The initiation reaction is relatively slow compared to the propagation reaction. Moreover, it is already accompanied by a considerable amount of propagation, which results in approximately 10% conversion at the end of the initiation period.

By combining time resolved ^{1}H NMR and SANS experiments we have shown, that there is a complex aggregation behavior in the polymerizing solution. We found at any polymerization stage two families of aggregated species having very different sizes. At the early reaction stages during the initiation period highly aggregated structures with $N_{ag\,g} \approx 100$ and $10 \text{Å} \leq R_g \leq 100 \text{Å}$ are found. As propagation progresses their aggregation number diminishes to a value of 4. These aggregates coexist with even larger, giant structures with $N_{agg} >> 1000$ and $R_g \approx 1000 \text{Å}$. The relative amount of giant aggregates decreases towards the end of the polymerization. It is not clear if both types of aggregates may consist of mixed aggregates containing both initiator and polymer chains.

[1] M. Morton, L. J. Fetters, *Rubber Chem. Technol.* **1975**, *A12*, 359.
[2] J. Roovers, S. Bywater, *Macromolecules* **1975**, *8*, 251.
[3] S. Bywater, D. J. Worsfold, *J. Organometal. Chem.* **1967**, *10*, 1.
[4] D. J. Worsfold, S. Bywater, *Can. J. Chem.* **1964**, *42*, 2884.
[5] J. Roovers, S. Bywater, *Macromolecules* **1968**, *1*, 328.
[6] H. Hsieh, *J. Polym. Sci.* **1965**, *A3*, 163.
[7] R. N. Young, R. P. Quirk, L. J. Fetters, *Adv. Polym. Sci.* **1984**, *56*, 1.
[8] H. L. Hsieh, R. P. Quirk, *"Anionic Polymerization"*, Marcel Dekker, New York 1996, p.135ff.
[9] D. J. Worsfold, S. Bywater, *Macromolecules* **1972**, *5*, 393.
[10] M. Szwarc, *Adv. Polym. Sci.* **1983**, *49*, 1.
[11] M. Van Beylen, S. Bywater, G. Smets, M. Szwarc, D. J. Worsfold, *Adv. Polym. Sci.* **1988**, *86*, 87.
[12] M. Morton, *"Anionic Polymerization. Principles and Practice"*, Academic Press, New York 1983, p.114.
[13] L. J. Fetters, N. P. Balsara, J. S. Huang, H. S. Jeon, K. Almdal, M. Y. Lin, *Macromolecules* **1995**, *28*, 4996.
[14] J. Stellbrink, L. Willner, O. Jucknischke, D. Richter, P. Lindner, L. J. Fetters, J. S. Huang, *Macromolecules* **1998**, *31*, 4189.
[15] J. Stellbrink, L. Willner, D. Richter, P. Lindner, L. J. Fetters, J. S. Huang, *Macromolecules* **1999**, *32*, 5321.
[16] J. Stellbrink, J. Allgaier, L. Willner, D. Richter, T. Slawecki, L. J. Fetters, *Polymer* **2002**, *43*, 7101.
[17] Ref. 8, p.157ff.
[18] J. S. Higgins and H. C. Benoit, *"Polymers and Neutron Scattering"* Oxford University Press, Oxford 1994.
[19] R_g denotes the radius of gyration.
[20] H. Benoît,. *J. Polym. Sci.*, **1953**, *11*, 507.
[21] H. S. Makowski and M. Lynn, *J. Macromol. Sci.*, **1966**, *1*, 443.
[22] M. Morton and L. J. Fetters, *J. Polym. Sci.*, Part A: **1964**, *2*, 3311.
[23] M. M. Al-Jarrah and R. N. Young, Polymer **1980**, *21*, 112.
[24] H. Watanabe, Y. Oishi, T. Kanaya, H. Kajii and F. Horii, *Macromolecules* **2003**, *36*, 220.
[25] A. L. Frischknecht and S. T. Milner J. Chem. Phys., **2001**, *114*, 1032.

Aluminate and Magnesiate Complexes as Propagating Species in the Anionic Polymerization of Styrene and Dienes

Alain Deffieux,* Larisa Shcheglova, Anna Barabanova, Jean Marc Maréchal, Stephane Carlotti

Laboratoire de Chimie des Polymères Organiques, ENSCPB-Université Bordeaux 1-CNRS, 16 avenue Pey Berland, 33607 Pessac-Cedex, France
E-mail: deffieux@enscpb.fr

Summary: The influence of MgR_2 and AlR_3 additives on alkyllithium initiators in the anionic polymerization of butadiene has been investigated in non polar solvents. A strong decrease of the diene polymerization rate in the presence of the two Lewis acids was observed, similarly to that observed in the retarded anionic polymerisation of styrene. With n,s-Bu$_2$Mg, the percentage of 1,2 vinyl units increases with the [Mg]/[Li] ratio. This behavior is specific to magnesium derivatives bearing secondary alkyl groups and likely results from the additional complexation of lithium species by free dialkylmagnesium and/or a 1,4- to 1,2- chain end isomerization process during chain exchanges between polybutadienyl active chains and dormant ones attached to magnesium species. These reversible exchanges also lead to the formation of one supplementary chain by initial dialkyl magnesium which acts as reversible chain transfer agent. On the contrary with the R_3Al/RLi systems the number of chains is only determined by the concentration of initial alkyllithium and no modification of the polybutadiene microstructure compared to lithium initiators (1,4 units = 80%) is noticed.
Dialkyl magnesiate complexes with alkali metal derivatives (i.e. alkoxide) are also able to influence the stereochemistry of the styrene insertion during the propagation reaction. Polystyrenes with different tacticities ranging from predominantly isotactic (85% triad iso) to syndiotactic (80% triad syndio) can be obtained with these initiators.

Keywords: aluminate and magnesiate complex; anionic polymerization; diene; stereochemistry; styrene

Introduction

We have recently reported the use of dialkylmagnesium (R_2Mg, $0 < r = [Mg]/[Li] \leq 20$) and of trialkylaluminum (R_3Al, $0 < r = [Al]/[Li] < 1$) as retarders of the styrene anionic polymerization

DOI: 10.1002/masy.200451102

initiated by alkyllithium derivatives at elevated temperature[1-3]. The formation of "ate" heterocomplexes between PSLi and the organometallic additives is the key for the kinetic retardation and the stabilization of active species at high temperature. However, the mechanism of monomer insertion and the nature of the propagating species has not been yet completely clarified. To get further informations on these elementary reactions, a study of the influence of the added Lewis acids on the stereochemistry of both the butadiene and styrene polymerizations has been conducted. These results are reported and discussed in term of polymerization mechanism.

Experimental

Materials: s-Butyllithium (1.3 M in cyclohexane from SAFC, France), ether-free n,s-dibutylmagnesium (n,s-Bu$_2$Mg ; 1.0 M in heptane from SAFC, France), triisobutylaluminum (*i*-Bu$_3$Al ; 1.0 M in cyclohexane from SAFC, France), lithium *tert*-butoxide (1.0M in hexane from SAFC, France), sodium *tert*-butoxide (powder 97% from SAFC, France) and potassium *tert*-butoxide (powder 95% from SAFC, France) were used as received. Cyclohexane and methylcyclohexane (99.5% from SAFC, France) were degassed over freshly crushed CaH$_2$, stored over polystyryllithium oligomers and distilled before use. Styrene (99% from SAFC, France) was degassed over freshly crushed CaH$_2$, stored over n,s-dibutylmagnesium and distilled before use. 1,3-butadiene (99% from SAFC, France) was purified one night over *s*-BuLi (1.3 M) at −30°C and distilled before use.

Polymerization: Polymerizations of 1,3-butadiene were carried out under dry nitrogen in cyclohexane at 40°C in glass flasks equipped with a quartz cell and fitted with PTFE stopcocks. PBLi seeds ([PBLi]= 5-8 10^{-3} M) were used as polymerization initiators. They were prepared by addition of 1,3-butadiene to a solution of s-BuLi in cyclohexane. A known amount of R$_2$Mg or *i*-Bu$_3$Al was then added to PBLi seed solutions to obtain the appropriate ratio [Mg]/[Li] ([Mg]/[Li] = 0 to 10) or [Al]/[Li] ([Al]/[Li] = 0 to 1). After addition of 1,3-butadiene, the polymerization kinetics were followed by dilatometry. The polybutadienes were then recovered by precipitation of the polymerization media into acidified methanol and their microstructure was determined by ^{1}H and ^{13}C NMR.

Styrene polymerizations were carried out under vacuum or argon atmosphere in cyclohexane (at 20°C) in glass flasks equipped with a quartz cell and fitted with PTFE stopcocks. The same kind of polymerizations were also carried out in methylcyclohexane (for temperatures below 5°C), but in sealed flasks. The initiating species n,s-Bu$_2$Mg / t-BuOMt (Mt = Li, Na, K) were prepared under argon by addition of the magnesium derivative and the alkoxide in the appropriate solvent and at the given temperature. The reactions were finally quenched by addition of degassed methanol in excess.

UV-Visible spectroscopy: The absorption spectra of the PBLi/R$_2$Mg and PBLi/i-Bu$_3$Al solutions were recorded on a UV-Vis spectrometer, Varian-Cary 3E, using a quartz cell (0.01 cm path-length) attached to the glass reactor ; ε_{PBLi} = 8300 L.mol^{-1}.cm^{-1} at 275 nm.

Polymer characterizations: The average molar masses and the molar mass distributions of polystyrene and polybutadienes were determined by size exclusion chromatography (SEC) on a Varian apparatus equipped with a JASCO HPLC pump, type 880-PU, refractive index/UV detectors and TSK Gel columns calibrated with polystyrene standards. According to this calibration, a corrective factor has been applied for the polybutadienes experimental $\overline{M}_n$'s[12].

The stereochemistry of the polystyrene was determined in *d*-chloroform by ^{13}C NMR on Brüker AC 250 and DPX 200 on the aromatic quaternary carbon. Calculation of tacticity was carried out by deconvolution of the signal between 145.5 and 147 ppm using the "WINNMR1D" software.

Fractionation attempts of the polystyrene samples into highly tactic (insoluble) and atactic (soluble) fractions were performed by solubilization in methylethyl ketone at 80°C. All of our samples were completely soluble under these conditions.

Results and Discussion

Butadiene polymerization

In terms of kinetics, copolymer structure, and stereochemistry, the random anionic copolymerization of styrene and 1,3-butadiene is highly dependent on the nature of the anionic initiator and on the polymerization conditions. In hydrocarbon media, alkyllithium used in very low concentration compare to monomer gives polydienes exhibiting a high cis-1,4 microstructure[4]. Conversely, in polar media or in the presence of polar additives or complexing agents, as well as with alkali metal counter-ions other than lithium, polydienes with higher amount of 1,2-vinyl structure are obtained. Only very limited work has dealt with the influence of Lewis acids additives[5] on the stereochemistry of polybutadiene initiated by alkyllithium in hydrocarbons. We investigate in the first part of this paper the role of R_2Mg, $(0 < r = [Mg]/[Li] \leq 20)$ and of i-Bu$_3$Al $(0 < r = [Al]/[Li] < 1)$ as additives in alkyllithium-initiated butadiene polymerization.

The UV-visible absorption spectra of polybutadienyllithium (PBLi) seeds in the presence of increasing amounts of n,s-Bu$_2$Mg and i-Bu$_3$Al are presented in Figures 1 and 2. The incremental addition of Lewis acids leads to a hypsochromic shift of the polybutadienyl band maximum ($\lambda = 275$ nm) to 259 nm for n,s-Bu$_2$Mg and 240 nm for i-Bu$_3$Al at r of about 1. These results are in agreement with the quantitative formation of 1:1 Lewis acid:PBLi complexes, see Scheme 1, as already reported for PSLi.

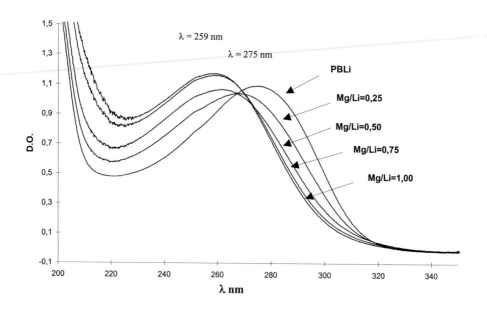

Figure 1. Influence of increasing amounts of n,s-Bu$_2$Mg on PBLi UV-visible spectrum ([Mg]/[Li] $\leq$ 1, T = 25°C, cyclohexane).

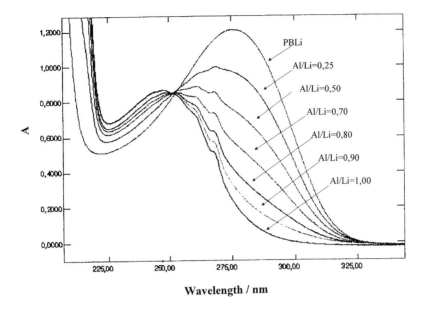

Figure 2. Influence of increasing amounts of i-Bu₃Al on PBLi UV-visible spectrum (T = 25°C, cyclohexane).

Scheme 1. Schematic representation of 1:1 Lewis acid:PBLi complexes.

The kinetics of 1,3-butadiene polymerization initiated by PBLi seeds in the presence of Lewis acid additives were investigated in cyclohexane at 40°C using the dilatometry technique. Since the aggregation state of Lewis acid:PBLi species is unknown, only apparent propagation rate constants have been determined using the kinetic equation

$$R_p = k_{p_{app}} [\text{"Li"}][\text{But}], \tag{1}$$

where ["Li"], the concentration of active PBLi species, is assumed equal to initial s-BuLi.

Both the initial butadiene and s-BuLi concentrations were kept approximately constant for the whole series of kinetic experiments. For the $(n,s\text{-}Bu_2Mg)_x$:PBuLi systems, results collected in Table 1 show that the polymerization rate and apparent propagation rate constant decrease drastically with increasing r (factor 270 from r = 0 to 4). The experimental PB molar masses also decrease with increasing r, as previously reported for styrene. This is in agreement with coinitiation, or reversible chain transfer, involving $n,s\text{-}Bu_2Mg$. At low [Mg]/[Li] ratios one supplementary PB chain is formed by $n,s\text{-}Bu_2Mg$ molecule, in addition to the one formed from PBLi seeds. However, at higher Mg/Li ratio the average number of PB chains per magnesium atom decreases suggesting incomplete coinitiation or reversible transfer processes with $n,s\text{-}Bu_2Mg$ moieties when present in large excess.

For triisobutylaluminum systems butadiene polymerization rate and apparent propagation rate constant, Table 2, also decrease drastically with increasing proportion of $i\text{-}Bu_3Al$ (factor 320 from r = 0 to 0.85) and no butadiene polymerization is observed for r equal or higher than one, as already described for styrene[2]. Experimental PB molar masses agree in this case with an initiation involving all the initial PBLi species, independently of r, whereas the contribution of the aluminum derivative to the formation of new PS chains is not observed.

Table 1. 1,3-Butadiene polymerization in the presence of $n,s\text{-}Bu_2Mg$/PBLi seeds at different [Mg]/[Li] ratios, cyclohexane, T = 40°C.

$[n,s\text{-}Bu_2Mg]$ /[RLi]	[PBLi] $mol.L^{-1}$ $\times10^3$	Rp/[M] min^{-1} $\times10^3$	$k_{P_{app}}$ (a) $L.mol^{-1}.min^{-1}$	$\overline{M}_n$ th[b] Li	$\overline{M}_n$ th[c] Li+Mg	$\overline{M}_n$ (SEC)[d]	$\dfrac{\overline{M}_w}{\overline{M}_n}$
0	8.0	15	1.9	5 200	-	4 600	1.06
1	7.0	2.7	0.4	19 700	10 000	10 000	1.08
2	5.6	0.24	0.04	20 400	7 200	9 500	1.40
4	5.3	0.038	0.007	10 000	2 400	5 000	1.60

(a) $k_{P_{app}} = (R_p/M) / [PBLi]$
(b) theoretical molar masses (g.mol^{-1}) calculated assuming the contribution of [PBLi] alone
(c) theoretical molar masses (g.mol^{-1}) calculated assuming the contribution of [PBLi + R$_2$Mg]
(d) experimental molar masses (g.mol^{-1})

Table 2. 1,3-Butadiene polymerization in the presence of i-Bu$_3$Al/PBLi seeds at different [Al]/[Li] ratios, cyclohexane, T = 40°C.

[i-Bu$_3$Al] /[RLi]	[PBLi] mol.L^{-1} $\times 10^3$	Rp/[M] min^{-1} $\times 10^3$	$k_{p_{app}}$[(a)] L.mol^{-1}. min^{-1}	$\overline{M_n}$ th[(b)] Li	$\overline{M_n}$ (SEC)[(c)]	$\dfrac{\overline{M_w}}{\overline{M_n}}$
0	8.0	15	1.9	5 200	4 600	1.06
0.70	6.2	0.51	0.08	14 400	13 500	1.16
0.85	6.1	0.034	0.006	6 400	6 600	1.18
1	6.0	0	0	-	-	-

[(a)] $k_{p_{app}} = (R_p/M) / [PBLi]$
[(b)] theoretical molar masses (g.mol^{-1}) calculated assuming the contribution of [PBLi] alone
[(c)] experimental molar masses (g.mol^{-1})

Hsieh and Wang[5] have reported that the association of n,s-Bu$_2$Mg to alkyllithium initiators (r lower than 1) does not affect the stereochemistry of the diene polymerization. However, at higher [Mg]/[Li] ratios, Table 3, increasing proportions of n,s-Bu$_2$Mg induce an important and continuous increase of the 1,2- vinyl unit content in the polymer; whereas no significant change in the 1,4- cis/trans ratio is observed. At r = 1, only 13% of the butadiene units exhibit a vinylic 1,2- structure, but this is more than 50% at r = 10. This would suggest at first that 1,3-butadiene insertion proceeds into both uncomplexed PBLi and mixed Mg:Li complexes, and the proportion of the latter increases with r. However the quite quantitative formation of the 1:1 heterocomplex at r ≈ 1, supported both by UV-visible spectroscopy and kinetic data, should not yield a linear variation of the PB microstructure in the whole Mg:Li range examined (0 to 10). This indicates the contribution of a second reaction mechanism which could be the additional complexation of lithium species by free dialkylmagnesium (Scheme 2) and/or a 1,4- to 1,2- chain end isomerization process during chain exchanges between polybutadienyl active chains and dormant ones attached to magnesium species, as illustrated in Scheme 3. An isomerization of PBLi ends through chain exchanges inside aggregates has been already discussed by Bywater[4]. The presence of these secondary processes does not allow to use the observed PB microstructure as a probe of the structure of polymerization active species.

For i-Bu$_3$Al/PBLi systems, both the 1,4- and 1,2- unit ratio (1,4- units = 87-90%) and the cis/trans ratio in 1,4- units (cis/trans = 0.76) remain constant and close to those found with s-BuLi, Table 3. A similar tendancy was reported by Arest-Yakubovitch in the case of AlR$_3$/sodium-based initiators[6]. Since no reversible chain exchanges between the two metals take place in *i*-Bu$_3$Al/PBLi systems, chain end isomerization, as in dialkylmagnesium based systems, is not possible. However, because the PB microstructures observed with *i*-Bu$_3$Al/PBLi systems is very close to those observed with PBLi alone, the question concerning butadiene insertion either directly into active AlR$_3$/PBLi complexes or into remaining uncomplexed PBLi, present in very low concentration, cannot be answered yet.

Table 3. Influence of [n,s-Bu$_2$Mg]/[Li] and [i-Bu$_3$Al]/[Li] ratios on the configuration of polybutadienes; polymerization solvent: cyclohexane; T = 40°C.

[n,s-Bu$_2$Mg] /[RLi]	[i-Bu$_3$Al] /[RLi]	1,2 %	1,4 %		cis % / trans %
			cis	trans	
0	0	10	38.5	51.5	0.75
-	0.5	9.5	39	51.5	0.76
-	0.7	10	39	51	0.76
-	0.9	13	36	51	0.70
1	-	13	37	50	0.75
2	-	20	35	45	0.76
4	-	31	29	40	0.73
10	-	49	20.5	30.5	0.67

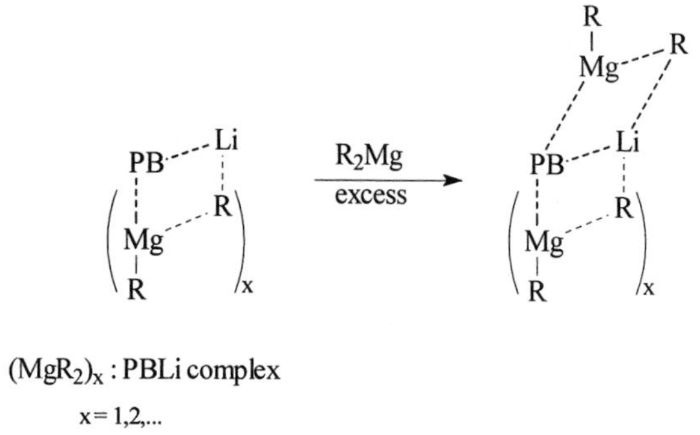

(MgR$_2$)$_x$: PBLi complex

x = 1,2,...

Scheme 2. Possible representation of the additional complexation of lithium species by free dialkylmagnesium.

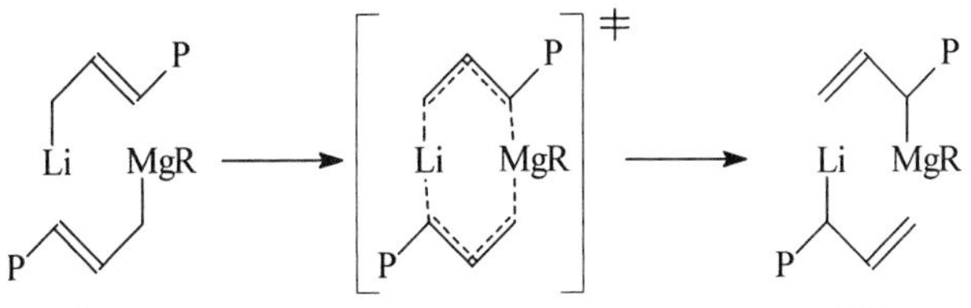

Scheme 3. Possible 1,4 to 1,2 isomerisation mechanism of polybutadienyl ends inside magnesiate complexes.

Stereoregulation in styrene anionic polymerization initiated by magnesiate complexes

Although alkali metal alkoxide as well as dialkylmagnesium are inactive alone in styrene polymerization, their combination yields an active initiating system, in agreement with the formation of alkali metal alkyl moieties through ligand exchanges in the complex[7]. Data concerning styrene polymerization initiated by alkali metal tert-butoxide/n,s-dibutylmagnesium "ate" complexes in hydrocarbon are given in Table 4. Besides the observed effect of temperature on both the reactivity and initiator efficiency, which both decrease when temperature goes down, magnesiate systems with different alkali metal yield polystyrene with different tacticities. At 20°C, with t-BuOLi/n,s-Bu$_2$Mg a PS with a predominantly syndiotactic microstructure, close to that observed with s-BuLi (i=0.10 ; h = 0.25 ; s = 0.65) is formed. No significant effect of the temperature on the syndioregulation is noticed. With the sodium cation the polystyrene shows an atactic structure. Conversely, a predominantly isotactic polymer is obtained with the potassium cation. The tacticity of polystyrenes prepared by the t-BuOK/n,s-Bu$_2$Mg system (r=1) in the range 20 to -40°C are given in Table 4.

As it may be seen, the isotactic character increases significantly when decreasing the temperature (85% of isotactic triads at –40°C) although in these conditions yields are low due in particular to a low initiation efficiency with respect to initial *t*-BuOK and the possibility of formation of alkylpotassium moieties. This can be explained by the insolubility of *t*-BuOK and the heterogeneous character of the polymerization system. These systems clearly differ from those described by Hogen-Esch[8],[9] and others[10],[11] obtained by the association of alkyllithium derivatives with lithium hydroxide as additive. No isotactic character is observed when lithium is used as alkali metal in the magnesiate complex. Fractionation of the highly isotactic PS into highly isotactic and atactic fractions could not be achieved in contrast to isotactic polystyrene obtained with alkyllithium/lithium hydroxide, suggesting one single type of propagating sites and the presence of stereodefects between isotactic polystyrene sequences.

Table 4. Polymerization of styrene initiated by the systems n,s-Bu$_2$Mg / t-BuOMt[a] (Mt = Li, Na, K ; Mg/Mt=1; cyclohexane ; 16 hours ; $\overline{Mn}_{th}$ =10 000 g.mol^{-1}).

Initiating system t-BuOM$_{et}$ / n,s-Bu$_2$Mg	T (°C)	Yield[b] (%)	$\overline{Mn}_{exp}$ (g/mol)	MMD	f[c] (%)	Tacticity[d]		
						iso	hetero	syndio
Li	20	76	18 500	1.3	41	0.15	0.22	0.64
Li[e]	-40	8	4 000	1.1	20	0.16	0.26	0.59
Na	20	72	21 500	1.3	33	0.28	0.30	0.42
K	20	91	25 000	1.4	36	0.61	0.29	0.10
K[e,f]	0	84	105 000	1.5	8	0.63	0.28	0.09
K[e,f]	-20	5	16 600	1.3	3	0.72	0.22	0.06
K[e,g]	-40	5	6 300	1.4	8	0.85	0.14	0.01

[a] [Initiator] = 5.10^{-3} M, [styrene] = 0.48 M

[b] Polymer yield determined gravimetrically

[c] initiator efficiency; f = (yield$_{wt}$. $\overline{Mn}_{th}$) / $\overline{Mn}_{exp}$

[d] determined by ^{13}C NMR of quaternary carbon in CDCl$_3$ at room temperature

[e] solvent = methylcyclohexane

[f] reaction for 24h

[g] reaction for 48h

In conclusion, the differences observed with "ate" complexes compared to the general behavior of alkyllithium systems, for styrene and 1,3 butadiene copolymerization in hydrocarbons yield some support to polymerization mechanisms involving monomer insertion into R$_2$Mg:PBLi and i-Bu$_3$Al:PBLi heterocomplexes. However the situation is not completely clear so far.

For magnesiate systems, secondary processes such as chain end isomerization or additional complexation of "ate" complexes by MgR$_2$ can explain the continuous change observed on the PB microstructure with increasing the proportion of dialkylmagnesium in the system. The mechanism yielding isotactic rich polystyrene in the presence of potassium metal alkoxide/ dialkylmagnesium initiating systems is also not completely elucidated. Isoregulation through insertion into magnesiate complexes or involving the potassium cation can still be postulated.

For alkylaluminum systems, the microstructure of butadiene units does not change with r (r < 1) and remains similar to PBLi in absence of any additives. Since 1,4- configuration of butadiene units is not altered, these results suggest that 1,4- insertion does not necessarily requires a preliminary strong diene coordination on the active species before insertion. The very low ionic character of the active species, ionic in ate complexes, might be in this case the determining factor for 1,4-insertion. Other results given by the study of styrene and butadiene copolymerization, which will be published soon, tend to show significant differences between the "ates" complexes and alkyllithiums alone systems.

Acknowledgements

BASF and Métaux Spéciaux are greatfully acknowledged for supporting these studies.

[1] P. Desbois, M. Fontanille, A. Deffieux, V. Warzelhan, S. Lätsch, C. Schade, *Macromol. Chem. Phys.* **1999**, *200*, 621.
[2] P. Desbois, M. Fontanille, A. Deffieux, V. Warzelhan, C. Schade, *Macromol. Symp.* **2000**, *157*, 151.
[3] S. Ménoret, S. Carlotti, M. Fontanille, A. Deffieux, P. Desbois, C. Schade, W. Schrepp, V. Warzelhan, *Macromol. Chem. Phys.* **2001**, *202*, 3219.
[4] D.J. Worsfold, S. Bywater, *Macromolecules* **1978**, *11*, 582.
[5] H.L. Hsieh, I.W. Wang, *Macromolecules* **1986**, *19*, 299.
[6] a) A. Arest-Yakubovich, *Macromol. Symp.* **1994**, *85*, 279 ; b) A. Arest-Yakubovich, *Chem. Reviews* **1994**, *19(4)*, 1.
[7] S. Ménoret, M. Fontanille, A. Deffieux, P. Desbois, *Polymer* **2002**, *43*, 7077.
[8] T. Makino, T.E. Hogen-Esch, *Polym. Prepr. (Am. Chem. Soc., Div. Polym. Chem.)* **1997**, *38(1)*, 164.
[9] T. Makino, T.E. Hogen-Esch, *Macromolecules* **1999**, *32*, 5712.
[10] J.E.L. Roovers, S. Bywater, *Macromolecules* **1975**, *8*, 251.
[11] L. Cazzaniga, R.E. Cohen, *Macromolecules* **1991**, *24*, 5817.
[12] M.F.K. Takahashi, M. De Lima, W.L. Polito, *Polymer Bulletin* **1997**, *38*, 455.

Macromol. Symp. **2004**, *215*, 29-40

Design and Synthesis of Functionalized Styrene-Butadiene Copolymers by Means of Living Anionic Polymerization

Mayumi Hayashi

Petrochemicals Research Laboratory, Sumitomo Chemical Co., Ltd., 2-1 Kitasode, Sodegaura-city, Chiba Pref. 299-0295, Japan
Tel: (+)-81-438-62-1276; Fax: (+)-81-438-63-6754

Summary: Styrene-butadiene copolymers (SBR) end-functionalized with dimethylamino groups at the initiating and terminating chain-ends were successfully synthesized by a one-step methodology based on living anionic polymerization using 1-(4-dimethylaminophenyl)-1-phenylethylene. The expected structures of the resulting copolymers were confirmed by ^{1}H-NMR, SEC, and SLS analyses, titration, and model reactions. Furthermore, the possible synthesis of tri-functionalized styrene-butadiene copolymer with dimethylamino groups in a chain by extending the methodology is described.

Keywords: anionic copolymerization; dimethylamino group; living polymerization; SBR; telechelics

Introduction

Chain-end-functionalized polymers such as telechelic polymers and macromonomers are industrially very important materials for preparing multi-block and graft copolymers and cross linked polymers with network structures.[1-4] Recently, these polymers have been utilized as precursory polymers for the synthesis of rod-coil block copolymers, specially shaped, branched polymers, cyclic polymers, and regular and asymmetric star-branched polymers.[5,6]

The methodology of living anionic polymerization is particularly suitable for the synthesis of well-defined chain-end-functionalized polymers with controllable molecular weights, narrow molecular weight distributions, and quantitative degrees of end-functionalizations.[4] Functional groups are generally introduced at either initiating or terminating chain-ends by using functionalized initiators or terminating agents. For this purpose, 1,1-diphenylethylene (DPE) and its derivatives are very useful and convenient compounds. They can react efficiently with carbanionic species like organolithium compounds as well as living anionic polymers of styrene

 DOI: 10.1002/masy.200451103

and 1,3-diene monomers in a monoaddition manner to quantitatively afford 1:1 adduct 1,1-diphenylalkyl anions (Scheme 1). Thus, they can be used as not only initiators but also terminators.

Scheme 1

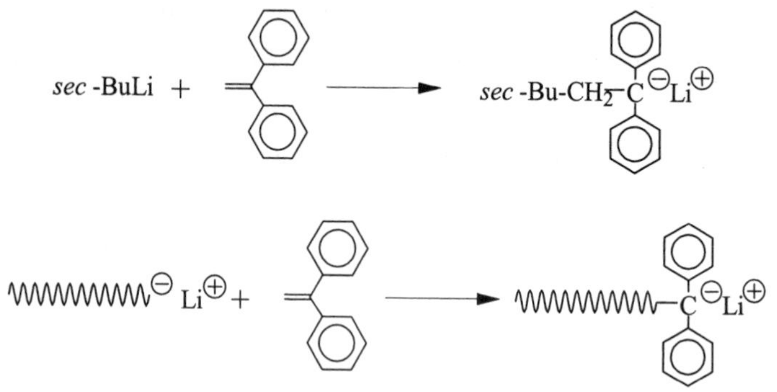

Quirk and his coworkers have greatly developed the methodology of using functionalized DPE derivatives to synthesize well-defined chain-functionalized polymers. [7-10] They demonstrated for the first time that 1,1-diphenylalkyl anions derived from functionalized DPE derivatives, often used as functionalized initiators in the polymerization of alkyl methacrylates and ethylene oxide, can also be used as effective initiators for the living anionic polymerization of less reactive styrene and 1,3-diene monomers. By this success, polystyrenes and poly(1,3-diene)s can be functionalized at both initiating and terminating chain ends. Asymmetric chain-end-functionalized polymers can also be synthesized by the use of different DPE derivatives at initiation and termination steps. Furthermore, in-chain-functionalized homopolymers and block copolymers have been synthesized by terminating living anionic polymers with functionalized DPE derivatives, followed by initiating the polymerization of the same or different monomers with the generated anions at the chain ends. Thus, functionalized DPE derivatives potentially offer a versatile methodology for the synthesis of various chain-functionalized polymers. However, one limitation of this methodology is that rigid stoichiometry is required in the addition reaction between DPE derivative and either initiator or propagating chain-end anion. If an excess of the DPE derivative is used, it may be incorporated into the polymer chain, resulting in the introduction of extra functional groups. If the anionic species is used in excess, unfunctionalized polymer chains are produced.

We herein report on a new, simple one-step methodology based on living anionic polymerization and DPE chemistry for the synthesis of chain-end-functionalized styrene-butadiene copolymers with dimethylamino groups at both initiating and terminating chain-ends. Although such functionalized polymers can be synthesized by the method of Quirk using two steps involving initiation and termination reactions, the same functionalized polymers can be synthesized by a one-step reaction in our new methodology. Furthermore, the possible synthesis of tri-functionalized styrene-butadiene copolymers with dimethylamino groups at desired positions in a chain by extending the methodology is described. Finally, the influence of number and position of dimethylamino groups on the physical properties of the resulting rubbery copolymers is discussed.

Experimental

Materials. Hexane, cyclohexane, styrene and 1,3-butadiene were purified to the level of anionic polymerization according to the literature procedures.[11] The functionalized 1,1-diphenylalkylanion used as an initiator was prepared by the reaction of appropriate amounts of 1-(4-dimethylaminophenyl)-1-phenylethylene (**1**)[7] with *sec*-BuLi in cyclohexane at 25 °C for 0.5 h under nitrogen.

α,ω-Functionalized Copolymers with Dimethylamino Groups. The anionic copolymerization of styrene and 1,3-butadiene was carried out in a mixture of hexane and cyclohexane at 60 °C in the presence of 15.2 mL of THF and 1.0 mL of 1,2-dimethoxyethane. The copolymerization was started by adding the functionalized initiator from a 2.1-fold excess of **1** (5.25 mmol) and *sec*-BuLi (1.0 M cyclohexane solution, 2.50 mL, 2.50 mmol) in cyclohexane (7.0 mL) to a mixture of styrene (58 g) and 1,3-butadiene (142 g) in hexane solution (3.5 L) in a 5L autoclave with stirring. Styrene (87 g) and 1,3-butadiene (213 g) were further fed for 1 h and 1.5 h, respectively throughout the polymerization. During stirring for 30 min after the copolymerization, an orange color characteristic to the 1,1-diphenylalkyl anion derived from **1** gradually developed. The polymer solution was stirred for an additional 1 h and quenched with degassed methanol. The resulting polymer was precipitated in methanol and purified by reprecipitation three times. It was finally dried under reduced pressure at 55 °C for 12 h.

Chain-Functionalized Copolymers with Three Dimethylamino Groups. The functionalized initiator prepared from a 3.1-fold excess of **1** (7.75 mmol), *sec*-BuLi (1.0 M cyclohexane solution, 2.50 mL, 2.50 mmol) and cyclohexane (7.0 mL) was used in this case. The copolymerization was first carried out under conditions identical to those described above. During the first stage of copolymerization, styrene (43.5 g) and 1,3-butadiene (106.5 g) were fed for 1 h and 1.5 h, respectively, and the polymerization mixture was stirred for additional 1.5 h. Then, the same amounts of styrene and 1.3-butadiene were again fed for 1 h and 1.5 h at the second stage of copolymerization, and the mixture was allowed to stir for an additional 1.5 h. The resulting polymer was precipitated in methanol and purified by reprecipitation three times. It was finally dried under reduced pressure at 55 °C for 12 h.

Polymer Characterization. Size-exclusion chromatography (SEC) was obtained at 40 °C with a TOSOH HLC 8020 instrument with UV (254nm) or refractive index detection. THF was used as a carrier solvent at a flow rate of 0.6 mL/min. A calibration curve was made using standard polystyrenes. ^{1}H NMR spectra were recorded on JEOL instrument at 270 MHz in CDCl$_3$. The measurement by static light scattering (SLS) was performed with an Otsuka Electronics SLS-600R instrument equipped with a He-Ne laser (633 nm) in THF at 25 °C. Infrared spectra were measured on a PERKIN ELMER FT-IR Spectrometer 1720X in order to determine the 1,2-vinyl content of butadiene unit. The styrene content was measured by the refractive index method.

Preparation of Vulcanized Rubber. The rubber composites listed in Table 1 were prepared from the resulting polymer (Mw .200kg/mol) by mechanical mixing. The mixing conditions were as follows: the dried copolymer was masticated in labo plastomill at 70 °C for 0.5 min, and then well-dried silica and the silane coupling agent were mixed with the resulting masticated polymer at 110 °C for 1.5 min. Furthermore, zinc oxide, stearic acid, and the antioxidant were mixed with the silica-filled rubber composites thus prepared at 120 °C for 3.5 min (first step). The master batch obtained at the first step was then mixed with sulfur and the accelerator at 70 °C for 3 min, followed by cooling to room temperature (second step). Finally, the resulting master batch was vulcanized at 160 °C for 45 min to prepare test pieces. Hardness, 300% modulus, and resilience of the resulting vulcanized polymers were measured according to ASTM STANDARD.[12]

Table 1. Composition of silica-filled rubber composite (weight per hundred rubber).

Copolymer	100	Antioxidant 6C	1.5
Silica	75	Accelerator CZ	1
Coupling agent	5	Accelerator DPG	1
Zinc Oxide	3	Sulfur	1.5
Stearic acid	2		

Results and Discussion

Synthesis of α,ω-Functionalized Copolymer with Dimethylamino Groups. We previously synthesized chain-end-functionalized styrene-butadiene copolymer with dimethylamino groups by the anionic copolymerization of styrene with 1,3-butadiene in a mixture of cyclohexane and hexane at 60 °C for 2.5 h with the functionalized initiator prepared from a 1.2-fold excess of **1** and *sec*-BuLi. Therefore, 0.2 equivalents of **1** remained after the preparation of the initiator. As described in experimental section, the copolymerization was performed by feeding two monomers throughout the polymerization. Since the weight ratio of 1,3-butadiene to styrene was 3/7 and 1,3-butadiene was fed longer, a pale yellow color of the poly(1,3-butadienyl) anion was always observed during the copolymerization, as expected. After the copolymerization, the polymerization mixture was allowed to stir for 30 min. Very surprisingly, the color gradually changed from pale yellow to bright orange color of the 1,1-diphenylalkyl anion presumably generated from **1** and the chain-end anion. This strongly suggests that the residual 0.2 equivalents of **1** are not incorporated into the copolymer, but remain intact during the course of the copolymerization and then react gradually with the propagating chain-end anion after the copolymerization.

This unexpected result prompted us to investigate the reaction pattern of **1** on the living anionic copolymerization of styrene and 1,3-butadiene in more detail with hope that a chain-end-functionalized styrene-butadiene copolymer having two dimethylamino groups at both initiating and terminating chain-ends can be synthesized as illustrated in Scheme 2. For this purpose, we prepared the dimethylamino-functionalized initiator from a 2.1-fold excess of **1** and *sec*-BuLi, and used it for the anionic copolymerization of styrene and 1,3-butadiene under the conditions

Scheme 2

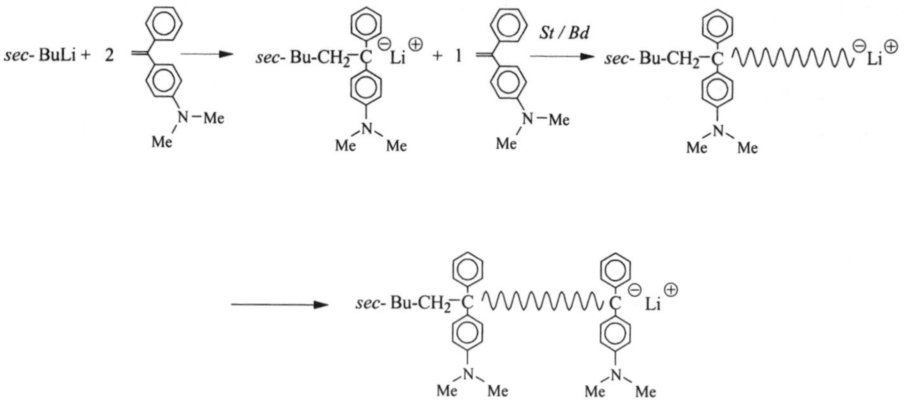

mentioned above.   Again, the pale yellow color was observed throughout the copolymerization, and the characteristic orange color gradually appeared after the copolymerization.

The resulting copolymer showed a sharp monomodal SEC distribution, the M_w/M_n value being 1.08.   The absolute M_w value of 205 kg/mol determined by SLS agreed well with that calculated (200 kg/mol).   The ^{1}H NMR spectrum of the reaction mixture showed the characteristic resonance at 2.90 ppm assigned to methyl protons of the dimethylamino groups introduced into the polymer chain.   In addition, a small resonance at 2.97 ppm corresponding to methyl protons of the dimethyl amino group of **1** was also present in this mixture.   The ratio of two peak intensities of 2.0/0.1 is exactly the same as the ratio assuming that two dimethylamino groups are introduced at both the initiating and terminating chain-ends and 0.1 equivalent of **1** remains unreacted as illustrated in Scheme 2.   The copolymer was purified by reprecipitation three times. The resonance at 2.90 ppm remained almost unchanged and the degree of dimethylamino-functionalization was 2.0 ± 0.1 using methyl benzoate (3.90 ppm) as an internal standard.   The functionalization degree was also determined by titration, using HClO$_4$ with crystal violet as an indicator, to be 2.0 ± 0.05.   These results showed that two dimethylamino groups were introduced into the polymer chain and the residual 0.1 equivalent of **1** remains intact after the copolymerization.

In our copolymerization system, one dimethylamino group is definitely introduced at the initiating chain-end *via* the initiation step with the functionalized 1,1-diphenylalkyl anion.   It is therefore indicated that the second dimethylamino group may be introduced at the terminating chain-end by the reaction of the propagating chain-end anion with **1** after the copolymerization. Since the resulting 1,1-diphenylalkyl anion cannot further react with **1**, it is reasonable that 0.1 equivalent of **1** remains unreacted.   Two more reactions have been performed to prove the fact that the resulting copolymer chain-end anion is completely converted from butadienyl anion to the 1,1-diphenylalkyl anion derived from **1**.

As the first reaction, we have reacted methyl 4-dimethylaminobenzoate (**2**) with a model poly(1,3-butadienyl) anion (A), the reaction product of (A) with **1** (B), and the polymer anion (C) prepared by our one-step methodology and compared the reaction products.   The polymer anion (A) was prepared by the anionic copolymerization of styrene and 1,3-butadiene with *sec*-BuLi under conditions similar to (C).   The polymer anion (B) was prepared by adding a 2.0-fold excess of **1** to the anion (A).   The three polymer anions thus prepared were reacted with **2** at 60 °C for 1 h. SEC profiles of the reaction mixtures are shown in Figure 1.   In all cases, there were only two sharp peaks. Lower molecular weight peaks represent the 1:1 ketonic products of **2** with polymer

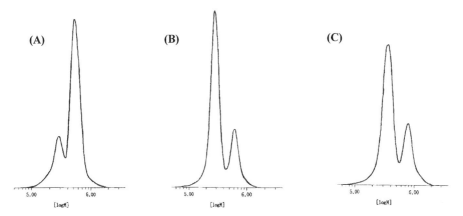

Figure 1. SEC profiles of the reaction mixtures of **2** with the model poly (1,3-butadienyl)anion **(A)**, the model 1,1-diphenylalkyl anion derived from **1 (B)** and the polymer anion **(C)** prepared by our one-step methodology.

anions, while higher ones at double the molecular weight may possibly be dimeric alcohols as illustated in Equation 1. The area ratios of the two peaks were 25/75, 75/25, and 74/26 in the reactions of **2** with polymer anions, (A), (B), and (C), respectively. Since the ratio obtained by the reaction of **2** with (C) was consistent with that obtained by the reaction of **2** with (B), the anion (C) prepared by our one-step methodology proved to be the 1,1-diphenylalkyl anion derived from **1**.

Equation 1

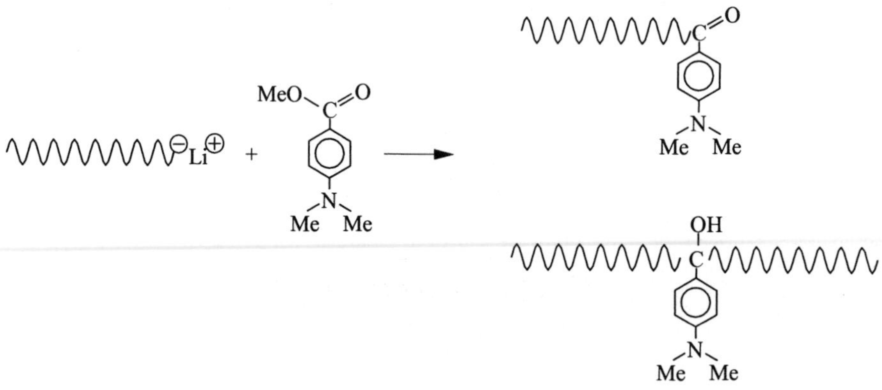

In order to further evaluate the structure of the terminal anion, we have reacted 1-(*N*-morpholinophenyl)-1-phenylethylene (**3**) with (A), (B), and (C) at 60 °C for 10 h. The introduction of morpholino group was observed in the copolymer obtained by the reaction of **3** with (A), while morpholino group was not present in the copolymers obtained by the reactions of **3** with (B) and (C). As it is well established that no addition reaction of DPE to 1,1-diphenylalkyl anion occurs under normal conditions, the result of the reaction of **3** with (B) is reasonable. Since the same result occurred for the reaction of **3** with (C), we conclude that (C) is terminated by the 1,1-diphenylalkyl anion from **1**.

All of these results confirm the expectation that styrene and 1,3-butadiene copolymerize with the dimethylamino-functionalized initiator and the resulting polymer anion reacts with the residual **1** after the copolymerization to afford an α,ω-functionalized copolymer with two dimethylamino groups at both the initiating and terminating chain-ends. It should be however mentioned that **1** undergoes copolymerization with styrene and 1,3-butadiene if a relatively large amount of **1** exists. Quirk and his co-workers[13] previously reported that **1** indeed copolymerized with either styrene or 1,3-butadiene to afford copolymers with many dimethylamino groups. These results suggest that the polymerization in the presence of **1** is very sensitive and depends significantly on changes of the applied concentration and reactivity of **1**, the reaction solvent, temperature and so on.

Synthesis of Chain-Functionalized Styrene-Butadiene Copolymer with Three Dimethyl-amino Groups. The copolymer chain-end anion prepared by the one-step methodology developed herein is still active and can reinitiate the copolymerization to extend a new polymer chain by the addition of monomers. If residual **1** is present in the system, it is expected that the new polymer chain-end anion reacts with **1** to introduce the third dimethylamino group at the chain-end as illustrated in Scheme 3.

Scheme 3

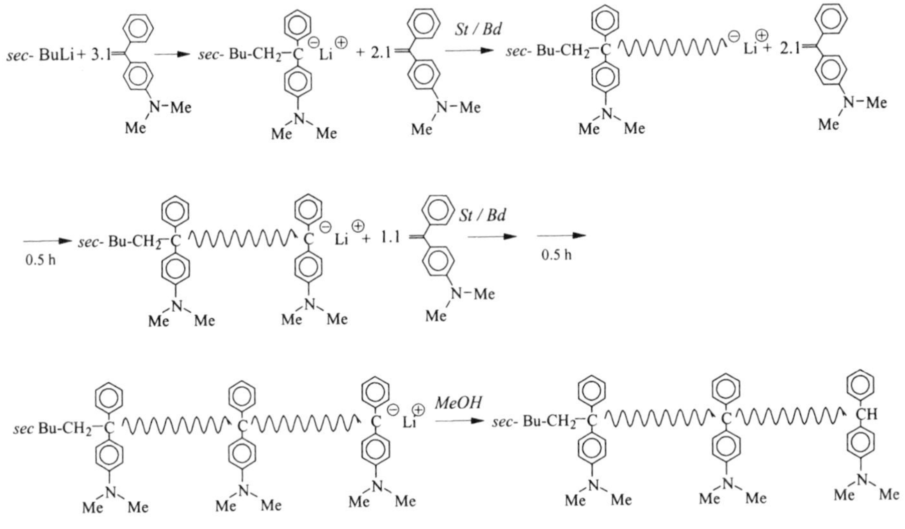

To test the possibility of this type of functionalized copolymer synthesis, the functionalized 1,1-diphenylethylene anion was newly prepared from a 3.1 molar ratio of **1** to *sec*-BuLi. The first copolymerization was similarly carried out with the initiator thus prepared under conditions identical to those employed for the one-step methodology. The color of the resulting polymer solution became gradually bright orange by allowing the polymerization mixture to stir for an additional 30 min after the copolymerization. The second copolymerization was started by feeding both styrene and 1,3-butadiene in the same manner. The color of the polymerization mixture instantaneously changed from bright orange to pale yellow and the pale yellow color remained stable as the copolymerization proceeded. After the polymerization, the reaction mixture was allowed to stir for 30 min and the color gradually changed again to bright orange with time.

The copolymers obtained at the first and second polymerization stages exhibit sharp monomodal SEC distributions, and the SEC peak of the first copolymer moves to a higher molecular weight side while maintaining a narrow molecular weight distribution, as shown in Figure 2. The characteristic resonance at 2.90 ppm for methyl protons of the dimethylamino group was observed in both copolymers. The degree of dimethylamino functionalization of the second copolymer was 3.0 ± 0.15. It was proved by the reactions with **2** and **3** that the copolymer chain-end anion after the second copolymerization is the 1,1-diphenylalkyl anion derived from **1**. These results in addition to the bright orange coloration in each copolymerization stage strongly indicate that the methodology works satisfactorily as illustrated in Scheme 3 to afford a special chain-functionalized styrene-butadiene copolymer with three dimethylamino groups, including

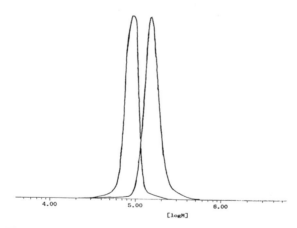

Figure 2. SEC profiles of the first and the second copolymers.

one each at the initiating and terminating chain ends as well as in the chain. Thus, we were successful in extending the one-step methodology to the synthesis of chain-functionalized styrene-butadiene copolymer with three dimethylamino groups. Although the influence of changes in molecular weight, copolymer composition, and reaction variables have to be studied in detail, the developed one-step methodology provides a promising tool for the design and synthesis of well-defined chain-functionalized styrene-butadiene copolymers with multiple dimethylamino groups. We are now investigating an extension of this one-step methodology to the introduction of more dimethylamino groups into copolymer chains.

Properties of Functionalized Styrene-Butadiene Copolymers

In this study, well-defined functionalized styrene-butadiene copolymers with one, two, or three dimethylamino groups were synthesized by one-step methodology. By testing physical properties of these polymers, the influence of the number and position of dimethylamino groups on the properties can be discussed. The samples and the results are listed in Table 2. As shown in the Table, the modulus of 300 % elongation becomes a bit higher with increasing number of dimethylamino groups. However, both hardness and resilience were independent of the number of dimethylamino groups. The influence of the number and position of dimethylamino groups on other properties is now under investigation.

Table 2. Properties of Functionalized Styrene-Butadiene Copolymers.

Polymer (number of -NMe$_2$)	Styrene (%)	Vinyl (%)	ML 1+4 100°C [1]	Hs [2] (Duro-A)	M300 [3] (Mpa)	Resilience 60°C [4] (%)
1	28.9	42.9	67	69	7.42	53
2	28.9	42.3	70	70	7.59	54
3	29.1	44.3	72	69	7.75	54

1) Measured according to ISO 289:1985
2) Measured according to ISO 48:1994
3) Measured according to ISO 37:1977
4) Measured according to ISO 4662:1986

Conclusions

We have developed a promising new one-step methodology for the synthesis of well-defined chain-functionalized styrene-butadiene copolymers with dimethylamino groups at both initiating and terminating chain-ends. By extending this methodology with the functionalized initiator prepared from 1 and sec-BuLi in a 3.1:1 molar ratio, a chain-functionalized styrene-butadiene copolymer with three dimethylamino groups could be synthesized. The simplicity and efficiency of this methodology makes it an attractive candidate for preparing multi-functionalized styrene-butadiene copolymers (SBR).

Acknowledgements

We would like to acknowledge Professor Akira Hirao and Dr. Naoki Haraguchi in Tokyo Institute of Technology for giving us some advice on our study.

[1] F. W. Harris, H. J. Spinelli, *Reactive Oligomers*, ACS Symposium Series, No. 282, Am. Chem. Soc., Washington D.C. **1985**

[2] E. J. Goethals, *Telechelic Polymers: Synthesis and Applications*, CRC Press, Boca Raton, Florida **1989**

[3] Y. Yamashita, *Chemistry and Industry of Macromonomers*, Marcel Dekker, Inc., Huthing & Wepf, Heidelberg **1993**

[4] H. L. Hsieh, R. P. Quirk, *Anionic Polymerization: Principles and Practical Applications*, Marcel Dekker, New York **1996**

[5] References are cited in A. Hirao, M. Hayashi, *Acta Polym.* **1999**, *50*, 219

[6] A. Hirao, M. Hayashi, Y. Tokuda, N. Haraguchi, T. Higashihara, S-W. Ryu, *Polymer J.* **2002**, *34*, 633

[7] R. P. Quirk, L-F. Zhu, *British Polym. J.* **1990**, *23*, 47

[8] R. P. Quirk, J. Yin, S-H. Guo, X-W. Hu, G. J. Summers, J. Kim, L-F. Zhu, J-J. Ma, T. Takizawa, T. Lynch, *Rubber Chem. Tech.* **1991**, *64*, 648

[9] R. P. Quirk, *Macromol. Chem., Macromol. Symp.* **1992**, *63*, 259

[10] R. P. Quirk, T. Yoo, Y. Lee, J. Kim, B. Lee, *Adv. Polym. Sci.* **2000**, *153*, 67

[11] N. Hadjichristidis, H. Iatrou, S. Pispas, M. Pitsikalis, *J. Polym. Sci. Part A: Polym. Chem.* **2000**, 38, 3211

[12] Annual Book of ASTM STANDARDS, Vol. 09.01, **1990**

[13] R. P. Quirk, L-F. Zhu, *Polymer International* **1992**, *27*, 1

Macromol. Symp. **2004**, *215*, 41-49

Synthesis of PS Star Polymers from Tetracarbanionic Initiators

Arnaud Lebreton, J. K. Kallitsis, Valérie Héroguez,* Yves Gnanou

Laboratoire de Chimie des Polymères Organiques (LCPO), ENSCPB, Avenue Pey-Berland, 33607 PESSAC Cedex, France
E-mail: heroguez@enscpb.fr; gnanou@enscpb.fr

Summary: A novel tetracarbanionic initiator as designed and used to obtain star polymers. Well-defined four-arm polystyrene stars exhibiting expected molar masses and low polydispersity indexes have been obtained in this way.

Keywords: anionic polymerization; core-first method; pluricarbanionic initiator; polystyrene; star polymers

Introduction

Star-type polymers attract interest because of their intrinsic features [1], namely, high segment density and low viscosity. In spite of severe experimental constraints, anionic polymerization is still one of the best routes to synthesize well-defined star-type polymers. Thanks to the long lifetime of active anionic species, features such as the functionality of the core, the polymolecularity and molar mass of branches can be precisely controlled in the stars obtained. There are two strategies [2] to produce star-type polymers, either the so-called "arm-first" or "core-first" methods. The arm-first approach entails the coupling of linear "living" chains to multifunctional agent such as SiCl$_4$. The second method involves the use of a plurifunctional core that can initiate polymerization in multiple directions. This approach is a major challenge since it requires the prior design of well-defined multifunctional initiators. Examples of such initiators are scarce, specially those that can trigger carbanionic polymerization [3-8].

In a recent addition to the field of anionic polymerization, Tsitsilianis et al. [9] described a novel difunctional initiator. Based on the well-known halogen-lithium exchange reaction that was first investigated by Gilman et al. [10] this initiator affords the synthesis of triblock polymers in a polar solvent.

DOI: 10.1002/masy.200451104

In this contribution we have also exploited the possibilities offered by this halogen-lithium chemistry to synthesize well-defined star-type polymers using the "core-first" method. To this end, the synthesis of a carbanionic compound from a dibromoaryl compound was first investigated and its behavior as a difunctional initiator studied. In a second step, a procedure similar to that used for the synthesis of this difunctional initiator was applied to obtain a tetracarbanionic initiator. The conditions that have been applied to obtain four-arm PS star polymers of targeted molar masses, low polydispersity indexes and accurate number of arms are discussed here.

Experimental part

Materials: Benzene was distilled over living polystyryllithium prior to its use. 2-Bromobutane, undecane, N,N,N',N''-tetramethylethylenediamine (TMEDA) were dried over calcium hydride for several days before being distilled. Styrene was dried over CaH_2 for 2 days, then distilled over dibutylmagnesium and redistilled just before use. Commercially available sec-butyllithium was titrated according to a literature method [18]. (1,2,3,4-tetra(p-bromophenyl)-5-phenyl)benzene (II) was synthesized as described [19]. 4, 4'-dibromobiphenyl (ACROS) and compound II were freeze-dried twice with benzene. Methanol was used as received.

All anionic polymerizations were performed under a slight argon overpressure using a tight reactor equipped with a argon inlet, magnetic stirring, a sampling device and burets meant to introduce solvent, sec-BuLi, monomer and deactivator.

Viscosimety measurements were realised using a Ubbelohde-type flow capillary viscosimeter. Intrinsic viscosities were obtained by extrapolation to zero concentration of the reduced and inherent viscosities. Measurements were performed in toluene at 35°C.

The size exclusion chromatography (SEC) equipment consisted of a JASCO HPLC pump type 880-PU, TOSOHAAS TSK gel columns, a Varian refractive index detector and a JASCO 875 UV/VIS absorption detector, THF being the mobile phase. The columns were calibrated with polystyrene standards. The actual molar masses of star samples were calculated from the response of the multiangle laser light scattering detector, Wyatt Technology, that was connected to SEC.

Synthesis of linear PS by pathway A: In a typical reaction, 4, 4'-dibromobiphenyl (0.25g, 0.8 mmol/L) was dissolved in 15 ml of benzene. 1.23 ml of sec-BuLi (1.6 mmol, [sec-BuLi]=1.3

mol/L) was added at once. After 20min at room temperature 1.23 ml of sec-BuLi (1.6 mmol) and 0.5 ml of TMEDA (3.2 mmol) were added. Then, 17ml (0.15 mol) of styrene was slowly added to the previous solution. Deactivation was performed after about 12 h of polymerization by addition of methanol. Precipitation in methanol yielded pure PS.

Synthesis of linear PS by pathway B: In a typical reaction, 4, 4'-dibromobiphenyl (0.25g, 0.8 mmol/L) was dissolved in 15 ml of benzene. Sec-BuLi (1.23 ml, 1.6 mmol) was added at once. After 20 min at room temperature the initiator formed was collected by filtration and then washed with 10 ml of benzene. Three successive filtration-washing processes were carried out. Afterward, initiator was dissolved in 30 ml of benzene prior to the addition of 0.25 ml (1.6 mmol) of TMEDA. Then, 8 ml (0.07 mol) of styrene was slowly added to this solution. Deactivation was performed after about 12 h of polymerization by addition of methanol. Precipitation in methanol yielded pure PS.

Synthesis of PS star: The same procedure as that used for the synthesis of linear PS was followed. Selective precipitation was carried out by slow addition of methanol into the benzene solution containing the star polymer.

Results and discussion

Synthesis of 4, 4'-dilithiobiphenyl from a 4, 4'-dibromobiphenyl:

4, 4'-dilithiobiphenyl was obtained by reaction of 4,4'-dibromobiphenyl with sec-butyllithium (Figure 1) through halogen-lithium exchange reaction [10] ; the latter gives rise to an equilibrium that lies towards the side giving the more stable organolithium compound. This chemistry is particularly useful for preparing aryl lithium. Under similar experimental conditions similar to those described in the literature [11-12], halogen-metal interconversion is complete when carried out in benzene. Likely due to its high degree of aggregation 4,4'-dilithiobiphenyl was found to be insoluble in benzene, but became soluble upon addition of trimethylethylenediamine (TMEDA). It could thus be used to initiate the polymerization of styrene.

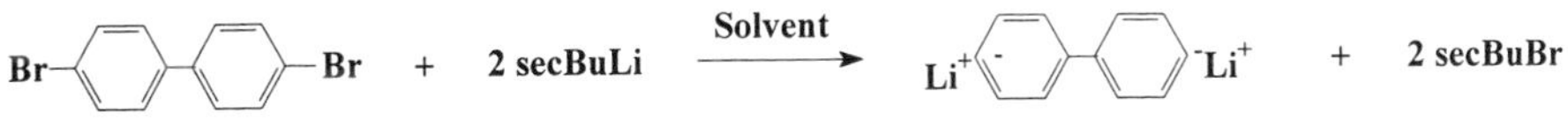

(1)

Figure 1. Synthesis of 4,4'-dilithiobiphenyl.

One feature of this halogen-lithium exchange reaction that caused complications was the formation of sec-butylbromide, an alkylbromide that was found to deactivate the growing polystyryllithium. In order to avoid such a deactivation, sec-butyl bromide had to be eliminated and two solutions have been explored (Figure 2) to this end :

A-   Before adding styrene an attempt was thus made to neutralize sec-BuBr with sec-BuLi : in fact, this compound reacted quantitatively with the bromide derivative in less than 1 minute in presence of TMEDA,

B-   As another means to eliminate sec-BuBr, insoluble 4, 4'-dilithiobiphenyl was isolated by filtration and then dissolved in benzene/TMEDA.

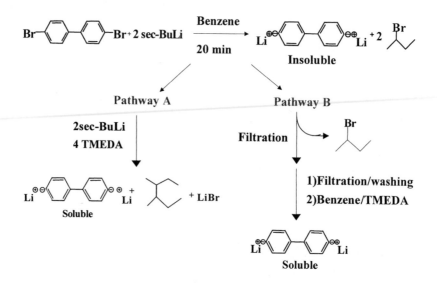

Figure 2. The two pathways used to purify 4,4'-dilithiobiphenyl initiator.

Styrene polymerization:

The anionic polymerization of styrene was triggered upon addition at room temperature of monomer into the flask containing the 4,4'-dilithiobiphenyl initiator obtained by the two pathways (A, B) described above. After complete conversion of styrene, methanol was added to the medium as quenching agent. All samples were characterized by size exclusion chromatography (SEC) (Table 1) and gas chromatography (GC).

Table 1. Characteristics of polystyrene samples obtained using routes A and B.

Pathways	$\overline{DP}_{n,targ.}$ a)	$\overline{M}_{n,targ.}$ b) g/mol	$\overline{M}_{n,exp}$ (SEC) g/mol	$\overline{M}_w/\overline{M}_n$
A	187	19500	19000/9200	bimodal
A	173	18000	16300/8300	bimodal
B	81	8500	9200	1.15
B	107	11200	12000	1.17
B	144	15000	16400	1.12

a) $\overline{DP}_{n,targ.} = 2 \times ([M]/2 \times [LiPhPhLi])$

b) $\overline{M}_{n,targ.} = 104 \times \overline{DP}_{n,targ.}$

Regardless of the experimental conditions used GC analysis showed that 100% of the dilithio compound had been consumed in the initiation of the polymerization of styrene. However, the samples obtained using initiator purified *via* pathway A, exhibited a bimodal molecular weight distribution (Figure 3).

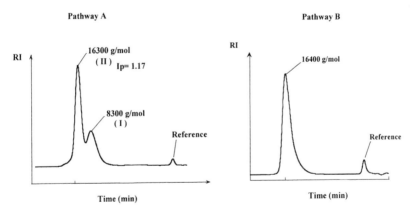

Figure 3. SEC chromatograms of linear polystyrene obtained from pathways A and B.

The population of lower molar mass (I), less than 10% in weight, was identified as resulting from the initiation by residual sec-butyllithium remaining in the reaction medium after neutralization of sec-BuBr. The molar mass of this minor population was about half that of the second population. Unlike the samples generated by route A, those obtained by the pathway B showed a low dispersity index and molar mass in excellent agreement with targeted values attesting to the efficiency of this method (Table 1, Figure 3).

Synthesis of tetracarbanionic initiator from a tetrabromoaryl compound:

The synthesis of tetrabromoaryl compound II was performed according to the procedure described in the literature [13-14]. Under the same conditions as those described above the tetralithioaryl compound III was synthesized in benzene. The yield of the halogen-metal reaction was checked by GC by measuring the quantity of sec-BuBr released. The value obtained was close to 100% attesting to the complete transformation of bromide atom into lithium. Due to its insolubility in benzene prior to the addition of TMEDA and its very good solubility in the presence of the latter additive, only the pathway B has been contemplated (Figure 4).

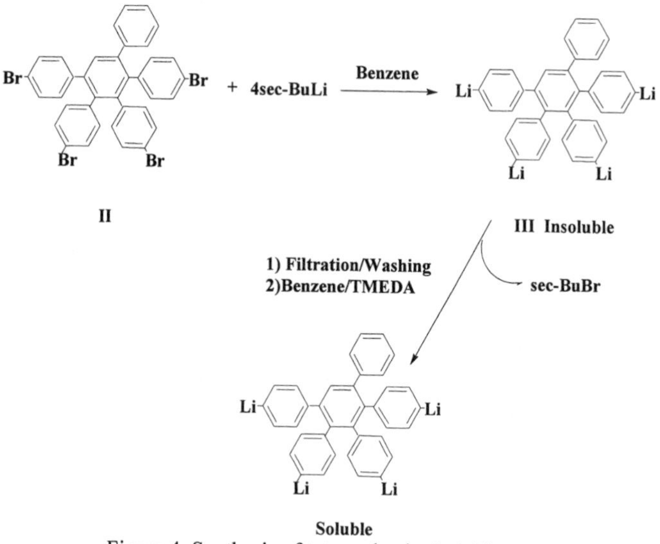

Figure 4. Synthesis of tetracarbanionic initiator.

Synthesis of four-arm star polymers:

The tetralithioaryl compound III obtained under these conditions was used to trigger the polymerization of styrene. After complete monomer conversion the carbanionic species were deactivated by addition of methanol. The resulting polymers were characterized by SEC coupled with laser light scattering so as to determine their actual molar masses. The chromatogram obtained attested to the complete consumption of the tetrafunctional initiator as no signal due to the protonated version of III could be detected in the region of lower masses. However, the SEC chromatograms revealed the presence of two populations of chains : that of largest amount (~90%) exhibited molar mass in agreement with a tetrafunctional initiation and a low molar mass distribution attesting to fast initiation step (Table 2). The second population represented hardly 10% by weight and its molar mass was four time lower than that of the main population. It was identified as linear polystyrene initiated by residual sec-BuLi which was entrapped inside aggregated tetracarbanionic species and could not be removed even after several washings and filtrations. Regardless of the number of washing steps with a dry solvent it was not possible to totally eliminate this second population. Nevertheless, pure polystyrene star polymers could be isolated as shown by SEC analysis upon selective precipitation using a methanol/benzene mixture(Figure 5).

Table 2. Characteristics of the four-arm polystyrene stars obtained.

$\overline{M}_{n,targ.}$ [a]	$\overline{M}_{w,DDL}$ [b] (B)	$\overline{M}_{n,SEC}$ (A)	$\overline{M}_w / \overline{M}_n$	$[\eta]_{star}$ [c]	$[\eta]_{linear}$ [c]	g'
g/mol	g/mol	g/mol		dl/g	dl/g	
39500	38400	9800	1.08	0.166	0.222	0.75
55000	49500	12500	1.09	-	-	-
120000	98600	25200	1.08	0.297	0.430	0.69

a) $\overline{M}_{n,targ.} = 104 \times ([M]/[PhLi]) \times 4$
b) Molar masses determined from SEC equipped with a laser light detector
c) Determined in toluene at 35°C

The branched nature of the samples obtained was demonstrated by viscometry and more particularly through the determination of g' which represents the ratio $[\eta]_{star}/[\eta]_{linear}$. As can be seen in Table 2, the g' values exhibited by all samples are very close to the theoretical values (0.71<g'<0.79) predicted by the theoretical models of Zimm [15] and Stockmayer [16]. Quite a bit large values had been previously reported by Roovers [17] for four-arm polystyrene stars in good solvent (g' = 0.86). These results substantiate the four-arm structure of the stars obtained.

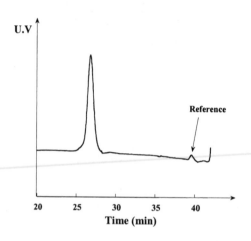

Figure 5. SEC chromatogram of a four-arm polystyrene star after purification.

Conclusion

In this study, it was demonstrated that phenyllithium species can be used to initiate the "living" anionic polymerization of styrene; a novel tetracarbanionic initiator carrying four phenyl lithium sites was synthesized and used to trigger the polymerization of styrene. Four-arm polystyrene star exhibiting the expected macromolecular architecture could be obtained in this way. The synthesis of PS-b-PB star block copolymers is currently in progress in our laboratory and will be described in a future contribution.

Acknowledgement

The authors wish to thank the Ministry of Higher Education (France) for providing a scholarship to Arnaud Lebreton.

[1]   J. Roovers, *Encyclopedia of Polymer Science and Engineering*, 2nd ed., Kroschwitz J. L., Ed., Wiley New York, **1985**, Vol.2, p. 478.
[2]   N. Hadjichristidis, M. Pitsikalis, S. Pispas, H. Iatrou, *Chemical Reviews* **2001**, *101*, 3747.
[3]   T. Fujimoto, S. Tani, K. Takano, M. Ogawa, M. Nagasawa, *Macromolecules* **1978**, *11*, 673.
[4]   W. Toreki, M. Takaki, T. E. Hogen-Esch, *Polym. Preprint* **1988**, *27(1)*, 355.
[5]   B. Gordon III, M. Blumenthal, J.E. Loftus, *Polym. Bull.* **1984**, *11*, 349.
[6]   R. Quirk, Y. Tsai, *Macromolecules* **1998**, *31*, 8016.
[7]   R. Quirk, T. Yoo, Y. Lee, J. Kim, B. Lee, *Adv. Polym. Sci.* **2000**, *153*, 66.
[8]   N.G. Vasilenko, E.A. Rebrov, A.M. Muzafarov, B. Esswein, B. Striegel, M. Möller, *Macromol. Chem. Phys.* **1998**, *199*, 889.
[9]   C. Tsitsilianis, G. Vogiagis, J.K. Kallitsis, *Macromol. Rapid Commun* **2000**, *21*, 1130.
[10]   H. Gilman, F.W. Moore, *J. Am. Chem. Soc.* **1940**, *62*, 1843.
[11]   H.J. Winkler, H. Winkler, *J. Am. Chem. Soc.* **1966**, *88*, 964.
[12]   D.E. Applequist, D.F. O'Brien, *J. Am. Chem. Soc.* **1963**, *85*, 744.
[13]   R.E. Lyle, E.J. DeWitt, N.M. Nichols, W. Cleland, *J. Am.Chem. Soc.* **1953**, *75*, 5959.
[14]   J.F. Wolfe, B.H. Loo, F.E. Arnold, *Macromolecules* **1981**, *14*, 915.
[15]   B.H. Zimm, R.W. Kilb, *J. Polym. Sci.* **1959**, *37*, 19.
[16]   W.H. Stockmayer, M. Fixman, *Ann. N.Y. Acad. Sci.* **1953**, *57*, 334.
[17]   J.E. Roovers, S.Bywater, *Macromolecules* **1972**, *5*, 384.
[18]   H. Gilman, F. Cartledge, *J. Organometal. Chem.* **1964**, *2*, 447.
[19]   A. Lebreton, V. Héroguez, Y. Gnanou, in progress.

Macromol. Symp. **2004**, *215*, 51-56

Fluorinated Amphiphilic Block Copolymers: Combining Anionic Polymerization and Selective Polymer Modification

Marc A. Hillmyer,[1] *Nathan W. Schmuhl,*[2] *Timothy P. Lodge*[1,2]

[1]Department of Chemistry, University of Minnesota, Minneapolis, MN 55455, USA
E-mail: hillmyer@chem.umn.edu; lodge@chem.umn.edu
[2]Department of Chemical Engineering and Materials Science, University of Minnesota, Minneapolis, MN 55455, USA

Summary: We report the preparation of novel fluorinated block copolymers using a two-step modification sequence. We first prepared model polyisoprene-poly-*tert*-butylmethacrylate block copolymers by anionic polymerization. Exposing these materials to difluorocarbene (generated by the thermolysis of hexafluoropropylene oxide) resulted in modification of the polyisoprene block to the corresponding difluorocyclopropane repeating unit without compromising the integrity of the poly-*tert*-butylmethacrylate block. Hydrolysis of the difluorocarbene-modified materials gave the corresponding difluorocarbene-modified polyisoprene-polymethacrylic acid diblock copolymers. These amphiphilic materials are expected to exhibit interesting self-assembly behavior in aqueous solution.

Keywords: anionic polymerization; block copolymers; fluoropolymers; modification; self-assembly

Introduction

The preparation of block copolymers containing a fluorinated block can be accomplished using a variety of synthetic strategies. These include sequential addition of monomers using a controlled/living polymerization scheme,[1] coupling of an end-functionalized fluorinated polymer with another end-functionalized material, and selective post-polymerization fluorination[2] of one segment of a parent block copolymer. Although all these methods have advantages, the last approach is particularly useful since, in principle, the level of fluorination can be controlled using either reaction time or stoichiometry. This allows for the systematic incorporation of fluorine, thereby generating block copolymer materials with tunable properties. The unique properties of fluorinated polymers motivate the preparation of such hybrid materials. Biocompatibility, low surface energy, unique solubility characteristics, and general incompatibility with most other polymers are a subset of the distinctive features of fluoropolymers that make their study particularly interesting. The combination of fluoropolymers with water-soluble macromolecules in one hybrid material (i.e., a block copolymer) yields "highly amphiphilic" materials that can

    DOI: 10.1002/masy.200451105

self-assemble in aqueous solution to give micellar structures at very low concentration (i.e., very low critical micelle concentrations) and can be used as dispersants, drug delivery vehicles, or viscosity modifiers. The preparation of such amphiphilic materials is described in this paper. We have taken a synthetic approach that involves the preparation of well-defined polyisoprene-poly-*tert*-butylmethacrylate diblock copolymers by sequential anionic polymerization.[3] Using a polydiene modification scheme that incorporates difluorocarbene in the polyisoprene backbone,[4] we have prepared the difluorocarbene-modified variants without significant degradation. Hydrolysis of the resultant fluorinated block copolymers gives the final amphiphilic materials. This work demonstrates that poly-*tert*-butylmethacrylate blocks are inert under the reaction conditions used for the difluorocarbene addition to polyisoprene, and that the hydrolysis of the poly-*tert*-butylmethacrylate segments can be accomplished without compromising the structural integrity of the difluorocyclopropane repeat unit in the difluorocarbene-modified polyisoprene block. The modification procedure is shown in Figure 1.

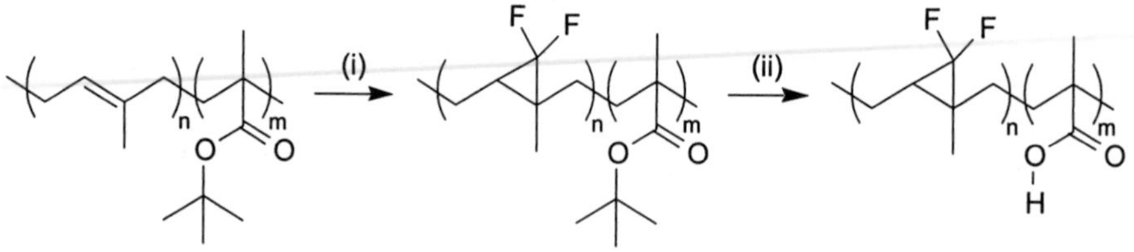

Figure 1. Sequential modification of polyisoprene-poly-*tert*-butylmethacrylate block copolymers. (i) Hexafluoropropylene oxide, 180 °C. (ii) Methanesulfonic acid / acetic acid, 100 °C.

Polyisoprene-poly-*tert*-butylmethacrylate block copolymers

The general synthetic protocol for the polyisoprene-poly-*tert*-butyl methacrylate (PI-PtBMA) diblock copolymers is as follows. The PI blocks were synthesized in cylcohexane (to obtain the desired 4,1 microstructure) at 40 °C using *sec*-butyl lithium as the initiator. A typical concentration of the polyisoprenyllithium chain ends was approximately 5 mmol·L^{-1}. After allowing 4 h for isoprene polymerization, an equal volume solution of LiCl predissolved in THF (typically having a concentration of 1.5 g of LiCl per liter of THF) was added to the reaction. As the THF solution was added, the temperature was slowly reduced to near −78 °C. The solution turned lemon yellow, indicating living polyisoprenyl anions in a polar solution. Once all of the THF solution was added and the reaction temperature was below −70 °C, the *tert*-butyl methacrylate was added. This resulted in a colorless solution, which was allowed to react for an additional 4 h. Termination was carried out by the addition of degassed methanol at the reaction temperature. Higher molecular weight PI-PtBMA polymers and those with large percentages of PtBMA were precipitated in deionized water. Precipitation in water removed some of the salt added in the reaction. The lower molecular weight polymers with a higher fraction of PI were concentrated, re-dissolved in hexane and filtered. The resulting solutions were again concentrated, diluted with THF and cast to dry.[5]

Table 1 summarizes the molecular characteristics for the PI-PtBMA block copolymers. The PI volume fractions (f_{PI}) were calculated using densities of PI = 0.91[3] and PtBMA = 1.022[6] $g \cdot cm^{-3}$ at 25 °C and the molar ratios from ^{1}H-NMR spectra. The molecular weights for these polymers were determined by SEC, both in comparison to polystyrene standards and using multi-angle light scattering (MALS) detection. The polydispersity indices (M_w/M_n) for the PI-PtBMA block copolymers were based on SEC using polystyrene standards. In all cases, polydispersities were less than 1.1.

Table 1. Molecular characteristics PI-PtBMA diblock copolymers.

Sample code	f_{PI}	M_n (kg·mol^{-1})^a	M_n (kg·mol^{-1})^b	M_n (kg·mol^{-1})^c	M_w/M_n^c
PI-PtBMA 9-2	0.74	10	10	29	1.04
PI-PtBMA 7-4	0.51	11	11	27	1.04
PI-PtBMA 4-7	0.25	11	10	28	1.05
PI-PtBMA 26-4	0.77	31	30	55	1.07
PI-PtBMA 22-10	0.55	32	29	59	1.03
PI-PtBMA 12-20	0.25	32	31	66	1.05

(a) Calculated based on quantitative monomer conversion and reaction stoichiometry.
(b) Determined by SEC with MALS detector using dn/dc calculated from dn/dc$_{PI}$ = 0.124 and dn/dc$_{PtBMA}$ = 0.079 mL·g^{-1} in THF.
(c) Determined by SEC with polystyrene standards.

Fluorination of PI-PtBMA Block Copolymers with Difluorocarbene

Cyclohexane solutions of the PI-PtBMA block copolymers were subjected to hexafluoropropylene oxide (HFPO) at 180 °C using a high-pressure reactor.[4] The thermolysis of HFPO generates difluorocarbene[7] and this adds selectively to the double bonds in the PI backbone. Prolonged reaction times were avoided since the PtBMA was somewhat sensitive to degradation at long reaction times. Additionally, 5 wt% 2,6-di-tert-butyl-4-methylphenol (BHT), a radical scavenger, relative to the weight of the PI block was also added to the mixture to prevent crosslinking. Reaction times necessary for complete conversion of the PI double bonds were determined in a separate study.[8] After the reaction the polymer solution was filtered, concentrated to dryness, redissolved in THF, and reverse precipitated in a 10:1 mixture of water/methanol. The precipitated polymer was collected and dried first at room temperature and then at 100 °C under vacuum < 100 mTorr for 48 h.

Using a combination of ^{1}H and ^{19}F NMR spectroscopies we determined that the PI block was selectively CF$_2$-modified to give FPI-PtBMA. The molecular characteristics of the difluorocarbene-modified FPI-PtBMA diblock copolymers are summarized in Table 2.[9] Molecular weights and FPI volume fractions (f_{FPI}) were calculated using the precursor PI-PtBMA molecular weights, assuming no degradation or side-reactions, and using RT densities (FPI = 1.26[10] and PtBMA = 1.022[6] g·cm^{-3}). Molecular weights were also determined with SEC using polystyrene standards and MALS and molecular weight distributions were calculated using polystyrene standards. Percent conversions of the polyisoprene double bonds were measured from the ^{1}H NMR spectra of the FPI-PtBMA block copolymers, and in all cases the conversions were > 99%.

These results presented in Table 2 are noteworthy since they are the first reported examples of the successful use of difluorocarbene for the modification of polydienes in the presence of methacrylates; the analogous difluorocarbene modification of polydienes in the presence of polystyrene blocks has been documented previously.[10] All polymers retained their narrow molecular weight distributions after CF$_2$ modification, with good agreement between the predicted molecular weights and those determined by using SEC/MALS. ^{1}H NMR spectra for these polymers indicated complete conversion of the PI blocks to FPI with little to no degradation of the PtBMA blocks. The volume fractions of the PtBMA blocks are significantly lower after fluorination due to the significant increase in the molecular weight of the FPI block from the addition of the two fluorine atoms and one carbon atom to each repeat unit.

Table 2. Molecular characteristics of the FPI-PtBMA diblock copolymers.

Sample code	f_{FPI}	M_n (kg·mol^{-1})^a	M_n (kg·mol^{-1})^b	M_w/M_n^{b}	Precursor
FPI-PtBMA 15-2	0.93	17	25	1.06	PI-PtBMA 9-2
FPI-PtBMA 12-4	0.83	19	30	1.06	PI-PtBMA 7-4
FPI-PtBMA 7-7	0.62	16	21	1.05	PI-PtBMA 4-7
FPI-PtBMA 46-4	0.94	47	58	1.07	PI-PtBMA 26-4
FPI-PtBMA 38-10	0.85	47	94	1.06	PI-PtBMA 22-10

(a) Determined by SEC with MALS detection using dn/dc calculated with dn/dc$_{FPI}$ = 0.04 and dn/dc$_{PtBMA}$ = 0.079 mL·g^{-1} in THF.
(b) Determined by SEC with polystyrene standards.

Hydrolysis of the FPI-PtBMA Block Copolymers

Hydrolysis reactions were performed in toluene using methanesulfonic acid (MSA) and acetic acid (AA) in the presence of BHT at 100 °C for 6h.[11] After the reaction, the solution was cooled to RT, concentrated to dryness, redissolved in THF, and precipitated in a 10:1 mixture of water/methanol. The precipitated polymer was collected, washed thoroughly with water, and dried first at room temperature under argon and then at 100 °C under reduced pressure (< 100 mTorr) for 48 h.

The results for the conversion of FPI-PtBMA to the corresponding polymethacrylic acid block copolymers (FPI-PMAA) are given in Table 3. While dramatic discrepancies were seen between predicted and measured M_n values for the FPI-PMAA 15-1 and 12-2 samples, FPI-PMAA 46-3 and 38-6 both exhibited M_n values fairly close to the predicted values based on complete conversion and no degradation. One possible reason for the discrepancy in the predicted and measured molecular weights for the FPI-PMAA 15-1 and 12-2 samples is some aggregation of the these materials in THF given their higher PMAA contents. However, the relatively narrow molecular weight distributions of these materials is not entirely consistent. Generally, molecular weight distributions were not observed to increase dramatically, except in the case of FPI-PMAA 12-2.

Table 3. Molecular characteristics of FPI-PMAA diblock copolymers.

Sample code	M_n (kg·mol^{-1})^a	M_n (kg·mol^{-1})^b	M_n (kg·mol^{-1})^c	$M_w/M_n{}^c$	Precursor
FPI-PMAA 15-1	16	80	53	1.04	FPI-PtBMA 15-2
FPI-PMAA 12-2	14	160	20	1.28	FPI-PtBMA 12-4
FPI-PMAA 46-3	49	48	57	1.07	FPI-PtBMA 46-4
FPI-PMAA 38-6	44	52	96	1.11	FPI-PtBMA 38-10

(a) Calculated value assuming 100% conversion and no FPI degradation or side reactions.
(b) Determined by SEC with MALS detector using dn/dc calculated with dn/dc$_{FPI}$ = 0.04 and dn/dc$_{PMAA}$ = 0.084 mL·g^{-1} in THF.
(c) Determined by SEC with polystyrene standards.

Conclusion

We have shown that model PI-PtBMA block copolymers prepared by sequential anionic polymerization can be selectively modified first by addition of difluorocarbene to the PI block and second by hydrolysis of the *tert*-butyl ester moieties in the PtBMA block. The self-assembly of these molecules in aqueous solution will be explored. These block copolymers have the potential to form interesting structures in water based on the high incompatibility of the fluorinated and water-soluble blocks.[12]

Acknowledgement

This work was supported primarily by the MRSEC Program of the National Science Foundation under Award Number DMR-980936.

[1] Jankova, K.; Hvilsted, S. *Macromolecules* **2003**, *36*, 1753.
[2] Reisinger, J. J.; Hillmyer, M. A. *Prog. Polym. Sci.* **2002**, *27*, 971.
[3] For related work on PI-PtBMA block copolymers see: Pochan, D. J.; Gido, S. P.; Zhou, J.; Mays, J. W.; Whitmore, M.; Ryan, A. J. *J. Polym. Sci. B, Polym. Phys.* **1997**, *35*, 2629.
[4] Ren, Y.; Lodge, T. P.; Hillmyer, M. A. *J. Am. Chem. Soc.* **1998**, *120*, 6830.
[5] The PI-PtBMA materials were contaminated with high molecular species as determined by SEC with a multi-angle light scattering detection. This material could be removed by simple filtration of a dilute heptane solution of these materials through a fine mesh ceramic filter. This procedure also removed residual lithium salts from the synthesis. See reference [8].
[6] *Aldrich Handbook* **2000**.
[7] Millauer, H.; Schwertferger, W.; Siegemund, G. *Angew. Chem. Int. Ed. Engl.* **1985**, *24*, 161.
[8] Schmuhl, N. W. "Fluorinated Amphiphilic Polymers by Sequential Modification of Polyisoprene-*b*-Poly(tert-butyl methacrylate) Block Copolymers" M.S. Thesis, University of Minnesota, 2002.
[9] Not performed on PI-PtBMA 12-20 due to lack of solubility in cyclohexane.
[10] Ren, Y.; Lodge, T. P.; Hillmyer, M. A. *Macromolecules* **2000**, *33*, 866.
[11] (a) DePorter, C. D.; Ferrence, G. M.; McGrath, J. E. *Polym. Prepr. (Am. Chem. Soc., Div. Polym. Chem.)* **1988**, *29(1)*, 343. (b) Yu, Y.; Zhang, L.; Eisenberg, A. *Langmuir* **1997**, *13*, 2578.
[12] Hillmyer, M. A.; Lodge, T. P. *J. Polym. Sci. Polym. Chem.* **2002**, *40*, 1.

Precise Synthesis of Star-Branched Polymers by Means of Living Anionic Polymerization Using 1,1-Bis(3-chloromethylphenyl)ethylene

Akira Hirao,* Tomoya Higashihara

Polymeric and Organic Materials Department, Graduate School of Science and Engineering, Tokyo Institute of Technology, 2-12-1, Ohokayama, Meguro-ku, Tokyo, 152-8552, Japan
Fax: +81-3-5734-2887; E-mail: ahirao@polymer.titech.ac.jp

Summary: The syntheses of three- and four-arm star-branched polymers by a new convenient methodology using 1,1-bis(3-chloromethylphenyl)ethylene are described. The methodology involves only two sets of reactions: an addition reaction of living anionic polymer to 1,1-bis(3-chloromethylphenyl)ethylene and a linking reaction of the intermediate polymer anion with ω-4-bromobutyl-functionalized polymer. All combinations of well-defined three- and four-arm stars comprised of polystyrene and polyisoprene segments were successfully synthesized by this methodology.

Keywords: anionic polymerization; benzyl halide; 1,1-diphenylethylene; star polymers; well-defined architectures

Introduction

In recent years, we have developed new methodologies for the synthesis of regular and, in particular, asymmetric star-branched polymers based on living anionic polymerization in combination with functionalized 1,1-diphenylethylene (DPE) derivatives.[1-12] The first example is 1,1-bis(3-*tert*-butyldimethylsilyloxymethylphenyl)ethylene (**1**).[9] Its *tert*-butyldimethylsilyloxymethylphenyl group is stable toward carbanionic species such as organolithiums and living anionic polymers of styrene and 1,3-diene monomers and quantitatively transformable into highly reactive benzyl halides. By reacting living anionic polymers with **1**, followed by transformation to benzyl halides, novel chain-functionalized polymers with a definite number of benzyl halide moieties have been successfully synthesized. A variety of star-branched polymers with well-defined structures could be synthesized by coupling such benzyl halide-chain-functionalized polymers with living anionic polymers. The second DPE derivative useful for the synthesis of star-branched polymer is 1-[4-(4-

DOI: 10.1002/masy.200451106

bromobutyl)phenyl]-1-phenylethylene (**2**).[6,10] The DPE function is readily introduced either at chain-ends or in chains by the reaction of **2** with living anionic polymers and polymer anions. The resulting DPE-chain-functionalized polymers have been utilized as precursor polymers to synthesize star-branched polymers. Furthermore, we have demonstrated that regular as well as asymmetric star-branched polymers can be successively synthesized by repeating the reaction of DPE-chain-functionalized polymer with living anionic polymer and the subsequent introduction of DPE function by the reaction of **2** with the generated anion as illustrated in Scheme 1.

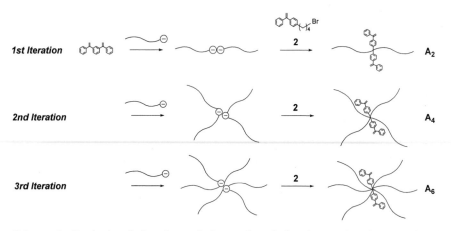

Scheme 1. Synthesis of A_2, A_4, and A_6 star-branched polymers based on an iterative methodology by using **2**.

Herein, we report on a new convenient methodology based on living anionic polymerization in combination with a new DPE derivative, 1,1-bis(3-chloromethylphenyl)ethylene (**3**), developed for the synthesis of three- and four-arm star-branched polymers comprised of two kinds of polymer segments.

Experimental Section

Materials. 1,1-Bis(3-chloromethylphenyl)ethylene (**3**) was synthesized according to the procedure previously reported by us.[13] All chemicals and solvents (> 98% from Aldrich, Japan) were purified as described elsewhere or used as received. ω-4-Bromobutyl-functionalized polystyrene and polyisoprene (designated as PS-Br and PI-Br) were synthesized by the procedure previously reported by us.[14]

Measurements. Both ^{1}H and ^{13}C NMR spectra were measured on a Bruker DPX300 in CDCl$_3$. Size-exclusion chromatograms (SEC) were generated in THF with a TOSOH HLC-8020 at 40 °C with UV (254 nm) or refractive index detection. Fractionation by SEC was performed at 40 °C using a TOSOH HLC-8020 type fully automatic instrument. The measurements for static light scattering (SLS) were performed with an Ohtsuka Electronics SLS-600R instrument equipped with a He-Ne laser (633 nm) in THF at 25 °C. Vapor pressure osmometry (VPO) measurements were made with a Corona 117 instruments in benzene at 40 °C with a highly sensitive thermoelectric couple and with equipment of very exact temperature control.

Procedures. Anionic polymerizations and reactions were carried out under high vacuum conditions (10^{-6} torr) using the usual break-seal technique. Polystyryllithium (PSLi) and polyisoprenyllithium (PILi) were prepared by the living anionic polymerization of styrene with *sec*-BuLi in THF at –78 °C for 0.5 h or isoprene with *sec*-BuLi in heptane at 40 °C for 2 h. In certain cases, PSLi and PILi were end-capped with a 1.5-fold excess of DPE in THF at –78 °C for 0.5 h and 12 h, respectively. They have been designated as PSD'Li and PID'Li. The reaction of living polymer with **3** was usually carried out in THF at –78 °C for 1 ~ 24 h. The resulting intermediate polymer anion was *in-situ* linked with ω-4-bromobutyl-functionalized polymer in THF at –78 °C for 12 h. Prior to the linking reaction, dibutylmagnesium (*ca.* 2 mol-%) was added to kill impurities in the ω-functionalized polymer. A 1.2 -fold excess of PSLi toward each reaction site of **3** was used in the reaction. A 1.2-fold excess of PS-Br or PI-Br was used toward total PSLi in the linking reaction. Each arm segment of star-branched polymers was designed to be *ca.*10 kg/mol in molecular weight. The resulting star-branched polymers were precipitated in methanol and isolated in more than 90% yields by fractional precipitation using cyclohexane-hexanes or fractionation with SEC. They were purified by twice reprecipitating and freeze-drying from their dry benzene solutions for 24 h.

Results and Discussion

Synthesis of Four-Arm Star-Branched Polymers. As illustrated in Scheme 2, we have developed a new convenient methodology using **3** for the synthesis of four-arm star-branched

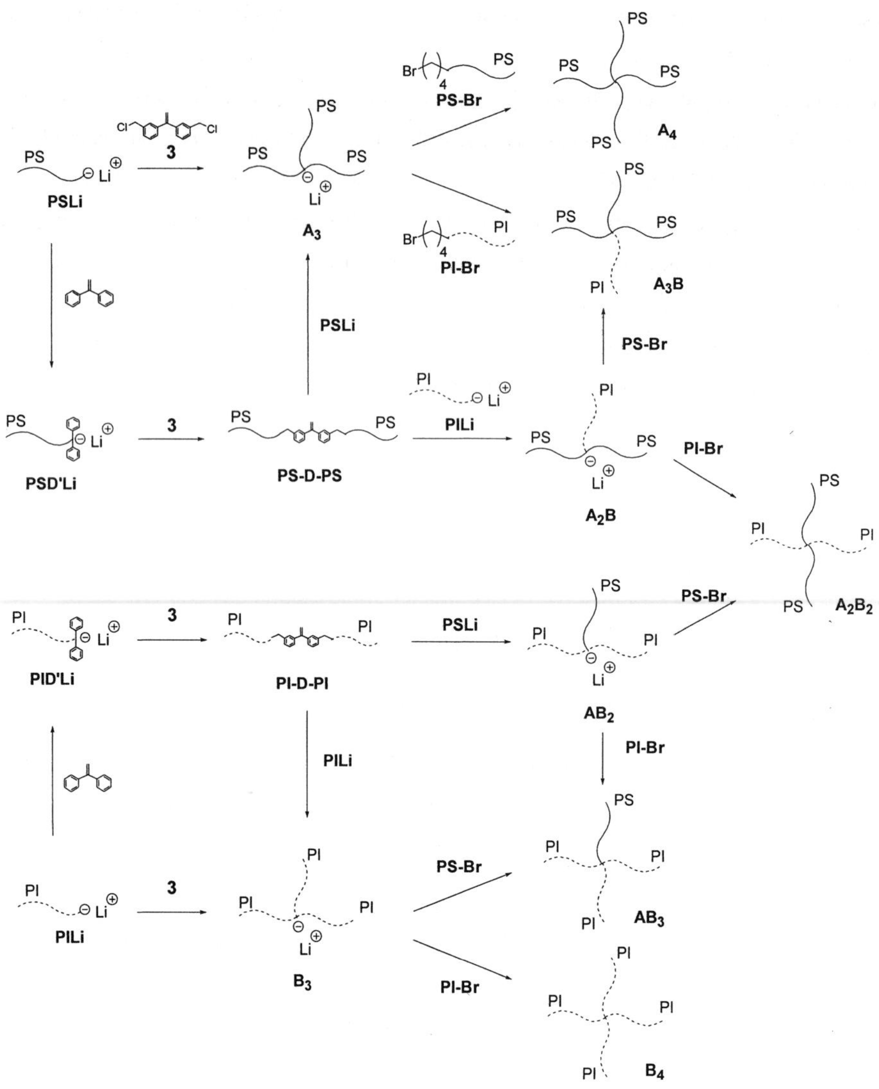

Scheme 2. Convenient synthesis of star-branched polymers by using **3**.

polymers comprised of polystyrene (PS) and polyisoprene (PI) segments. Either PSLi or PSD'Li was reacted with **3** in THF at –78 °C in order to examine the reactivity of **3**. PSLi underwent reactions with both the chloromethyl and vinylene groups of **3** to afford an intermediate three-armed star-branched polymer anion, while PSD'Li reacted only with the chloromethyl group of **3** as expected. In-chain-functionalized PS with a DPE moiety, PS-D-PS, was quantitatively obtained in the latter reaction. Thus, the differentiation between these two reaction sites of **3** is possible by using PSLi and PSD'Li.

A four-arm A_4 regular star-branched polystyrene was synthesized by a one-pot reaction involving the reaction of three equivalents of PSLi with **3** followed by the *in-situ* linking with PS-Br. Similarly, an asymmetric A_3B star was obtained by the use of PI-Br instead of PS-Br in the *in-situ* linking reaction. The SEC trace of the reaction mixture in the A_4 star synthesis shown in Figure 1(A) consisted of three peaks. They appear to correspond the objective A_4 star-branched polymer, a coupled product from PS-Br and the unreacted PSLi, and the unreacted PS-Br used in excess.

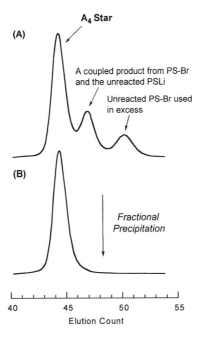

Figure 1. SEC RI traces of the polymers obtained by the addition reaction of three equivalents of PSLi with **3** followed by the in-situ linking with PS-Br. (A): before fractionation, (B): after fractionation.

A very similar SEC profile was also observed in the synthesis of an A_3B star. Overall yields of the star-branched polymers were estimated to be *ca.* 95% by comparing SEC peak areas (UV detector). These stars were isolated in *ca.* 90% yields using fractionation (see Figure 1(B)) and characterized by ^{1}H and ^{13}C NMR, FT-IR, SEC, and SLS. The results are summarized in Table 1.

Table 1. Synthesis of A_4, A_3B, A_2B_2, AB_3, and B_4 Star-Branched Polymers.

| type | M_n (kg/mol) | | | | M_w (kg/mol) | | | M_w/M_n | Composition (wt-%) PS/PI | |
	calcd	SEC[a]	^{1}H NMR	VPO	Calcd[b]	SLS[c]	dn/dc[c] (mL/g)	SEC[a]	calcd	^{1}H NMR
A_4	43.3	35.4	42.5	41.3	44.6	46.8	0.188	1.03	-	-
A_3B	43.5	37.9	43.0	43.3	45.7	44.2	0.170	1.05	79/21	81/19
A_2B_2	40.8	42.8	40.0	41.2	41.6	42.8	0.162	1.02	50/50	51/49
AB_3	40.8	44.1	41.0	40.0	42.4	42.3	0.144	1.04	24/76	26/74
B_4	42.1	38.5[d]	42.7	41.9	43.4	44.0	0.137	1.03	-	-

[a] Estimated from the caribration using polystyrene standards.
[b] Calculated from M_n (calcd) and M_w/M_n (SEC).
[c] Measured in THF at 25 °C.
[d] Estimated from the caribration using polyisoprene standards.

Both A_4 and A_3B stars exhibited sharp monomodal SEC distributions, the M_w/M_n values being 1.03 and 1.05, respectively. Their molecular weights determined by VPO (M_n) and SLS (M_w) were in good agreement with those predicted. The expected structure of the A_3B star was also confirmed by agreement of the [A]/[B] ratio measured by ^{1}H NMR with that calculated. Thus, two architecturally different stars could be synthesized from the same intermediate A_3 star-branched polymer anion. These stars were also obtained in more than 95% yields by the reaction of PSLi with PS-D-PS, followed by *in-situ* linking with PS-Br and PI-Br, respectively.

In order to synthesize A_2B_2 asymmetric star-branched polymer, PILi was reacted with PS-D-PS, followed by *in-situ* linking with PI-Br. The yield of the objective A_2B_2 star was however only 25% yield even after 168 h because the reaction of PILi with the DPE moiety of PS-D-PS proceeded very sluggishly in THF at −78 °C. On the other hand, PSLi reacted virtually quantitatively with PS-D-PS after 24 h under the same conditions as mentioned above. Thus, the reactivity difference between PSLi and PILi toward the DPE moiety of PS-D-PS is evident.

We have succeeded in synthesizing A_2B_2 star-branched polymer by changing the reaction route. At first, an in-chain-functionalized polyisoprene with DPE moiety, PI-D-PI, was prepared by the reaction of **3** with PID'Li. PSLi efficiently reacted with the PI-D-PI to afford an intermediate AB_2 star-branched polymer anion. By *in-situ* linking of the intermediate polymer anion with PS-Br, an A_2B_2 asymmetric star-branched polymer was successfully obtained in 97% yield. With use of PI-Br instead of PS-Br in the *in-situ* reaction, an AB_3 star could be obtained in 99% yield. As was seen in Table 1, they were well defined in architecture and precisely controlled in chain length.

Again, difficulty arose in the synthesis of B_4 regular star-branched polyisoprene due to the extremely low reactivity of PILi toward PI-D-PI. The reaction was therefore examined in more detail to increase the yield under various conditions: in THF at -78 °C for 14 days and in either heptane or *tert*-butylbenzene in the absence and presence of a small amount of THF at 30 °C for $24 \sim 48$ h. Unfortunately, the yields were always low ($< 25\%$). There was no further increase in reaction yield by raising the reaction temperature from -78 °C to -40 °C followed by -20 °C in THF. Instead, undesirable high molecular weight polymers were significantly produced presumably by the attack of PILi on the vinyl side chain of PI. Thus, all attempts were not successful to improve the yield in the reaction of PILi with PI-D-PI.

Surprisingly, PILi reacted efficiently with **3** to afford an intermediate B_3 star-branched polymer anion in *ca.* 85% yield in THF at -78 °C for 24 h. This suggests that PILi reacts predominately with the vinylene group of **3** prior to the reaction of PILi with the two chloromethyl groups that yields less reactive PI-D-PI. A regular B_4 star-branched polyisoprene was obtained by the reaction of PILi with **3** followed by the *in-situ* reaction with PI-Br.[15] Thus, all combinations of well-defined four-arm stars comprising of PS (A) and PI (B) segments could be successfully synthesized by developing the methodology using **3**.

Synthesis of Three-Arm Star-Branched Polymers. Three-arm star-branched polymers were synthesized only by reacting properly living anionic polymers or the DPE-end-capped polymer anion with **3**. For example, a regular A_3 or B_3 star-branched polymer was readily synthesized by the reaction of PSLi or PILi with **3**. The same A_3 star could also be obtained by the reaction of two equivalents of PSD'Li with **3**, followed by reacting with PSLi. Similarly, an AB_2 star was synthesized by reacting successively two equivalents of PID'Li and PSLi with **3**. These results are listed in Table 2.

Table 2. Synthesis of A$_3$, A$_2$B, and B$_3$ Star-Branched Polymers.

	M_n (kg/mol)				M_w (kg/mol)			M_w/M_n	Composition (wt-%) PS/PI	
type	calcd	SEC^a	^{1}H NMR	VPO	Calcd^b	SLS^c	dn/dc^c (mL/g)	SEC	calcd	^{1}H NMR
A$_3$	32.6	31.1	32.0	31.5	33.6	34.1	0.187	1.03	-	-
A$_2$B	30.5	29.7	31.2	30.0	31.4	31.7	0.170	1.03	67/33	69/31
B$_3$	31.1	28.4^d	31.7	30.9	31.7	32.5	0.138	1.02	-	-

^{a} Estimated from the caribration using polystyrene standards.
^{b} Calculated from M_n (calcd) and M_w/M_n (SEC).
^{c} Measured in THF at 25 °C.
^{d} Estimated from the caribration using polyisoprene standards.

Unfortunately, an A$_2$B star was synthesized only in 25% yield due to the low reactivity of PILi toward PS-D-PS prepared from PSD'Li and **3**. We have therefore proposed an alternative procedure for the synthesis of A$_2$B star-branched polymer as illustrated in Scheme 3. An A$_2$B star could be synthesized in 90% yield by this procedure involving the addition reaction of *sec*-BuLi with PS-D-PS, followed by *in-situ* linking with PI-Br.

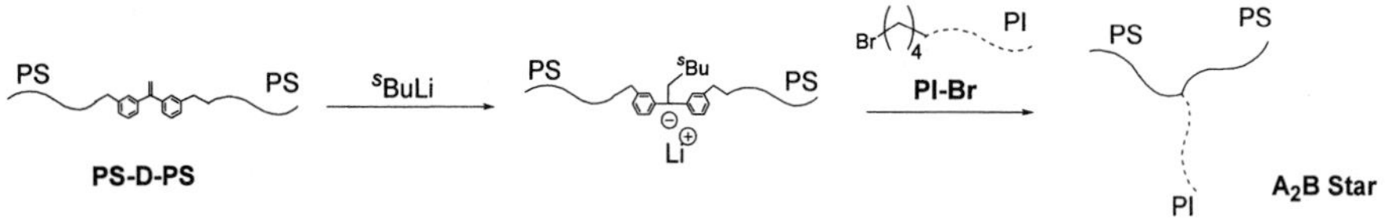

Scheme 3. Synthesis of A$_2$B star-branched polymer.

Conclusions

A simple and convenient methodology for the synthesis of well-defined three- and four-arm star-branched polymers comprising of PS and PI segments has been developed by using living anionic polymers and a new functional DPE derivative, **3**, as a core compound. All possible star-branched polymers, A$_4$, A$_3$B, A$_2$B$_2$, AB$_3$, B$_4$, A$_3$, A$_2$B, AB$_2$, and B$_3$ could be successfully synthesized. However, one limitation of this methodology is the relative lack of reactivity of PILi toward both PS-D-PS and PI-D-PI. Reaction routes and yields of certain star-branched polymers were indeed influenced by the limitation. Therefore, in order to establish the present methodology using **3** as a general procedure, further experiments should be needed by changing B segment from PILi to another living anionic polymer.

In this study, arm segments of all star-branched polymers synthesized herein were *ca.* 10 kg/mol in molecular weight. If molecular weight of the arm segment is different, a series of asymmetric star-branched polymers whose arms differ in molecular weight such as A_3A', $A_2A'_2$, and even $A_2A'A''$ may possibly be synthesized. In the same sense, the synthesis of four-arm star-branched polymers comprised of three kinds (A, B, and C) of polymer segments is now under investigation.

[1] M. Hayashi, K. Kojima, A. Hirao, *Macromolecules* **1999**, *32*, 2425.
[2] M. Hayashi, Y. Negishi, A. Hirao, *Proc. Japan Acad., Ser B* **1999**, *75*, 93.
[3] A. Hirao, M. Hayashi, N. Haraguchi, *Macromol. Rapid Commun.* **2000**, *21*, 1171.
[4] A. Hirao, M. Hayashi, Y. Tokuda, *Macromol. Chem. Phys.* **2001**, *202*, 1606.
[5] A. Hirao, A. Matsuo, K. Morifuju, Y. Tokuda, M. Hayashi, *Polym. Adv. Technol.* **2001**, *12*, 680.
[6] A. Hirao, M. Hayashi, T. Higashihara, *Macromol. Chem. Phys.* **2001**, *202*, 3165.
[7] T. Higashihara, M. Hayashi, A. Hirao, *Macromol. Chem. Phys.* **2002**, *203*, 166.
[8] A. Hirao, M. Hayashi, Y. Tokuda, N. Haraguchi, T. Higashihara, S. W. Ryu, *Polym. J.* **2002**, *34*, 633.
[9] A. Hirao, N. Haraguchi, *Macromolecules* **2002**, *35*, 7224.
[10] A. Hirao, T. Higashihara, *Macromolecules* **2002**, *35*, 7238.
[11] A. Hirao, Y. Tokuda, *Macromolecules* **2003**, *36*, 6081.
[12] T. Higashihara, M. Kitamura, N. Haraguchi, K. Sugiyama, A. Hirao, J. H. Ahn, J. S. Lee, *Macromolecules* **2003**, *36*, 6730.
[13] M. Hayashi, S. Loykulnant, A. Hirao, S. Nakahama, *Macromolecules* **1998**, *31*, 2057.
[14] A. Hirao, M. Tohoyama, S. Nakahama, *Macromolecules* **1997**, *30*, 3484.
[15] In addition to a B_3 star-branched polymer anion, PI-D-PI ($\sim$ 10%) and high molecular weight polymers ($\sim$ 5%) were also produced in the reaction of PILi with **3**. Therefore, the resulting regular B_4 star-branched polyisoprene was carefully isolated in *ca.* 60% yield by SEC fractionation.

Macromol. Symp. **2004**, *215*, 67-79

Synthesis and Properties of Macrocyclic Vinylaromatic Polymers Containing a Single 1,4-Benzylidene or 9,10-Anthracenylidene Group

Rong Chen, Gennadi G. Nossarev, Thieo E. Hogen-Esch*

Loker Hydrocarbon Institute and Department of Chemistry, University of Southern California, Los-Angeles, California 90089-1661, USA

Summary: The synthesis is reported of well-defined and narrow molecular weight distribution macrocyclic poly(2-vinylnaphthalene) (P2VN) and poly(9,9-dimethyl-2-vinylfluorene) (PDMVF) containing a single 1,4-benzylidene or 9,10-anthracenylidene unit. The synthesis involves the potassium naphthalide (K-Naph) initiated polymerization of 2VN or DMVF in THF at -78 °C followed by end-to-end coupling of the resulting P2VN dianions under high dilution conditions (10^{-6}–10^{-4} M) with 1,4-bis(bromomethyl)benzene (DBX) or 9,10-bis(chloromethyl)anthracene (BCMA). The molecular characterization was carried out by size exclusion chromatography (SEC), NMR and MALDI-TOF.

Keywords: anionic polymerization; anthracene; macrocyclic polymers; polyvinylfluorene; polyvinylnaphthalene

Introduction

There has been a revival of interest in recent years in the synthesis of macrocyclic vinyl aromatic polymers.[1-17] We have successfully synthesized and characterized macrocyclic and matching linear polystyrene (PS),[11,12] poly(2-vinylpyridine) (P2VP),[3,12] poly(α-methylstyrene) (PAMS)[13] and poly(2-vinylnaphthalene) (P2VN),[14] and vinylaromatic block copolymers[15-17] by end-to-end cyclization of the corresponding dianions with bifunctional electrophiles, such as 1,4-bis(bromomethyl)benzene (DBX) and dibromomethane in highly dilute solutions (10^{-4}-10^{-6} M). Vinyl aromatic polymers containing 2-naphthyl or 2-fluorenyl pendent groups have interesting photoluminescent properties and good chemical and thermal stabilities.[12,14,18-21] The molar absorptivity of fluorene in the near UV region is in the

                DOI: 10.1002/masy.200451107

order of 10^4 and its fluorescence quantum efficiency is relatively high (about.80 percent).[20] Such polymers or copolymers are of interest in energy transfer studies in light harvesting polymers.[22-26]

Here we describe the synthesis, characterization, and properties of the potassium naphthalide mediated electron transfer initiated polymerization of 2-vinylnaphthalene (2VN) and 9,9-dimethyl-2-vinylfluorene [(DMVF) and the intramolecular cyclization of the resulting polymer dianion precursors in THF with DBX or 9,10-bis(chloromethyl)anthracene (BCMA) under high dilution conditions to give the corresponding macrocyclic P2VN and PDMVF (Scheme 1). The incorporation of anthracene in such polymers is of interest as this is a useful fluorescent acceptor probe in the study of energy migration.[25,26]

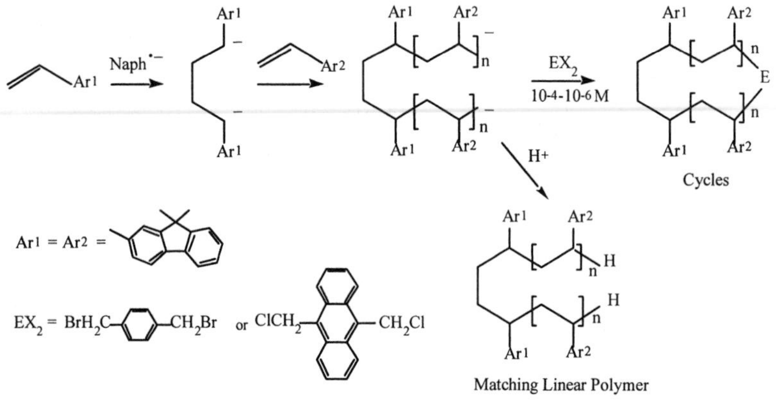

Scheme 1. Synthesis of macrocyclic and matching linear vinylaromatic polymers.

Experimental Section

Materials. The procedures for the preparation and characterization of macrocyclic vinylaromatic polymers using break-seal techniques have been reported.[11a,14] The DMVF was prepared by bis-methylation of 2-vinylfluorene[27] (2VF) at the 9-position using a modified procedure.[28] UV-Vis absorption on the P2VN samples was run on a Varian Cary 50 spectrometer with a baseline correction using 50 mg/L P2VN solutions in cyclohexane/THF (90/10 v/v) mixtures and 1 mm quartz cells. UV-Vis measurements of the PDMVF samples were carried out in cyclohexane at 5 mg/L on a Varian Cary 50 spectrometer using a 1 cm quartz cell with a baseline set using solvent. Fluorescence experiments were run in cyclohexane at 5 mg/L in a 1 cm fluorescence cell on a PTI Quanta Master TM Model C-60SE spectrofluorimeter with a 3 nm bandpass.

Results and Discussion

A number of narrow MW distribution macrocyclic P2VN and PDMVF with DP_n's from 7 to 142 were synthesized in THF at -78 °C through the potassium naphthalide initiated polymerization and coupling at anion concentrations that fluctuated typically between 10^{-4} and 10^{-6}M using 1,4-bis(bromomethyl)benzene (DBX) or 9,10-bis(chloromethyl)anthracene (BCMA) (Tables 1 and 2).

Table 1. Formation of macrocyclic P2VN by coupling of P2VN-K_2 and DBX or BCMA in THF at -78 °C[a]

#	$M_n^{calc} \cdot 10^{-3}$ (g/mole) [f]	P2VN linear [b,d]			P2VN cyclic [b,d]			
		M_p^b 10^{-3} (g/mole)	$M_n \cdot 10^{-3}$ (g/mole)	PDI	M_p^b 10^{-3} (g/mole)	$M_n \cdot 10^{-3}$ (g/mole)	PDI	<G> [c]
1	5.70	6.50	5.60	1.12	5.24	11.5	3.86	0.80
2	5.70	7.00	6.00	1.12	5.45	11.7	5.22	0.78
3	0.55	0.95	0.74	1.34	0.91	0.94	2.65	0.95
4[e]	0.85	1.10	1.00	1.18	0.98	1.15	2.70	0.89
5	1.10	1.25	1.15 (1.25)	1.16	1.07	1.15 (1.00)	1.13	0.86
6	1.50	1.80	1.60 (1.75)	1.16	1.53	1.67 (1.58)	1.12	0.85
7[e]	2.30	2.89	2.50	1.15	15.0	4.70	-	-
8	3.40	4.40	3.80 (4.05)	1.12	3.35	4.00 (2.90)	1.11	0.76
9[e]	3.70	4.43	4.00 (4.00)	1.11	3.48	4.40 (3.15)	1.13	0.79
10	11.0	14.8	13.2 (12.3)	1.16	10.6	15.0 (11.1)	1.12	0.71

a. Coupling agent was DBX except as indicated. *b.* MW's determined by SEC using polystyrene standards Apparent peak molecular weight. *c.* Ratio of apparent peak molecular weights of macrocyclic and linear P2VN. *d.* The values in parentheses are Mn's of fractionated P2VN. *e.* BCMA was used as coupling reagent. *f.* Calculated MW's

The first reaction (Table 1, #2) was carried out at a relatively high anion concentration of $5.10^{-3}M$ in order to determine the coupling efficiency (Table 1, #2, Figure 1a). The SEC of the unfractionated product shows a peak at 12 mL that corresponds to the expected high MW "polycondensation"[29-31] product (M_p=112,000) (Figure 1a). The presence of a significant fraction of the intended P2VN macrocycle at 15.5 mL with an apparent peak MW (M_p) of about 5,500 macrocyclic polymer at high anion concentrations is consistent with predictions.[32]

The apparent MW corresponding to the peak at 14.5 mL (M_p=11,200) is almost exactly double of that of the primary macrocycle, and is attributed to the macrocyclic dimer formed by one inter- and one intramolecular coupling. The linear P2VN (M_p =7000) is shown for comparison. At the typically much lower anion concentrations the SEC yields of the cycles are much higher (Table 1, # 6, Figure 1b).

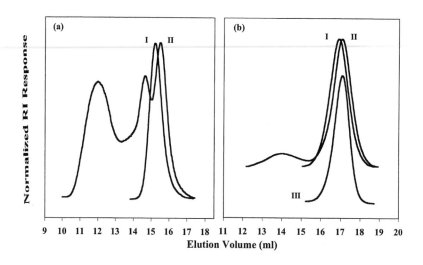

Figure 1. SEC (RI) traces of (a) linear P2VN (I) and (DBX) cyclization (Table 1, # 2) and (b) precursor P2VN (I), product of coupling with DBX under high dilution (II) (Table 1, # 6). and fractionated macrocyclic P2VN (III).

The M_n of the highest MW fraction in Figure 1a indicates a high coupling efficiency as it is 17 times that of the linear precursor giving an effective degree of step polymerization of 34. The corresponding apparent average functionality of the corresponding P2VN-K$_2$ calculated from the SEC data, at least at 5.10^{-3}M, is equal to 1.94 indicating that approximately three percent of total anion might have been inadvertently terminated.[33] Thus, the expected fraction of P2VN chains terminated at *both* chain-ends is negligible (about 0.001) at least at this concentration.

The linear analogues with the same MW were prepared by protonation of the P2VN dianion using methanol. Although the presence of a 1,4-benzylidene unit in macrocyclic PS affects its spectroscopic properties,[11a] this should not be the case for P2VN as absorption and emission of 1,4-benzylidene is negligible in this case.[20]

Table 2. Cyclization of DMVF Coupled with DBX or BCMA at −78 °C in THF[a,]

#	Mn^{calc} (g/mole)	Linear PDMVF			Cyclic PDMVF			SEC Yield[d]	<G>[e]
		Mn[b] (g/mole)	Mp[b] (g/mole)	PDI	Mn[b] (g/mole)	Mp[b,c] (g/mole)	PDI[b]		
1	2200	2620	2920	1.12	2560	2600	1.16	72%	0.91
2	3300	3930	4420	1.12	3470	3930	1.09	65%	0.89
3[f]	3530	4210	4970	1.13		4420		50%	0.89
4	6180	6560	7810	1.12	5860	6600	1.14	50%	0.84
5[g]	8820	9710	11420	1.10	8080	8930	1.12	66%	0.77
6	9240	10320	11990	1.09	7910	8820	1.09	70%	0.73
7	26440	31240	34420	1.07	20840	24670	1.08	38%	0.71

a. [Monomer] = 0.15 M; cyclization carried out over about 20 minutes at anion concentration of ~ 10^{-5}-10^{-6} M. *b.* determined by SEC using polystyrene standards.. *c.* SEC maxima of unfractionated cyclic PDMVF. *d.* estimated from SEC. *e.* <G>: see text. *f.* Cyclization carried out at [anion] =10^{-2} M. *g.* BCMA was used as coupling reagent.

The apparent functionality for the PDMVF under comparable conditions (Table 2, #3) is 1.90, indicating about 5 percent anion termination at each chain end and thus the presence of about 0.25 percent PDMVF inadvertently terminated at both ends. However the presence of somewhat higher amounts of linear impurities at the much lower anion concentrations employed during most coupling reactions can not be excluded.

The efficient formation of P2VN and PDMVF macrocycles is confirmed by decreases in the values of <G> (= M_{cyclic}/M_{linear}) with increasing DP's from 0.95 to 0.71 for P2VN[14] and from 0.91 to 0.71 for PDMVF (Tables 1 and 2). We have observed such trends for all macrocyclic polymers we have studied.[3,11-16] This may be due to the severe conformational restraints in the cycles as the number of monomer units decreases.

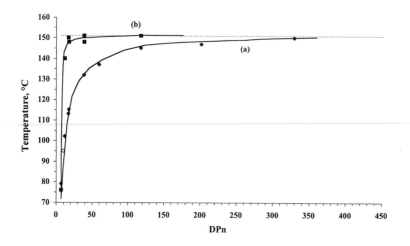

Figure 2. Relationship between glass transition temperature (T_g) and degree of polymerization (DP_n) of linear (a) and macrocyclic (b) PDMVF.

The PDMVF cyclization yields estimated from the SEC chromatograms varied from 38 to 72 percent and generally decreased with increasing DP_n, consistent with predictions.[32] For the case of the PDMVF cycles MALDI-TOF measurements of the fractionated macrocyclic PDMVF and the matching linear polymer (Table 1, # 1) showed convincing evidence for ring formation. Thus the interval between the peaks, corresponding to the molar mass of DMVF unit, is 220 Da in both cases and the masses of the cycles were 102 Da higher than the matching linear polymers due to the 1,4-benzylidene linkages.

Glass transitions. The dependence of the glass transition temperature (T_g) of linear and cyclic P2VN on DP_n is similar to that observed for PS[12] and PAMS.[13] Thus, as shown in Figure 2, the T_g's of the P2VN macrocycles above a DP_n of about 20 are equal to that of the highest MW linear P2VN (about 150 °C) and independent of MW.[14] Below a DP_n of 20, the P2VN cycles decrease sharply with decreasing MW. However, as shown in Figure 3, the glass transition temperatures of the macrocyclic PDMVF did not follow this trend. Thus, the T_g values of the cycles actually *increased*

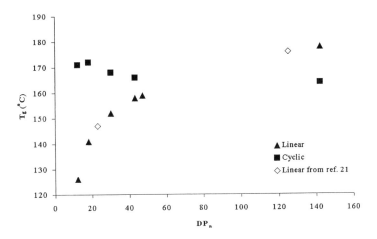

Figure 3. Relationship between glass transition temperature (T_g) and degree of polymerization (DP_n) of macrocyclic and linear PDMVF.

with decreasing DP_n and the differences with the linear PDMVF were larger than for any other vinylaromatic macrocycle (as large as 45 °C, Table 2, # 1). Furthermore, the T_g values of the linear and cyclic PDMVF's at high MW's did not converge at the highest MW, indicating that the T_g's of the cycles did not yet approach their high MW limit. These differences are tentatively attributed to the increasing rigidity of the conformationally encumbered PDMVF macrocycles (see below).

UV spectra. PDMVF exhibits a complicated UV spectrum that resembles that of free fluorene (except for a 6 nm red shift observed for the substituted fluorene), for which a relatively complete assignment has been proposed (Figure 4).[37] Compared to the model compound, both linear and cyclic polymers have lower molar absorptivities with some bands being affected more than others. Furthermore, at equal fluorenyl

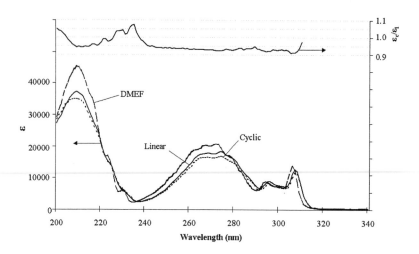

Figure 4. Extinction coefficients of cyclic (ε_C) and linear (ε_L) PDMVF (DP$_n$ = 30). and their ratio ($\varepsilon_C/\varepsilon_L$) and that of 9,9-dimethyl-2-ethylfluorene; concentration = 5 mg/L in cyclohexane.

concentrations, the cycles have lower UV absorptivities than the matching linear polymers at almost all wavelengths, except that at 229 nm and 235 nm where the absorptivities are higher. As shown more clearly in the figure (top), the ratio of extinction coefficients of cyclic and linear PDMVF ($\varepsilon_C/\varepsilon_l$) indicate a clearly structured progression of bands that correspond to distinct fluorene transitions with similar $\varepsilon_C/\varepsilon_l$ spectra being observed for each MW. The intensity increases at 229 and

235 nm for the rings relative to the linear polymers are due to "borrowing" from the other transitions.[38]

As number average degrees of polymerization (DP_n's) increase from 12 to 142, the UV absorptivities of the linear and cyclic polymers at 307 nm decrease dramatically, by 15 and 30 percent respectively, the absorptivities of the linear polymers being 4-22 percent higher than those of the matching macrocycles depending on MW (Figure 5).

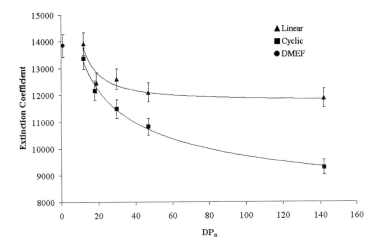

Figure 5. Extinction coefficients of 9,9-dimethyl-2-ethylfluorene (DMEF) model compound (at 306 nm), cyclic(■) and linear (▲) PDMVF (at 307 nm) as a function of degree of polymerization (DP_n) (trend-lines added).

The absorptivity of the lowest MW linear PDMVF (DP_n = 12, $\varepsilon \sim 13900$) is nearly the same as that of the 9,9-dimethyl-2-ethylfluorene (DMEF) model compound ($\varepsilon \sim 13800$). Absorptivities of linear PDMVF decrease about 12 percent when the DP_n increases from 12 to 18 but decreases little at higher DP_n's. The large hypochromic effects for PDMVF cycles suggest the presence of stronger chromophore stacking in the more congested higher MW rings leading to weak electronic interactions between π-chromophores.[39,40] Similar decreases in absorptivity of the P2VN cycles compared to the linear P2VN chains were observed.[41]

9,10-Anthracenylidene macrocycles. The formation of macrocyclic P2VN and PDMVF containing a single 9,10-anthracenylidene unit by reaction of P2VN,K$_2$ or PDMVF,K$_2$ with 9,10-bis(chloromethyl)anthracene proceeded well under high dilution conditions as indicated by a symmetrical SEC peak and clear shifts to a higher elution volumes compared with the linear polymers and high SEC yields (50-70%). In addition, the <G> values follow the same trends obtained for the above DBX mediated P2VN cycles being slightly higher for the same MW's, most likely on account of the larger 9,10-anthracenylidene spacers.

The uniform incorporation of a single AN into macrocyclic P2VN was confirmed by identical RI and UV SEC traces of the fractionated macrocycles run at both 405 nm and 295 nm, the wavelength where only naphthalene absorbs (data not shown). The anthracene content in the fractionated macrocycles calculated using the DP$_n$ from the linear precursor and absorption at 405 nm using the extinction coefficient for 9,10-dimethylanthracene[20] (ε =10,000) was about 90 percent, confirming a high degree of anthracene incorporation. Studies on the photophysics of these labeled macrocycles are being carried out.[36]

Fluorescence. As shown in Figure 6, cyclic and linear PDMVF's with a DP$_n$ of 18 exhibit strong and characteristic fluorene ("monomer") emissions at 311 and 322 nm (λ_{ex} = 307 nm). There is a weaker emission between 360 and 370 nm that is likely due to excimer formation, consistent with the absence of a long red-shifted tail in the fluorescence spectrum of the DMEF model.[41]

Like the absorption spectra, the shape and peak position of the monomer emission band of cyclic and linear PDMVF are the same. However, after correction for absorptivity differences, the emission intensities of the macrocycles at 311 nm and 322 nm are increased by 19 and 16 percent respectively, compared to the matching linear polymers. The total emission quantum yield of the DP$_n$ = 18 macrocycle is the same as the DMEF model but all linear polymers have lower emission yields.

Enhanced monomer emission polymers has also been observed for cyclic PS [11] and cyclic P2VN.[42]

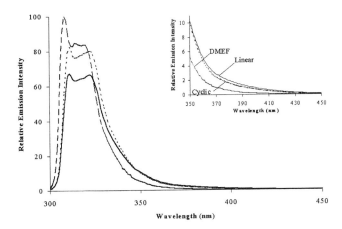

Figure 6. Fluorescence emission spectra of 9,9-dimethyl-2-ethylfluorene (DMEF) model compound (dashed line, l_{ex} = 306 nm), cyclic (dotted line) and linear (solid line) PDMVF with DP_n = 18 (l_{ex} = 307 nm). Concentration: 5 mg/L in cyclohexane. Counts normalized by optical density at excitation wavelength.

Compared to the linear PDMVF, the macrocycle shows a decrease in the excimer band at 360 – 400 nm, as shown in the inset of Figure 6, and although the excimer emission increases with increasing DP_n, the macrocycles always have lower excimer emission yields than the linear polymers. The decrease in monomer emission and concomitant increase in excimer emission observed for both cyclic and linear polymers with increasing MW is consistent with the increased number of excimer traps available on a given chain or cycle.

Compared to the linear PDMVF, the formation of excimer sites on the macrocycles may be hindered due to bond angle and torsional strains especially for the low MW cyclic PDMVF. This is consistent with the much higher glass transition temperatures of the cyclic- compared to matching linear PDMVF. Low MW linear polymers also have been shown to have smaller excimer site densities, presumably due to entropic effects that are consistent with their lower glass transition temperatures.[43]

Acknowledgements

This work was supported in part by NSF-DMR 9810283, NSF-STC 594-608 and the Loker Hydrocarbon Research Institute. We thank Dr. W. Weber for the use of TGA and DSC instrumentation in his lab at USC. We also acknowledge the Mass Spec Facility at UC, Riverside for their help with the MALDI-TOF mass spectrometry. Finally we wish to thank one of the referees for the constructive comments.

[1]   (a) Keul, H.; Höecker, H. *Large Ring Molecules* 1st ed.; Semlyen, J. A., Ed.; John Wiley & Sons: New York, **1996**; Chapter 10. (b) Ederle, Y.; Naraghi, K.; Lutz, P. J., Eds. *Synthesis of Cyclic Macromolecules*; Wiley-VCH, Weinheim **1999**.

[2]   Roovers, J. *Macromolecules* **1985**, 18, 1359.

[3]   Hogen-Esch, T. E.; Sundararajan, J. Toreki, W. *Makromol. Chem., Macromol. Symp.* **1991**, 47, 23.

[4]   (a) Rique-Lubet, L.; Schappacher, M.; Deffieux, A. *Macromolecules* **1994**, 27, 6318. (b) Deffieux, A.; Schappacher, M. *Macromolecules* **2001**, 34, 5827.

[5]   Deffieux, A.; Schappacher, M.; Rique-Lubet, L. *Polymer* **1994**, 35, 4562.

[6]   Ishizu, K.; Kanno, H. *Polymer* **1996**, 37, 1487.

[7]   Pasch, H.; Deffieux, A.; Ghahary, R.; Schappacher, M.; Rique-Lubet, L. *Macromolecules* **1997**, 30, 98.

[8]   Kubo, M. Hayashi, T. Kobayashi, H. Tsubo, K. Itoh, T. *Macromolecules* **1997**, 30, 2805.

[9]   (a) Oike, H.; Imaizumi, H. Mouri, T.; Yoshioka, Y. Uchibori, A.; Tezuka, Y. *J. Am. Chem. Soc.* **2000**, 122, 9595. (b) Oike, H.; Hamada, M.; Eguchi, S.; Danda, Y.; Tezuka, Y. *Macromolecules* **2001**, 34, 2776.

[10] Lepoittevin, B.; Dourges, M. A.; Masure, M.; Hemery, P.; Baran, K.; Cramail, H. *Macromolecule* **1999**, 32, 8218.

[11] (a) Alberty, K. A.; Tillman, E. Carlotti, S.; King, K.; Bradforth, S. E.; Hogen-Esch, T. E.; Parker, D.; Feast, W. J. *Macromolecules* **2002**, 35, 3856. (b) Gan, Y. D.; Dong, D. H.; Carlotti, S.; Hogen-Esch, T. E. *J. Am. Chem. Soc.* **2000**, 122, 2130.

[12] Gan, Y.; Dong, D. H.; Hogen-Esch, T. E. *Macromolecules* **1995**, 28,383.

[13] Dong, D. H.; Hogen-Esch, T. E. *E-Polymers* **2001**, 007.

[14] Nossarev, G. G.; Hogen-Esch, T. E. *Macromolecules* **2002**, 35, 1604.

[15] Yin, R.; Hogen-Esch, T. E. *Macromolecules* **1993**, 26, 6952.

[16] Gan, Y.; Zoeller, J.; Yin, R.; Hogen-Esch, T. E. *Makromol. Chem., Makromol. Symp.* **1994**, 77, 93.

[17] Lescanec, R. L.; Hajduk, D. A.; Kim, G. Y.; Gan, Y.; Yin, R.; Gruner, S. M.; Hogen-Esch, T. E.; Thomas, E. L. *Macromolecules* **1995**, 28, 3485.

[18] Zhang, X.; Hogen-Esch, T. E. *Macromolecules* **2000**, *33*, 9176.

[19] Zhang, X.; Hogen-Esch, T. E. Manuscript submitted.

[20] (a) Birks, L. B. Photophysics of Aromatic Molecules; John Wiley: London, 1970. (b) Berlman, I. B. Handbook of Fluorescence Spectra of Aromatic Molecules; Academic Press: London, 1971. (c) Berlman, I. B. Energy Transfer Parameters of Aromatic Compounds; Academic Press: New York, 1973.

[21] Semarak, S. N.; Frank, C. W. Adv. Polym. Sci. **1983**, 54, 33. ; Webber, S. E. *Chem. Rev.* 1990, **90**, **1469.**

[22] Guillet, J. E. *Polymer Photophysics and Photochemistry*; Cambridge University Press: Cambridge, **1985.**

[23] (a) Guillet, J. E. *Trends Polym. Sci.* **1996**, 4, 41. (b) Nowakowska, M.; Strorsberg, J.; Zapotoczny, S. Guillet, J. E. *New J. Chem.* **1999**, 23, 617.

[24] Webber, S. E. *Chem. Rev.* **1990**, 90, 1469.

[25] (a) Pokorno, V.; Vyprachticky, D.; Pecka, J. Mikes, F. *J. Fluoresc.* **1999**, 9, 59. (b) Tong, J.-D.; Ni, S.; Winnik, M. A. *Macromolecules* **2000**, 33, 1482.

[26] Ushiki, H.; Horie, K. Okamoto, A.; Mita, I. *Polym. J.* **1981**, 13, 191.

[27] Wong, K-H.; Ambroz, L.; Smid, J. *Polymer Bulletin* **1982**, *8*, 411.

[28] Xia, C.; Advincula, R. C. *Macromolecules* **2001**, 34, 5854.

[29] (a) Carothers, W.H. *Trans. Faraday Soc.* **1936**, 32, 39. (b) Odian, G. *Principles of Polymerization* John Wiely & Sons: New York, **1991**, Chapter 5.

[30] Nossarev G. G.; Tillman E.; Hogen-Esch T. E. *J. Polym. Sci. Part A: Polymer Chemistry*, **2001**,*39*, 3121; Tillman, E. S.; Hogen-Esch, T. E. *J. Polym. Sci., Part A: Polym. Chem.* **2002**, 40, 1081.

[31] Dong, D.; Hogen-Esch, T. E.; Shaffer, J. S. *Macromol. Chem. Phys.* **1996**, 197, 3397.

[32] Jacobson, H.; Stockmayer, W. H. *J. Chem. Phys.* **1950**, 18, 1600.

[33] Tillman, E. S.; Hogen-Esch, T. E. *Macromolecules* **2001**, *34*, 6616.

[34] Lin, J.; Fox, M.A. *Macromolecules* **1994**, *27*, 902.

[35] Jagur-Grodzinski, J.; Szwarc, M. *J. Am. Chem. Soc.* **1969**, *91*, 7594.

[36] Nossarev, G.G.; Bradforth, S.E.; Hogen-Esch, T.E. In preparation.

[37] Sudipati, M. G.; Daverkausen J.; Maus, M.; Hohlneicher, G. *Chem. Phys.* **1994**, 181, 289.

[38] (a) Tinoco, I. J. *J. Am. Chem. Soc.* **1960**, 82, 4785. (b) Rhodes, W.; Chase, M. W. *Rev. Mod. Phys.* **1967**, 39, 348. (c) Cantor, C. R.; Schimmel, P. R. *Biophysical Chemistry*; W. H. Freeman & Co. San Francisco, **1980**; Part II, Chapter 7.

[39] Vala, M. T. Jr.; Rice, S. A. *J. Chem. Phys.* **1963**, 39, 2348.

[40] (a) Okamoto,K.; Itaya, A.; Kusabayashi, S. *Chem. Lett.* **1974**, 3, 1167. (b) Kowal, J. *Macromol. Chem. Phys.* **1995**, 196, 1195.

[41] Horrocks, D. L.; Brown, W. G. *Chem. Phys. Lett.* **1970**, 117.

[42] Nossarev, G. G.; Bradforth, S. E.; Hogen-Esch, T. E. Manuscript in preparation.

[43] Ishii, T.; Handa, T.; Matsunage, S. *Macromolecules* **1978**, 11, 40.

(Star PS)-*block*-(Linear PI)-*block*-(Star PS) Triblock Copolymers – Thermoplastic Elastomers with Complex Branched Architectures

*Tianzi Huang, Daniel M. Knauss**

Chemistry Department, Colorado School of Mines, Golden, CO 80401, USA
E-mail: dknauss@mines.edu

Summary: Polystyrene-*block*-polyisoprene-*block*-polystyrene triblock copolymers were synthesized with star-shaped branching in the polystyrene phase. The block copolymers were formed through sequential anionic polymerization by first synthesizing linear polystyrene, followed by star coupling using 4-(chlorodimethylsilyl)styrene, then the polymerization of isoprene, followed by difunctional coupling with dichlorodimethylsilane. The polymerization was followed by gel permeation chromatography and the resulting copolymers were characterized by ^1H NMR spectroscopy to examine the polyisoprene microstructure.

Keywords: anionic polymerization; branched; copolymerization; elastomers; star polymers

Introduction

Polystyrene-*block*-polyisoprene-*block*-polystyrene (SIS) triblock copolymers where the inner polyisoprene block is the major component are well known thermoplastic elastomers.[1,2] The triblock copolymers are phase separated at room temperature and the hard and rigid polystyrene domains that are distributed in the polyisoprene matrix act as thermally reversible multi-functional junction points to give a physically crosslinked elastomeric network.[1,3,4] The block copolymers are also soluble in many solvents (such as toluene, THF, dichloromethane, etc.), but can regain their phase-separated morphology and mechanical properties when the solvent is evaporated. These thermoplastic elastomers are therefore reprocessible by heating or solvent casting, but their morphology and resulting mechanical properties are influenced by their process history.[5,6]

SIS triblock copolymers have been synthesized by living anionic polymerization initiated by organolithium compounds in a hydrocarbon solvent (see references in review articles[2,7-9]). Three basic methods have been described for the synthesis of these triblock copolymers: the sequential addition of monomers to an alkyllithium initiator, the sequential addition of monomers to a di-functional initiator, and the sequential addition of monomers to an alkyllithium initiator followed by coupling reactions.

We have been investigating a one-pot convergent living anionic polymerization procedure for introducing star-branching into polystyrene.[10-12] Star-shaped polystyrene with a living site at the core can be synthesized by slowly introducing a dual functional coupling agent to living polystyryllithium.[10-12] The resulting living star can initiate the polymerization of styrene monomer to obtain a star-*block*-linear diblock polystyrene.[13-15] The living diblock polystyrene can be coupled by adding a difunctional coupling agent to obtain star-*block*-linear-*block*-star triblock (pom-pom) polystyrene,[14,16] or it can be coupled with a multifunctional coupling agent to yield a more complex architecture.[17]

Star-branching can also be introduced into the polystyrene block of SIS triblock copolymers by this convergent living anionic polymerization technique. The resulting SIS triblock copolymer with the pom-pom structure can be used as a model sample to study the influence of the structural modification on the morphology, mechanical properties, and processibility of SIS thermoplastic elastomers. While other researchers have begun investigation of thermoplastic elastomers with other types of complex architectures through controlled, multistep syntheses,[18-20] our synthetic method can potentially yield unique structures in a single reaction container.

Experimental details

Materials Styrene (99%, inhibited with ~ 10 - 15 ppm 4-*tert*-butylcatechol), *p*-chlorostyrene (97%, inhibited with ~ 500 ppm 4-*tert*-butylcatechol), 1,2-dibromoethane(99%), dichlorodimethylsilane (99%) and trichloromethylsilane (99%) were obtained from Aldrich Chemical Company. Reagents were dried over calcium hydride and distilled under argon or under reduced pressure immediately prior to use. Isoprene (99%, inhibited with ~ 100 ppm 4-*tert*-butylcatechol) was distilled twice over calcium hydride and distilled once over a dibutylmagnesium solution (1.0 M in heptane, Aldrich Chemical Co.) just prior to use. *sec*-Butyllithium in a mixture of cyclohexane and heptane was kindly donated by FMC, Lithium

Division. The effective molarity of the solution was determined to be 1.10 M by repeated initiation and polymerization of styrene and the subsequent analysis of the molecular weights by GPC-MALLS. HPLC grade tetrahydrofuran (99.9+%) was obtained from Fisher Scientific, dried over sodium metal, and distilled from sodium benzophenone ketyl under argon immediately prior to use. HPLC grade cyclohexane (99%) from Fisher Scientific was purified by repeated washings with H_2SO_4 and water and distilled from sodium metal. 4-(Chlorodimethylsilyl)styrene (CDMSS) was synthesized as reported,[11] and was distilled from calcium hydride under reduced pressure immediately prior to use. All glassware, glass syringes, and needles were oven dried at 150 °C for at least 24 hours, and cooled under argon. The glassware was further flame dried under an argon purge after assembly. Gastight syringes were prepared by washing with a dilute *sec*-butyllithium solution followed by washing with dry cyclohexane.

Polymerization Polymerizations were conducted at room temperature. A typical polymerization procedure is as follows (Experiment 1): Styrene (2.5 mL, 21.8 mmol) was charged to 100 mL of cyclohexane in a 250 mL round bottom flask sealed with a rubber septum under an argon atmosphere. *sec*-BuLi was added dropwise to titrate impurities until a pale yellow color indicative of the polystyryllithium was observed. The calculated charge of *sec*-BuLi initiator (2.5 mL, 2.75 mmol) was then added. After two hours, an aliquot was removed and precipitated into argon-purged methanol. A solution of CDMSS in cyclohexane (approximately 0.50 M) was slowly added by using a gastight syringe with the addition rate controlled by a syringe pump. The addition rate of the CDMSS solution was 1.0 mL/hour over a course of 4.2 hours to introduce a total of 4.2 mL (2.10 mmol) of the CDMSS solution. Thirty minutes after the complete addition, an aliquot (5 mL) was removed and precipitated in argon-purged methanol. Isoprene (6.0 mL, 60 mmol) was then added to the solution from a syringe, and the solution was allowed to stir for three hours at room temperature. An aliquot (5 mL) was removed and THF (3 mL) was added to the solution. A dichlorodimethylsilane mixture in cyclohexane (0.20 M) was added slowly into the living anion solution using a gastight syringe. The addition rate was controlled to a rate of 0.20 mL/hour by a syringe pump and the addition was continued until the reaction solution turned colorless. The reaction mixture was then precipitated into methanol, filtered, washed with methanol, and dried to a constant weight at room temperature in a vacuum oven. 2,6-di-*tert*-butyl-4-methylphenol (0.02 wt %) was added to the methanol to prevent the oxidation of the polyisoprene block. The reaction yield was quantitative after considering the sampled aliquots.

Characterization

Molecular weights and molecular weight distributions were characterized by gel permeation chromatography (GPC) performed on a Hewlett-Packard model 1084B liquid chromatograph equipped with two Hewlett-Packard Plgel 5μ Mixed-D columns (linear molecular weight range: 200 – 400,000 g/mol), a calibrated RI (Waters R401) detector, and a Wyatt Technology miniDAWN multi-angle laser light scattering (MALLS) detector (λ = 690 nm, three detector angles: 45°, 90° and 135°), using Astra 1.5.0b2 molecular weight characterization software. Elutions were carried out at an ambient temperature with THF as the solvent and a flow rate of 0.70 mL/min. The refractive index increment (dn/dc) used for the polystyrene samples was 0.193 mL/g, which was obtained for the polystyrene stars[11] and which also corresponds to linear polystyrene in THF.[21] The dn/dc of polyisoprene was determined to be 0.117 mL/g in THF at room temperature by using an Abbe Refractometer (Bausch & Lomb). The refractive index increments of the SIS copolymers were calculated based on the summation of the products of the dn/dc of each block times its weight fraction.[22] ^{1}H NMR spectroscopy was performed on samples dissolved in deuterated chloroform on a Chemagnetics CMX Infinity 400 instrument.

Results and Discussion

Synthesis Method The synthesis of (star PS)-*block*-(linear PI)-*block*-(star PS) triblock is depicted in Scheme 1. Living star-shaped polystyrene anions were synthesized via the convergent living anionic polymerization method by introducing a less than stoichiometric amount of CDMSS to living polystyrene anions in a procedure similar to that previously described.[16,17] Isoprene monomer was added into the living star-shaped polystyrene anions to make a (star PS)-*block*-(linear PI) living diblock. The syntheses of the living star and the (star PS)-*block*-(linear PI) were performed in cyclohexane without any added polar modifiers to maintain the predominately *cis*-1,4 structure of the polyisoprene block, which is important for the formation of thermoplastic elastomers.[1,2] To the living diblocks was added the difunctional coupling agent dichlorodimethylsilane (DCDMS) to obtain the (star PS)-*block*-(linear PI)-*block*-(star PS) triblock. A small amount of THF was added into the solution before the addition of DCDMS to increase the coupling reaction rate.[16] The addition of reactants was conducted using syringe transfer techniques, and the rate of addition was controlled by using a syringe pump.

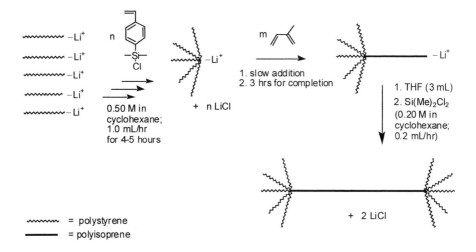

Scheme 1. Reaction sequence for the synthesis of (star PS)-*block*-(linear PI)-*block*-(star PS) triblock copolymer.

The structural variations, such as the length of the initial chain and the average number of arms, can be readily controlled by the molar ratio of styrene monomer to *sec*-BuLi and the amount of CDMSS added, respectively.[16] The molecular weight of the linear polyisoprene block can also be controlled by varying the amount of isoprene monomer relative to the star-shaped macroinitiator.

Synthesis of Living Stars In the convergent process, CDMSS is slowly introduced to the living polystyrene solution using a gastight syringe, with the addition rate controlled by a syringe pump. The CDMSS reacts with the living polymer by first terminating one chain through the substitution of the silyl chloride in the formation of a macromonomer, and then another living chain quickly adds to the vinyl group of the macromonomer to form a polymer with two coupled chains and a living site at the junction point. The series of reactions are repeated as more CDMSS is added, and a star-shaped polymer with a hyperbranched core is finally obtained.[11]

We have studied this convergent process in cyclohexane solutions with different amounts of added THF.[11,16] The THF was added because the rates of both the substitution reaction and addition reaction can be increased.[23,24] The maximum number of arms of star-shaped polystyrene under given reaction conditions, and the relationship between the desired number of arms of a living star and the amount of CDMSS added, have been reported.[16] In this work, it was desired to synthesize living star-shaped polystyrenes in a hydrocarbon solvent without any added THF. An experiment was designed to compare star-shaped polystyrene after stoichiometric CDMSS addition, synthesized with and without added THF. The results are compared in Figure 1. Convergent living anionic polymerization was determined to proceed favorably in cyclohexane to synthesize star-shaped polystyrenes

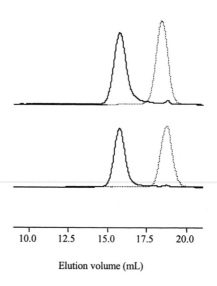

10.0   12.5   15.0   17.5   20.0

Elution volume (mL)

Figure 1. GPC chromatograms of star-shaped polystyrenes (... initial linear polystyrene, — star-shaped polystyrene). (a) Synthesized in cyclohexane only; linear polystyrene (M_n = 1210 g/mole), star-shaped polystyrene (M_n = 12,900, PDI = 1.16, no. of arms = 9.5) (b) synthesized in cyclohexane with 3% v/v THF; linear polystyrene (M_n = 940 g/mole), star-shaped polystyrene (M_n = 12,100, PDI = 1.14, no. of arms = 11).

The number of coupled arms in the living star is related to the stoichiometric amount of CDMSS added. To obtain an f-armed star, $(1-1/f)$ of the stoichiometric amount of CDMSS is needed and the resulting living anion concentration is $1/f$ of the initial concentration of living anions.[16] This relationship was used to control the number of arms of living polystyrene stars in this work.

Living polystyrene chains were prepared in cyclohexane by initiating styrene monomers with *sec*-BuLi. An aliquot of the reaction solution was sampled and subsequently characterized by

GPC-MALLS. The desired amount of CDMSS was then added slowly by using a gastight syringe with the addition rate controlled by a syringe pump. After complete addition, an aliquot of the living star-shaped polystyrene solution was removed and isolated by precipitating in argon-purged methanol, then characterized by GPC-MALLS. Using the molecular weights of the living star and the initial living chains, the number of arms, f, can be readily calculated based on Equation 1 as previously reported,[11,16,17]

$$ G = \frac{\log(M_{star} + M_{branch}) - \log(M_{initial} + M_{branch})}{\log 2}; \qquad f = 2^G \qquad \text{Eq. 1} $$

where G equals the number of generations of growth, M_{star} equals the number average molecular weight of the star-shaped polystyrene, M_{branch} equals the number average molecular weight of the residue from the coupling agent, and $M_{initial}$ equals the number average molecular weight of the initial linear polystyrene chain. The molecular weight characterization results of living polystyrene chains and stars from aliquots removed during the experiments are shown in Table 1. In this work, different arm lengths of the living polystyrene star, from ~ 1,000 g/mol to ~ 6,300 g/mol were used. The average number of arms of the living star was determined to be around 5 (4.4 – 5.3) by controlling the amount of CDMSS added.

Table 1. Synthesis and characterization for initial chains and stars

Exp.	St/BuLi (mmol/ mmol)	[a]M_n of initial chain (g/mol)/PDI	Addition rate of CDMSS(mmol/hr) /BuLi(mmol)	[a]Mn of PS star (g/mol)/PDI	[b]f: Avg. no. of arms
1	8.0	1, 060/1.05	0.50/2.75	6,260/1.35	5.3
2	20	2,210/1.04	0.50/2.75	11,700/1.30	5.0
3	50	5,530/1.01	0.02/1.10	28,600/1.27	4.7
4	30	3,470/1.04	0.50/2.08	18,400/1.27	5.1
5	55	6,300/1.02	0.25/1.16	28,200/1.22	4.4
6	30	3,130/1.02	0.50/2.08	15,300/1.24	4.7

All reactions were done at room temperature in cyclohexane and the addition of CDMSS (approximately 0.50M in cyclohexane) was completed in the time range of 4-5 hours. [a]Determined by GPC-MALLS; [b]Determined by Eq. 1, the number of arms $f = 2^G$.

Synthesis of Triblock Copolymers The star-shaped polystyrene contains a living chain end at the focal point of the hyperbranched core; thus, it can be used as a macroinitiator to polymerize sequentially added isoprene monomer. In a hydrocarbon solvent, the crossover rate of the polystyryllithium to the polyisoprenyllithium is approximately 150 times faster than the polymerization rate of isoprene.[25] The initiation rate of the living polystyrene star should therefore be fast despite any steric hindrance at the living site.

Experiments were performed to polymerize isoprene from the polystyrene star initiator. Based on the expected quantitative initiation and subsequent propagation of living anionic polymerization, the length of the linear polyisoprene block was controlled by the stoichiometric amount of isoprene monomer to the living polystyrene star. The mass ratio of the polystyrene block to the polyisoprene block was fixed at 35/65 by controlled monomer feed ratios.

Table 2. Synthesis and characterization for (star PS)-*block*- (linear PI) diblock and SIS triblock copolymers

Exp.	[a]Mass ratio PS/PI	[b]M_n of diblock (g/mol)/PDI	Addition rate of DCDMS (mmol/hr)	[b,c]M_n of product (g/mol)/PDI	[d]Diblock content (wt %)
1	35/65	16,000/1.12	0.04	26,400/1.12	21
2	35/65	31,200/1.08	0.04	54,000/1.05	15
3	35/65	79,900/1.08	0.02	131,000/1.05	22
4	35/65	42,400/1.07	0.04	74,400/1.06	14
5	35/65	62,900/1.05	0.02	107,000/1.05	18
6	35/65	34,100/1.07	0.04	58,200/1.07	17

All reactions were done at room temperature in cyclohexane with 3% v/v THF added prior to the addition of DCDMS (approximately 0.20 M in cyclohexane). [a]By feed ratio; [b]Determined by GPC-MALLS with dn/dc = 0.144 mL/g; [c]Before fractionation, peak includes residual diblock; [d]Calculated based on the theoretical molecular weight of triblock (i.e., 2 x M_n of diblock); wt% diblock = 100 x (2 x M_n of diblock – M_n of product)/(M_n of product).

After sampling for GPC-MALLS characterization, the living (star PS)-*block*-(linear PI) diblocks with a 35/65 mass ratio were coupled with dichlorodimethylsilane (DCDMS) to form the triblock. To increase the coupling efficiency, 3% v/v THF was added before the addition of DCDMS.[16,24] A dilute solution of DCDMS in cyclohexane (0.20 M) was added at a slow rate to the living diblock anions. The slow addition technique maintains a deficiency of coupling agent to allow the diblock anions to couple efficiently.[16,26]

The molecular weights and molecular weight distributions for the diblock and triblock samples were characterized by GPC-MALLS. The refractive index increments (dn/dc) used for the (star PS)-*block*-(linear PI) diblock and the (star PS)-*block*-(linear PI)-*block*-(star PS) triblock, were calculated based on the summation of the products of the dn/dc of each block multiplied by its weight fraction.[22] The dn/dc of the polyisoprene block was determined to be 0.117 mL/g in THF at room temperature by using an Abbe Refractometer (Bausch & Lomb), and the dn/dc of the polystyrene block was 0.193 mL/g.[11,21] The dn/dc is calculated to be 0.144 mL/g for the diblock and triblock with a polystyrene to polyisoprene mass ratio of 35/65.

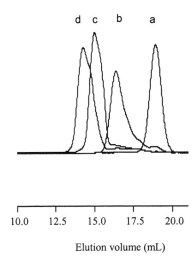

Figure 2. GPC chromatograms for Experiment 1. (a) Initial chain; (b) star; (c) (star PS)-*block*-(linear PI) diblock; (d) (star PS)-*block*-(linear PI)-*block*-(star PS) triblock copolymer.

The progression of the molecular weight of samples during the synthesis is demonstrated by the representative GPC-MALLS chromatograms of Experiment 1 (Figure 2). From high elution volume to low elution volume, the corresponding initial linear chains, the polystyrene star, the (star PS)-*block*-(linear PI) diblock, and the (star PS)-*block*-(linear PI)-*block*-(star PS) triblock are observed as relatively narrow peaks.

The GPC results for the (star PS)-*block*-(linear PI) diblock and the (star PS)-*block*-(linear PI)-*block*-(star PS) triblock are presented in Table 2. The molecular weight of the diblock was controlled from 16,000 g/mol to 79,900 g/mol. The molecular weight of the final product containing the triblock is not exactly twice that of the corresponding diblock because of the co-existence of diblock residue. The weight percentage of residual diblock was calculated based on the difference in average molecular weight of the final sample (diblock plus triblock) and the theoretical triblock molecular weight (twice the molecular weight of the diblock),[16,27] and is presented in Table 2. The amount of diblock in each sample varies from 14 to 22 wt% by the calculation. It is reported that the commercial Kraton SIS triblock copolymers possess bimodal molecular weight distributions with 15 to 20 wt % of apparently diblock copolymers with one-half of the molecular weight of the triblock copolymers.[28] The diblock residue has been demonstrated to lower the mechanical properties of the final products.[29,30]

Microstructure Characterization   The (star PS)-*block*-(linear PI)-*block*-(star PS) triblock samples were characterized by ^{1}H NMR spectroscopy. The ^{1}H NMR spectrum of the triblock copolymer from Experiment 1 is shown in Figure 3. Signals from the linear polyisoprene block and the star-shaped polystyrene block are observed. The peaks are readily identified by comparison with ^{1}H NMR spectra of polystyrene and polyisoprene.[31-34]. The small peak at 0.1 ppm represents the silyl methyl groups from the CDMSS residue in the polystyrene star block. The small peaks at 0.6-0.8 are due to the residue of the intiator (*sec*-butyl group). The peaks at chemical shifts from 1.3 ppm to 2.3 ppm, are attributed to signals from the residue of the initiator and from the polymerized vinyl groups of polyisoprene and polystyrene, together with the typical signals from the methylene groups at 2.1 ppm and the methyl group from the 1,4-structures of polyisoprene (1.7 ppm for the *cis*-1,4-structure and 1.6 ppm for the *trans*-1,4-structure). The double peak at 4.7 ppm represents the two olefinic protons from the 3,4-structure of polyisoprene and the single peak at 5.1 ppm represents the olefinic proton from the 1,4-structure of polyisoprene. The peaks with a chemical shift of 6.4 ppm to 7.2 ppm are signals from the aromatic ring of the polystyrene block.

The microstructure of the polyisoprene block can be analyzed based on this spectrum. The ratio of the peaks at 4.7 ppm (two olefinic protons from the 3,4-structure) and 5.1 ppm (one olefinic proton from 1,4-structure) is approximately 1:9 by integration, which indicates that there is only a small amount of polyisoprene units with the 3,4-structure co-existing with the polyisoprene units with1,4-structure. The ratio of the *cis*-1,4-structure relative to the *trans*-

1,4-structure can be estimated based on the peak intensity at 1.7 ppm and 1.6 ppm. The precise ratio is hard to obtain because these two peaks are mixed with signals from polymerized vinyl groups of the polystyrene block, but it is easy to tell that the *cis*-1,4-structure is the predominant microstructure.

Preliminary property characterization Films have been made by solution casting from toluene for all of the triblock samples. The films are transparent and demonstrate the typical mechanical properties of thermoplastic elastomers. Complete examination of the morphology, melt viscosity, and tensile properties, which should be influenced by introduction of branching, is being done and more detailed results will be forthcoming. Other samples are also being produced to better control the extent of branching, composition, and the efficiency of coupling.

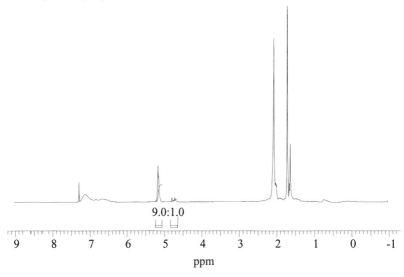

Figure 3. ^{1}H NMR spectrum for (star PS)-*block*-(linear PI)-*block*-(star PS) triblock copolymer (Experiment 1). Solvent: CDCl₃.

Conclusions

Styrene based thermoplastic elastomers with a novel structure, (star PS)-*block*-(linear PI)-*block*-(star PS) triblock copolymers, were synthesized by the sequential addition of isoprene monomer and DCDMS to living polystyrene stars obtained by convergent living anionic polymerization. GPC-MALLS characterization demonstrated the progression of the molecular weight during the synthesis. Samples with controllable arm lengths, numbers of arms, molecular weights of the diblock, and polystyrene to polyisoprene mass ratios were obtained. The microstructure of the resulting copolymers was characterized by ^{1}H NMR spectroscopy. Preliminary mechanical property tests demonstrate that these copolymers possess elastomeric properties as expected based on their phase separated morphologies.

Acknowledgements

The authors acknowledge support for this work from the NSF (Grant DMR-9985221) for a CAREER award to D.M.K. Support from Kraton Polymers is also gratefully acknowledged. Acknowledgement is made to the donors of the Petroleum Research Fund, administered by the ACS, for partial support of this research.

[1] G. Holden; N. R. Legge. In *Thermoplastic Elastomers, A Comprehensive Review*; Legge, N. R.; Holden, G.; Schroeder, H. E., Eds.; Hanser Publishers: New York, 1987.
[2] G. Holden. *Rubber World* **1993**, *208*, 25-30.
[3] G. Holden; R. Milkovich. U.S. Patent, 1964.
[4] L. H. Sperling. *Polymeric Multicomponent Materials: An Introduction*; John Wiley & Sons, Inc.: New York, 1997.
[5] M. Morton. *Encyclopedia of Polymer Science and Technology*; John Wiley and Sons: New York, 1971; Vol. 15.
[6] J. F. Beecher; L. Marker; R. D. Bradfod; S. L. Aggarwal. *J. Polym. Sci., Part C* **1969**, *26*, 117-134.
[7] M. Morton. In *Thermoplastic Elastomers, A Comprehensive Review*; Legge, N. R.; Holden, G.; Schroeder, H. E., Eds.; Hanser Publishers: New York, 1987.
[8] H. L. Hsieh; R. P. Quirk. *Anionic Polymerization: Principles and Practical Applications*; Marcel Dekker, Inc.: New York, 1996.
[9] J. E. McGrath, Ed. *Anionic Polymerization. Kinetics, Mechanics, and Synthesis*; American Chemical Society: Washington, D.C., 1981; Vol. 166.
[10] H. A. Al-Muallem; D. M. Knauss. *Polym. Prepr. Am. Chem. Soc., Div. Polym. Chem.* **1997**, *38(1)*, 68-69.
[11] D. M. Knauss; H. A. Al-Muallem; T. Huang; D. T. Wu. *Macromolecules* **2000**, *33*, 3557-3568.
[12] H. A. Al-Muallem; D. M. Knauss. *J. Polym. Sci.: Part A: Polym. Chem.* **2001**, *39*, 3547-3555.
[13] H. A. Al-Muallem; D. M. Knauss. *J. Polym. Sci.: Part A: Polym. Chem.* **2001**, *39*, 152-161.
[14] D. M. Knauss; T. Huang. *Polym. Prepr. Am. Chem. Soc., Div. Polym. Chem.* **2000**, *41(2)*, 1232-1233.
[15] D. M. Knauss; T. Huang. *Polym. Prepr. Am. Chem. Soc., Div. Polym. Chem.* **2000**, *41*, 1397-1398.
[16] D. M. Knauss; T. Huang. *Macromolecules* **2002**, *35*, 2055-2062.
[17] D. M. Knauss; T. Huang. *Macromolecules* **2003**, *36*, 6036-6042.
[18] H. Iatrou; J. W. Mays; N. Hadjichristidis. *Macromolecules* **2000**, *31*, 6697-6701.
[19] D. Uhrig; J. W. Mays. *Macromolecules* **2002**, *35*, 7182-7190.
[20] R. Weidisch; S. P. Gido; D. Uhrig; H. Iatrou; J. Mays; N. Hadjichristidis. *Macromolecules* **2001**, *34*, 6333-6337.
[21] M. B. Huglin. In *Polymer Handbook*, 3rd ed.; Bandrup, J.; Immergut, E. H., Eds.; John Wiley & Sons, Inc: New York, 1989; p V445.
[22] S. Michielsen. In *Polymer Handbook*, 4th ed.; Brandrup, J.; Immergut, E. H.; Grulke, E. A., Eds.; John Wiley & Sons, Inc: New York, 1999; p VII/548.
[23] F. J. Welch. *J. Am. Chem. Soc.* **1960**, *82*, 6000-6005.
[24] M. O. Hunt, Jr.; A. M. Belu; R. W. Linton; J. M. DeSimone. *Macromolecules* **1993**, *26*, 4854-4859.
[25] D. J. Worsfold. *J. Polym. Sci.: Polym. Chem. Ed.* **1967**, *5*, 2783-2789.
[26] D. N. Schulz; A. F. Halasa. *J. Polym. Sci.: Polym. Chem. Ed* **1977**, *15*, 2401-2410.
[27] H. L. Hsieh. *Rubber Chem. Technol.* **1976**, *45*, 1305-1310.
[28] L. J. Fetters; B. H. Meyer; D. McIntyre. *J. Appl. Polym. Sci.* **1972**, *16*, 2079-2089.

[29] H. R. Lovisi; L. F. Nicolini; A. A. Ferreria; M. L. S. Martins. In *ACS Symposium series 696, Applications of Anionic Polymerization Research*; Quirk, R. P., Ed.; American Chemical Society: Washington D.C., 1998.

[30] T. Matsubara; M. Ishigura. In *ACS Symposium series 696, Applications of Anionic Polymerization Research*; Quirk, R. P., Ed.; American Chemical Society: Washington D.C., 1998.

[31] N. Ishihara; T. Seimiya; M. Kuramoto; M. Uoi. *Macromolecules* **1986**, *19*, 2464-2465.

[32] H. Sato; Y. Tanaka. *J. Polym. Sci.: Polym. Chem. Ed* **1979**, *17*, 3551-3558.

[33] H. Sato; A. Ono; Y. Tanaka. *Polymer* **1977**, *18*, 580-585.

[34] Y. Tanaka; H. Sato. *Polymer* **1976**, *17*, 113-116.

Synthesis of Poly(1,3-cyclohexadiene) Containing Star-Shaped Elastomers

*D. T. Williamson, T. E. Long**

Virginia Polytechnic University and State University, Department of Chemistry, Polymer Materials and Interfaces Laboratories, Blacksburg, VA 24061-0212, USA

Summary: The synthesis of high molecular weight star-shaped polymers comprising poly(1,3-cyclohexadiene-*block*-isoprene) diblock arms coupled to a divinyl benzene (DVB) core is reported. The number average molecular weights of the diblock arms were varied from 30000 to 50000 and the ratio of DVB to n-butyllithium (nBuLi) was systematically varied from 3:1 to 12:1. Size exclusion chromatography coupled with light scattering detection was utilized to detect the formation of star-shaped polymers and the presence of star-star coupling. The molecular weight distribution ($<M_w>/<M_n>$) of the star polymers ranged from 1.25 to 1.50. The effect of poly(1,3-cyclohexadiene) content on the mechanical properties of these novel elastomers is reported. The elastic modulus, elongation at break, and tensile strength of these elastomers were all found to be a function of the percentage of poly(1,3-cyclohexadiene). The glass transition temperatures were determined using both differential scanning calorimetry (DSC) and dynamic mechanical analysis (DMA). Atomic force microscopy was performed in the tapping mode (TMAFM) to verify the presence of microphase separation.

Keywords: anionic polymerization; atomic force microscopy; 1,3-cyclohexadiene; star-shaped polymer; thermoplastic elastomer

Introduction

Poly(styrene-*block*-butadiene-*block*-styrene) (SBS) and poly(styrene-*block*-isoprene-*block*-styrene) (SIS) triblock copolymers are well known to exhibit an excellent combination of elasticity and thermoplasticity. These triblock copolymers are multiphase materials, wherein the polystyrene blocks microphase separate from the polydiene blocks. The microphase-separated domains are responsible for the remarkable elastomeric properties of these macromolecules. This elastomeric behavior persists over a wide temperature range with the lower temperature limit dictated by the T_g of the poly(1,3-butadiene) (-90 °C) or polyisoprene (-60 °C) domains and the upper temperature limit dictated by the T_g of the polystyrene (100 °C) domains. The polystyrene domains lose physical integrity and thus these elastomers are unable to maintain their elastomeric properties, when the temperature approaches 100 °C.[1]

Numerous studies were performed to increase the upper use temperature of thermoplastic elastomers in order to compete with vulcanized rubber.[2] Jerome and coworkers recently synthesized triblock copolymers of poly(isobornyl methacrylate)-*block*-poly(1,3-butadiene)-*block*-poly(isobornyl methacrylate).[2] These elastomers exhibited improved tensile strength (30 MPa), high percent elongation (1000 %), and a high upper use temperature of 160 °C. Earlier efforts involved the preparation of thermoplastic elastomers containing poly(α-methylstyrene)[3], poly(ethylene sulfide),[4] and poly(*p*-methylstyrene).[5] While these polymer blocks exhibited a higher glass transition temperature, the resulting elastomers exhibited either a low ceiling temperature or poor ultimate mechanical properties.

In contrast to other polydienes, poly(1,3-cyclohexadiene) PCHD exhibits a remarkably high T_g (150 °C), 50 °C higher than the T_g of polystyrene.[1] Unfortunately, the polymerization of 1,3-cyclohexadiene is plagued by a number of molecular weight limiting side reactions. While a wide variety of polymerization techniques were studied in the mid-1960's and early 1970's to identify a viable polymerization methodology for 1,3-cyclohexadiene, the resulting polymers exhibited unpredictable molecular weights and broad molecular weight distributions. [6-10]

In 1997, Natori reported the first living anionic polymerization of 1,3-cyclohexadiene in the presence of the ligating agent N,N,N'N'-tetramethylethylenediamine (TMEDA). As an indication of the living nature of the reaction, the polymerization exhibited a linear relationship between number average molecular weight and percent conversion. A series of studies was reported, in which the synthesis and characterization of poly(1,3-cyclohexadiene)-*block*-poly(1,3-butadiene)-*block*-poly(1,3-cyclohexadiene) triblock copolymers were examined. [11-14] These triblock copolymers exhibited relatively narrow molecular weight distributions with varying percentages of 1,3-cyclohexadiene. Increased incorporation of 1,3-cyclohexadiene broadened the molecular weight distributions. The presence of TMEDA during the polymerization increased the T_g of the poly(1,3-butadiene) block.

In later studies, Natori et al. altered the synthetic strategy and utilized both a difunctional initiator and sequential addition.[14] A difunctional initiator was utilized, in the presence TMEDA, to initiate the polymerization of the poly(1,3-butadiene) block. Following the synthesis of the poly(1,3-butadiene) blocks, 1,3-cyclohexadiene was added to synthesize the PCHD blocks. In a difunctional system such as this, the outer blocks of the copolymer were the same molecular weight. However, in a sequential addition, the last block of a triblock copolymer does not typically exhibit the same molecular weight as the first block.[1] It is therefore a concern that the asymmetric triblock structure will affect the elastomeric properties and corresponding morphologies.[15] An alternative approach to PCHD containing thermoplastic elastomers involves coupling strategies which permit the synthesis of highly branched polymeric structures, such as star-shaped polymers.[16] Star-shaped polymers exhibit useful rheological properties and are used as additives to improve polymer viscosity. [17,18] The synthesis of PCHD star-shaped polymers was recently reported. [19,20]

Star-shaped polymers, wherein the arms of the star are nearly identical in chain length, are readily prepared using living anionic polymerization methodologies.[1,16] Typically, two methods are utilized for the synthesis of star-shaped macromolecules. These approaches are the "core-first" and the "arm-first" methods.[16,21-25] The "core-first" and "arm-first" methods may also be combined in a single synthesis to generate stars with two different types of polymer arms via the "in-out" method. In the "arm-first" method, a living monofunctional arm of known length is reacted with a plurifunctional compound. "Arm-first" methods can be employed with various linking agents such as silicon tetrachloride and p- and m-divinylbenzene.[16] When chlorosilane based linking agents are used, the functionality of the chlorosilane dictates the number of arms in the star polymer. When using divinyl linking agents, the divinyl compounds undergo homopolymerization and form star polymers with a DVB core.[1] These star-shaped polymers exhibit a greater number of arms than the functionality of a single divinyl compound.[26] An advantage of the "arm-first" method is the facile synthesis of star-shaped polymers with narrow molecular weight distributions. In contrast to the "arm-first" method, "core-first" methods generate a reactive core prior to the addition of arm forming monomers.[16] The arm forming monomers then polymerize in a radial fashion from the core in a living manner. This enables the outer chain ends of the star-shaped macromolecule to be functionalized using various termination agents.[16] The outer chain ends of a star can be functionalized when the "arm-first" route is used, however, this approach requires the use of protected functionalized initiators.[1] The functionalized periphery can subsequently be utilized for the preparation of networks or star-shaped polymers containing copolymeric branches.[16] The disadvantage of the "core-first" method is the generation of polymers with relatively large molecular weight distributions.[16]

Herein, the synthesis of poly(1,3-cyclohexadiene-*block*-isoprene) DVB coupled, star-shaped elastomers via the arm-first method is reported. The molecular weight characteristics of these novel elastomers were examined using size exclusion chromatography in combination with multiple angle laser light scattering (MALLS). The effect of poly(1,3-cyclohexadiene) content on the elastomeric properties was examined and the presence of microphase separation was verified using atomic force microscopy.

Experimental Section

Materials. 1,3-cyclohexadiene (Aldrich, 98%) and isoprene (Aldrich, 99%) were degassed several times and distilled at reduced pressure (0.10 mmHg, 10 °C) from dibutylmagnesium (DBM, 0.84 M). Divinylbenzene (DVB) (Aldrich, 80% divinylbenzene comprising a mixture of the meta and para isomers and 20 wt% ethylvinylbenzene) was distilled under reduced pressure (0.10 mm, 25 °C) from calcium hydride (Aldrich, 95%) and dibutylmagnesium immediately prior to use. The ethylvinylbenzene was not removed and all further references to DVB assume the presence of the ethylvinylbenzene (20 mol%). n-Butyllithium (FMC Corporation, Lithium Division, 1.35 M in n-hexane) was used without further purification. TMEDA (Aldrich, 99%) was distilled at reduced pressure (0.13-0.16 mmHg, 10 °C) from calcium hydride and stored under nitrogen at –25 °C until ready for use. Cyclohexane (Burdick-Jackson, HPLC grade) was stirred over sulfuric acid (10:1 cyclohexane:sulfuric acid) for 7-10 days, decanted, and distilled from a sodium dispersion under nitrogen. The cyclohexane was then vacuum distilled from poly(styryllithium) immediately prior to use. All reagents were transferred using syringe and cannula techniques under ultrapure (99.999%) nitrogen.

Synthesis of a Pre-formed TMEDA/nBuLi Adduct. A 25-mL round-bottomed flask was charged with nBuLi (5 mL, 8 mmol) and cooled to -20 °C. The tertiary diamine TMEDA (0.59 mL, 4 mmol) was added drop-wise. Upon addition of the tertiary diamine, the white TMEDA/nBuLi adduct formed rapidly. The adduct was re-dissolved by heating to 67 °C and used to initiate the polymerization.

Synthesis of Poly(1,3-cyclohexadiene)-*block*-Poly(isoprene)-*block*-DVB. A 500-mL round-bottomed flask containing anhydrous cyclohexane (60 mL, 0.54 mol) and 1,3-cyclohexadiene (6.70 mL, 62.5 mmol) was maintained at 25 °C. The preformed TMEDA/nBuLi adduct (0.56 mmol) was added using a syringe and the polymerization was maintained at 25 °C for 120 minutes. Following the polymerization of 1,3-cyclohexadiene, isoprene (17 mL, 78 mmol) and cyclohexane (300 mL, 1.71 mol) were added to the reaction. Upon addition of isoprene, the polymerization immediately changed to a light green color, indicative of the poly(isoprenyllithium) species in the presence of TMEDA. The isoprene was polymerized for 2 h prior to the addition of DVB (0.72 mL, 3.51 mmol). Upon addition of DVB, the polymerization immediately changed to a deep red color, indicative of the divinylbenzyllithium species. The DVB coupling reaction was allowed to proceed for 12 h at 40 °C. The polymerization was terminated using degassed methanol (Burdick-Jackson, HPLC grade). The resulting polymer was precipitated into isopropanol (2.5 L), filtered, and dried at 50 °C in vacuo for 12-18 h. An antioxidant such as Irganox 1010 (0.10 weight % compared to the polymer) was added to the precipitation solvent to retard oxidative degradation during subsequent storage.

Polymer Characterization. ^{1}H NMR spectra were determined in CDCl$_3$ at 400 MHz with a Varian Spectrometer. Glass transition temperatures were determined using a Perkin-Elmer Pyris 1 DSC at a heating rate of 10 °C/min under nitrogen. Glass transition temperatures are reported as the midpoint of the change in heat capacity during the second heat. Molecular weights were determined using size exclusion chromatography with a Waters 2410 refractive index detector (SEC-DRI) in combination with a Wyatt Technology Minidawn MALLS detector (SEC-MALLS) for absolute molecular weight measurements. The dn/dc values were determined on-line using the calibration constant for the RI detector and the mass of the polymer sample. For all samples, it was assumed that 100% of the polymer eluted from the column during the measurement. SEC measurements were performed at 40 °C in chloroform at a flow rate of 1.0 mL/min.

Film Preparation. Film samples for viscoelastic measurements were prepared by dissolving the block copolymers in chloroform (10 wt%) in the presence of an antioxidant (Irganox 1010, Ciba-Geigy Group). The polymer solution was poured into a Petri dish and the solvent was allowed to slowly evaporate over 5-7 days at 25 °C. The films were then dried under vacuum at 40 °C for 4 h and finally dried at 80 °C under vacuum for 24 h to a constant weight. These films were clear, colorless and elastomeric. The average film thickness (0.25 mm, n = 5) was measured using a metric micrometer.

Dynamic Mechanical Analysis. Dynamic mechanical analysis (DMA) was performed using a TA-instruments 2988 dynamic mechanical analyzer. The storage modulus (G′) and loss modulus (G″) were measured at a fixed frequency and strain (110%). Samples were cut (1 x 8 mm) from the 0.25 mm solution cast films and deformed at a constant rate (1Hz) while heating at a rate of 5 °C/min. The glass transition temperature for the polyisoprene block was determined by measuring the peak in the loss modulus. The glass transition temperature of the poly(1,3-cyclohexadiene) block was estimated by measuring the temperature where the elastomer mechanically failed. It was assumed that at this elevated temperature the onset of the poly(1,3-cyclohexadiene) glass transition resulted in the failure of the poly(1,3-cyclohexadiene) domains and subsequent mechanical failure of the elastomer.

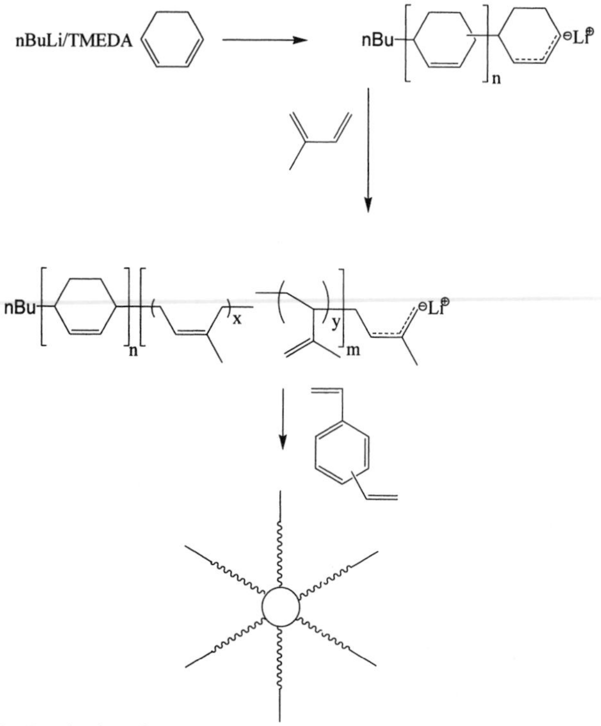

Scheme 1. Synthesis of poly(1,3-cyclohexadiene-*block*-isoprene) DVB coupled star-shaped elastomers.

Tensile Measurements. The tensile properties of the polymers were determined using a Texture Analyzer, (Texture Analysis Inc). Films samples were solution cast as described above, cut into microdumbells (10 x 2.70 x 0.25 mm), and characterized as outlined in ASTM D 882-01. Data are reported as the average of 3 or more measurements.

Atomic Force Microscopy. Tapping mode atomic force microscopy (TMAFM) was performed on a Nanoscope III microscope from Digital Instruments Inc. at room temperature in air using microfabricated cantilevers provided by the manufacturer (spring constant of 30 N m^{-1}).[27] Digital Instruments image processing software was used for image analysis. The set point and tapping amplitude were 1.80 and 2.80 V, respectively. Silicon wafers were prepared by sonication in methanol and subsequently dried under nitrogen. Polymer samples were prepared for AFM analysis by dissolving the polymer in toluene (4 wt%) and spin coating the solution onto prepared silicon wafers at 2000 rpm. The samples were annealed at 120 °C for 12 h.

Results and Discussion

Star-shaped polymers with poly(1,3-cyclohexadiene-*block*-isoprene) arms were prepared using the "arm-first" method with divinylbenzene as the coupling agent (Scheme 1). The poly(1,3-cyclohexadiene-*block*-isoprene) arms were prepared via sequential addition of the monomers and the viscosity of the reaction solution remained low during the synthesis of the arms. The poly(1,3-cyclohexadiene) block remained soluble during the polymerization and the crossover to the isoprene was efficient, as indicated by the rapid color change (yellow to green). It was previously demonstrated that monomer concentration effects the degree of livingness of 1,3-cyclohexadiene polymerizations.[28] As such, the synthesis of the poly(1,3-cyclohexadiene) block was performed at a monomer concentration of 10 wt%. To minimize the reaction viscosity during the synthesis of the diblocks, the reaction was diluted to a total polymer concentration of 8 wt% when the isoprene charge was added. Upon addition of DVB to the poly(1,3-cyclohexadiene)-*block*-poly(isoprenyllithium) solution, the reaction color immediately changed from green to a deep red color indicative of the formation of a highly delocalized benzylic anion resulting from the rapid crossover from the poly(1,3-cyclohexadiene)-*block*-poly(isoprenyllithium) anion to the DVB monomer. The SEC molecular weight characterization of a star-shaped polymer with poly(1,3-cyclohexadiene-*block*-isoprene) arms is shown in Figure 1. The degree of coupling was estimated from the SEC-DRI measurements, although in some cases the peak from the residual uncoupled diblock could not be deconvoluted from the star polymer peak and no attempts to isolate the star-shaped elastomers by fractionation were made. For all samples synthesized in this study, the degree of coupling of the poly(1,3-cyclohexadiene-*block*-isoprene) arms to the DVB core was approximately 90-95%.

To study the effect of the DVB/nBuLi ratio on the coupling efficiency and number of arms per star, a series of star- shaped polymers was synthesized using a constant diblock arm length of 40000 g/mol (10000 PCDH-30000 PI) with varying molar ratios of DVB to nBuLi (Table 1). The resulting polymers exhibited high number average molecular weights (80000 to 192000) and

Table 1. Molecular weights, molecular weight distribution, coupling efficiency of poly(1,3-cyclohexadiene-*block*-isoprene) DVB coupled star-shaped polymers synthesized with varying ratios of DVB/nBuLi.

Polymer	Block size PCHD-b-PI	DVB/nBuLi	$<M_n>$**	$<M_w/M_n>$**	Coupling Efficiency (%)*	Number of Arms (f)
30TPE31	10K-30K	3:1	80000	1.27	98	2.7
30TPE61	10K-30K	6:1	120000	1.27	98	4.0
30TPE81	10K-30K	8:1	192000	1.25	95	6.4
30TPE121	10K-30K	12:1	140000	1.31	96	4.7

SEC Conditions: 40 °C, Chloroform, *SEC-DRI, **SEC-MALLS

narrow molecular weight distributions (1.25 to 1.31). As expected, an important factor for the control of the star polymer number average molecular weight ($<M_n>$) was the DVB/nBuLi ratio. Star-shaped polymers comprising arms with a number average molecular weight of 40,000 g/mol were prepared at a linking efficiency of approximately 90-95%. The resulting polymers all exhibited relatively narrow molecular weight distributions and a high percent of conversion of the living anionic arms to the DVB core (Table 1).

At a low DVB/nBuLi ratio of 3:1 (sample # 30TPE31), the average number of arms per star (f) was 2.7, which indicated that approximately 30% of the coupled diblocks were dimers. The remaining 70% was comprised of three arm stars. As the DVB/nBuLi ratio was increased in the range of 3:1 to 8:1, the core size increased and a greater number of arms attached to the core. When the DVB/nBuLi ratio was further increased from 8:1 (sample # 30TPE81) to 12:1 (sample # 30TPE121), the average number of arms decreased from 6.4 to 4.7. Previous studies have demonstrated that increased ratios of DVB/nBuLi do not necessarily result in more highly branched star-shaped polydienes, rather that at higher DVB/nBuLi ratios, star formation is more quantitative, i.e., more arm polymers are coupled, but does not always increase the branch functionality.[26] It was also shown that at higher DVB/nBuLi ratios, there was a higher propensity for star-star coupling, increasing the higher molecular weight fraction of the polymer sample

(sample # 30TPE121). Characterization of the star-shaped polymers using SEC-MALLS enabled the identification of these higher molecular weight species. As shown in Figure 1, the relative amount of higher molecular weight species increased at higher molar ratios of DVB/nBuLi. This was attributed to star-star coupling, which was shown to occur in polyisoprene based star-shaped polymers using DVB as the coupling agent.[26] The chromatograms in Figure 1 are depicted in the descending order that reflects a qualitative improvement in the uniformity of the molecular weight distribution, and the 8:1 DVB:nBuLi molar ratio was preferred based on this qualitative analysis.

In spite of the introduction of these higher molecular weight species, the molecular weight distribution of these star-shaped, elastomeric polymers did not increase significantly when the DVB/nBuLi ratio was increased. A ratio of DVB/nBuLi of 8:1 (sample # 30TPE81) resulted in elastomeric, star-shaped materials with the highest molecular weight (192000) and narrowest molecular weight distribution (1.25). As such, a ratio of DVB/nBuLi of 8:1 was used to synthesize a series of polymers containing varying compositions of 1,3-cyclohexadiene.

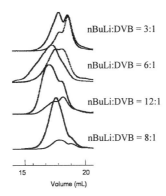

Figure 1. Molecular weight characterization of the series of star-shaped polymers with varying ratios of DVB/nBuLi. The light scattering data (top curve) is superimposed on the refractive index data (bottom trace).

The relative volume fraction of the component blocks in a block copolymer was previously shown to determine the morphology of the block copolymers in both the solid state and in solution.[29] The morphology in the solid state impacts the thermal and mechanical properties of the elastomeric material. A series of poly(1,3-cyclohexadiene) star-shaped elastomers were synthesized with varying volume fractions of poly(1,3-cyclohexadiene) repeat units (15% to 50%) and a constant 8:1 ratio of DVB/nBuLi to investigate the effect of polymer composition on morphology and mechanical properties (Table 2). The composition of the resulting polymers was determined using ^{1}H NMR spectroscopy and agreed well with the relative monomer amounts charged to the reactions. With the exception of sample # 15TPE81, the resulting polymers exhibited high molecular weights and narrow molecular weight distributions. The broad molecular weight distribution and reduced coupling efficiency for sample 15TPE81 was attributed to an increase in viscosity and loss of stirring efficiency upon DVB addition

Table 2. Molecular weight characterization of poly(1,3-cyclohexadiene-*block*-isoprene) DVB coupled star-shaped polymers synthesized with varying percentages of poly(1,3-cyclohexadiene).

Polymer	Block Size PCHD-b-PI	DVB/nBuLi	$<M_n>$**	$<M_w/M_n>$**	Coupling Efficiency (%)*
15TPE81	10K-40K	8:1	350000	1.50	75
30TPE81	10K-30K	8:1	192000	1.25	95
50TPE81	10K-10K	8:1	120000	1.28	85

SEC Conditions: 40 °C, Chloroform, *SEC-DRI, **SEC-MALLS

The glass transition temperatures of the polymers were characterized using both differential scanning calorimetry (DSC) and dynamic mechanical analysis (DMA) (Table 3). The glass transition temperature (T_g) of the polyisoprene block ranged from -15 °C to -13 °C. These elevated glass transition temperatures relative to polyisoprene containing a high percentage of 1,4-addition (T_g = -65 °C) were attributed to the high percentage of 3,4-addition that resulted from the presence of TMEDA in the polymerization solution. The average percentage of 3,4-addition was estimated as 53% using ^{1}H NMR spectroscopy. As expected, the volume fraction of the poly(1,3-cyclohexadiene) block was too low to accurately measure a T_g using DSC analysis. A poly(1,3-cyclohexadiene) homopolymer control was synthesized ($<M_n>$ = 10000, $<M_w>/<M_n>$ = 1.05) and the glass transition temperature of the control was measured. The T_g for this polymer was 138 °C, in good agreement with our earlier studies.[30]

DMA was used to effectively measure the T_g of both the poly(1,3-cyclohexadiene) and

Table 3. Thermal analysis of poly(1,3-cyclohexadiene-*block*-isoprene) DVB coupled star-shaped polymers synthesized with varying ratios of DVB/nBuLi.

Polymer	T_g polyisoprene*	T_g polyisoprene**	T_g PCHD*	T_g PCHD**
PCHD(10K)			138 °C	Not Determined
30TPE31	-15 °C	-5 °C	Not Detected	146 °C
30TPE61	-14 °C	-2 °C	Not Detected	148 °C
30TPE81	-13 °C	-3 °C	Not Detected	145 °C
30TPE121	-14 °C	-1 °C	Not Detected	151 °C

*DSC Conditions: 20 °C/min, second heat, N_2
**DMA Conditions: 5 °C/min, second heat, N_2, 1Hz

polyisoprene blocks. The glass transition temperature for the polyisoprene block (-5 to -1 °C) was measured as the peak in the loss modulus and was within the error of the measurement. These T_gs varied from -5 to -1 °C, which was within the error of the measurement. The glass transition temperature for the poly(1,3-cyclohexadiene) block was measured using the DMA as approximately 148 °C and agreed well with the T_g measured by DSC analysis for the poly(1,3-cyclohexadiene) homopolymer control ($<M_n> = 10000$, $T_g = 138$ °C). The presence of two different glass transition temperatures indicated that the polyisoprene blocks phase separated from the poly(1,3-cyclohexadiene) blocks.

Elastomeric properties were expected to arise from the phase separation between the high T_g poly(1,3-cyclohexadiene) domains and the low T_g polyisoprene domains. Tensile tests were performed to determine the effect of the DVB/nBuLi ratio on the elastomeric properties of the polymers. The tensile properties of samples that were synthesized with a DVB/nBuLi ratio of 3:1, 6:1, 8:1, and 12:1 were compared. All four polymers exhibited an average elongation at break of 633% and an average tensile strength of 17.0 MPa. As expected, the ratio of DVB/nBuLi did not affect the elastic modulus, elongation at break, or the tensile strength of the polymers.

A number of previous studies have shown the impact of composition on the tensile properties of the polymer.[29] Increasing the percentage of the high T_g component was shown to alter the polymer morphology and impact the tensile properties of the polymer. Tensile tests were performed on polymers with varying volume fractions of poly(1,3-cyclohexadiene) (15%, 30%, 50%, and 100%) to examine the effect of polymer composition on the elastomeric properties of the polymers (Table 4). As the percentage of poly(1,3-cyclohexadiene) increased, the elongation at break decreased and the tensile strength increased. The elastic modulus also increased as the percentage of 1,3-cyclohexadiene increased. Previous studies by Natori et al. reported the synthesis of poly(1,3-cyclohexadiene-*block*-butadiene-*block*-1,3-cyclohexadiene) that were prepared via sequential addition with approximately 30% poly(1,3-cyclohexadiene) content.[14] These polymers typically exhibited a tensile strength of 20-25 MPa and a percent elongation of 700%. The star-shaped elastomers prepared in this study exhibit similar mechanical properties to linear polymers with a similar composition.

Table 4. Mechanical analysis of poly(1,3-cyclohexadiene-*block*-isoprene) DVB coupled star-shaped polymers synthesized with varying percentages of poly(1,3-cyclohexadiene).

Polymer	Elastic Modulus (MPa)	Elongation at Break (%)	Tensile Strength (MPa)
15TPE81	0.5 ± 0.1	745 ± 81	7.2 ± 0.2
30TPE81	2.2± 0.1	650 ± 24	15.1 ± 0.1
50TPE81	9.1± 0.1	212 ± 39	17.0 ± 0.1
Poly(1,3-CHD)	23.0± 0.1*	1.8 ± 0.1	Not Determined

Tensile testing performed according to ASTM D-882-01
Youngs Modulus

Multiple glass transition temperatures in combination with excellent elastomeric properties strongly indicated that the poly(1,3-cyclohexadiene) domains phase separated from the polyisoprene domains.[31] Tapping mode atomic force microscopy (TMAFM) was used to examine the microphase separation of the poly(1,3-cyclohexadiene) and the polyisoprene blocks (Figure 2). Silicon wafers were spin coated with a 4 wt% solution of the elastomer (30TPE81) in toluene and the samples were annealed at 120 °C for 12 h under vacuum. The presence of microphase separation between the poly(1,3-cyclohexadiene) block and the polyisoprene block was readily apparent (Figure 2A and B). The image on the left is a topographical image of the surface (Figure 2A) and the image on the right is the phase contrast image of the surface (Figure 2B). Previous studies described the utility of the phase contrast image for the detection of microphases based on the differences in the viscoelastic response of the various blocks.[27] The reduced interaction time of the AFM tip with the poly(1,3-cyclohexadiene) domains resulted in near zero phase lag values causing the poly(1,3-cyclohexadiene) domains to appear darker (Figure 2B). The polyisoprene domains remained in contact with the AFM tip forcing a delay in the tip motion and an increase in the phase lag, giving rise to the brighter areas in the image (Figure 2B). The topographical image of the surface was also used to characterize the polymer surface (Figure 2A). Previous studies suggested that the height difference between the different domains arises from the ability of the lower T_g domains to relax by protruding out of the surface of the film. Accordingly, the lighter regions in the image were attributed to the protruding domains of the polyisoprene blocks, which constituted approximately 70% of the TMAFM image. The relative percentages of the two domains in the AFM image (30% poly(1,3-cyclohexadiene) and 70% polyisoprene) agreed well with the composition of the star-shaped elastomer, which was 30% poly(1,3-cyclohexadiene) and 70% polyisoprene. To our knowledge, this is the first report verifying the presence of microphase separation in poly(1,3-cyclohexadiene-*block*-isoprene) block copolymers using atomic force microscopy.

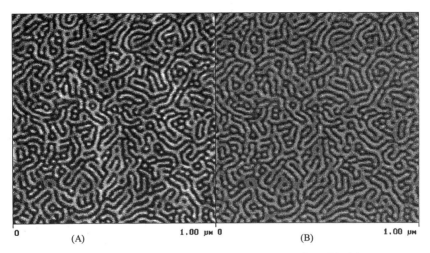

0   (A)   1.00 μм 0   (B)   1.00 μм

Figure 2. Microphase separation present in a poly(1,3-cyclohexadiene-*block*-isoprene) DVB coupled star-shaped elastomer by TMAFM (Sample # 30TPE81).

Conclusion

A novel series of poly(1,3-cyclohexadiene-*block*-isoprene) star-shaped elastomers was successfully synthesized using a convergent "arm-first" approach with DVB. These star-shaped elastomers exhibited controlled molecular weights and narrow molecular weight distributions indicative of a controlled living anionic polymerization. As expected, increased ratios of

DVB/nBuLi resulted in elevated levels of star-star coupling. The tensile strength, elongation at break, and elastic modulus were dependent on the percentage of poly(1,3-cyclohexadiene) present in the elastomer. Microphase separation of the poly(1,3-cyclohexadiene) and polyisoprene blocks was verified using both DMA and TMAFM analysis. TMAFM was performed and the presence of both poly(1,3-cyclohexadiene) and polyisoprene domains was verified. The domains in the TMAFM image were assigned based on the percent composition of the polymer sample and the relative differences in the viscoelastic response between the poly(1,3-cyclohexadiene) and polyisoprene blocks.

Acknowledgments

The authors would like to acknowledge the financial support of Kraton™ Polymers, the Petroleum Research Fund (ACS-PRF# 35190-AC7) as administered by the American Chemical Society, and the NASA-Langley GSRP program. In addition, the FMC Lithium Division is acknowledged for their gracious donation of alkyllithium reagents.

(1)   Hsieh, H. L.; Quirk, R. P. Anionic Polymerization: Principles and Practical Applications; Marcel Dekker: New York, 1996.
(2)   Ming Yu, J.; Dubois, P.; Jerome, R. *Macromolecules* **1996**, *29*, 7316.
(3)   Fetters, L. J.; Morton, M. *Macromolecules* **1969**, *2*, 453.
(4)   Morton, M.; Mikesell, S. L. *J. Macromol. Sci., Chem.* **1993**, *A7*, 1391.
(5)   Quirk, R. P. *Polym. Prepr. (Am. Chem. Soc., Polym. Chem. Div.)* **1985**, *26(2)*, 14.
(6)   Sharaby, Z.; Jagur-Grodzinski, J.; Martan, M.; Vofsi, D. *J. Polym. Sci., Polym. Chem. Ed.* **1982**, *20*, 901.
(7)   Lefebvre, G.; Dawans, F. *J. Polym. Sci.* **1964**, *A2*, 3277.
(8)   Sharaby, Z.; Martan, M.; Jagur-Grodzinski, J. *Macromolecules* **1982**, *15*, 1167.
(9)   Mango, L. A.; Lenz, R. W. *U.S. Nat. Tech. Inform. Serv.* **1972**, *12*, 402.
(10) Mango, L. A.; Lenz, R. W. *Polym. Prepr. (Am. Chem. Soc., Polym. Chem. Div.)* **1971**, *12(2)*, 402.
(11) Natori, I.; Inoue, S. *Macromolecules* **1998**, *31*, 982.
(12) Natori, I.; Inoue, S. *Macromolecules* **1998**, *31*, 4687.
(13) Natori, I.; Imaizumi, K.; Yamagish, H.; Kazunori, M. *J. Polym. Sci., Part B: Polym. Phys.* **1998**, *36*, 1657.
(14) Imaizumi, K.; Ono, T.; Natori, I.; Sakuri, S.; Takedo, K. *J. Polym. Sci., Part B: Polym. Phys.* **2001**, *39*, 13.
(15) Matsen, M. W. *Polymer Science* **2000**, *113*, 5539.
(16) Meneghetti, S. P.; Lutz, P. J.; Rein, D. Star-Shaped Polymers via Anionic Polymerization Methods ; Mishra, M., Kobayashi, S., Eds.; Marcel Dekker: New York, 1999, pp 27.
(17) Fetters, L. J.; Kiss, A. D.; Pearson, D. S. *Macromolecules* **1993**, *25*, 647.
(18) Knischka, R.; Lutz, P. J.; Sunder, A.; Mulhaupt, R.; Frey, H. *Macromolecules* **2000**, *33*, 315.
(19) Hong, K.; Mays, J. W. *Macromolecules* **2001**, *34*, 782.
(20) Williamson, D. T.; Elman, J. F.; Madison, P. H.; Pasquale, A. J.; Long, T. E. *Macromolecules* **2001**, *34*, 2108.
(21) Hadjichristidis, N.; Fetters, L. J. *Macromolecules* **1980**, *13*, 193.
(22) Ishizu, K.; Sunahara, K. *Polymer* **1995**, *36*, 4155.
(23) Hadjichristidis, N. *J. Polym. Sci., Part A: Polym. Chem* **1999**, *37*, 857.
(24) Storey, R. F.; Shoemake, K. A. *J. Polym. Sci., Part A: Polym. Chem* **1999**, *37*, 1629.
(25) Ishizu, K.; Kitana, H.; Ono, T.; Uchida, S. *Polymer* **1999**, *40*, 3229.

(26) McGrath, J. E., Ed. *Anionic Polymerization: Kinetics, Mechanisms, and Synthesis*; American Chemical Society: Washington D.C., 1981; Vol. 166.

(27) Leclere, P.; Lazzaroni, R.; Bredas, J. L.; Yu, J. M.; Dubois, P.; Jerome, R. *Langmuir* **1996**, *12*, 4317.

(28) Williamson, D. T.; Long, T. E. *Polym. Prepr. (Am. Chem. Soc., Polym. Chem. Div.)* **2001**, *42(2)*, 634.

(29) McGrath, J. E. An Introductory Overview of Block Copolymers; Meier, D. J., Ed.; Harwood Academic Publishers: New York, 1979; Vol. 3.

(30) Williamson, D. T.; Hudelson, C. L.; Long, T. E. *Polym. Prepr. (Am. Chem. Soc., Polym. Chem. Div.)* **2002**, *43(1)*, 502.

(31) Stroble, G. The Physics of Polymers; 2nd ed.; Springer-Verlag: Berlin, 1997.

(32) Hashimoto, T.; Mitsuhiro, S.; Fujimura, M.; Kawai, H. Microphase Separation and the Polymer-Polymer Interphase in Block Copolymers; Meier, D. J., Ed.; Hardwood Academic: New York, 1979; Vol. 3.

Synthesis and Structure – Property Relationships for Regular Multigraft Copolymers

Jimmy W. Mays,[1,2] *David Uhrig,*[3] *Samuel Gido,*[4] *Yuqing Zhu,*[4] *Roland Weidisch,*[5] *Hermis Iatrou,*[6] *Nikos Hadjichristidis,*[6] *Kunlun Hong,*[2] *Frederick Beyer,*[7] *Ralf Lach,*[5,8] *Matthias Buschnakowski*[8]

[1]Department of Chemistry, University of Tennessee, Knoxville, TN 37996, USA
[2]Chemical Sciences Division, Oak Ridge National Laboratory, Oak Ridge, TN 37831, USA
[3]School of Chemistry, Physics, and Earth Sciences, Flinders University, GPO Box 2100, Adelaide SA 5001, Australia
[4]Department of Polymer Science and Engineering, University of Massachusetts at Amherst, Amherst, MA 01003, USA
[5]Polymer Research Institute, 01068 Dresden, Germany
[6]Department of Chemistry, University of Athens, Athens 157 71, Greece
[7]U.S. Army Research Laboratory, Polymers Research Branch, Aberdeen Proving Ground, MD 21005, USA
[8]Dept. of Materials Science, University Halle-Wittenberg, 06099 Halle, Germany

Summary: Multigraft copolymers with polyisoprene backbones and polystyrene branches, having multiple regularly spaced branch points, were synthesized by anionic polymerization high vacuum techniques and controlled chlorosilane linking chemistry. The functionality of the branch points (1, 2 and 4) can be controlled, through the choice of chlorosilane linking agent. The morphologies of the various graft copolymers were investigated by transmission electron microscopy and X-ray scattering. It was concluded that the morphology of these complex architectures is governed by the behavior of the corresponding miktoarm star copolymer associated with each branch point (constituting block copolymer), which follows Milner's theoretical treatment for miktoarm stars. By comparing samples having the same molecular weight backbone and branches but different number of branches it was found that the extent of long range order decreases with increasing number of branch points. The stress-strain properties in tension were investigated for some of these multigraft copolymers. For certain compositions thermoplastic elastomer (TPE) behavior was observed, and in many instances the elongation at break was much higher (2-3X) than that of conventional triblock TPEs.

Keywords: anionic polymerization; graft copolymer; mechanical properties; molecular architecture; morphology

© 2004 International Union of Pure and Applied Chemistry

DOI: 10.1002/masy.200451110

Introduction

Graft copolymers are branched block copolymers having a backbone composed of one type of polymer with pendant side chains that are chemically different from the backbone.[1] Graft copolymers are complex materials due to potential variations in: 1. backbone length and polydispersity, 2. branch length and polydispersity, 3. number of branch points per molecule, 4. location of branch points along the backbone, and 5. functionality of the branch points, i.e. the number of branches attached at each branch point. A synthetic approach for producing true model graft copolymers would have all five of these variables under tight control. Unfortunately, the conventional strategies of "grafting from", "grafting onto", and "copolymerization of macromonomers" are usually only effective at controlling the first two parameters, even when living polymerizations are used.

For example, Figure 1 shows a recent example of grafting from chemistry.[2] The backbone is polystyrene (PS) incorporating some 4-methylstyrene as comonomer. Living radical polymerization was used to synthesize this copolymer backbone, so its polydispersity is low. Treatment of the backbone with potassium superbase produced benzylpotassium anionic active sites that are used for the polymerization of 4-methoxystyrene. This method thus controls backbone and branch lengths and polydispersities. On the other hand, the number of branch points per molecule and their placement along the backbone is a statistical process, controlled mostly by the reactivity ratios of styrene and 4-methystyrene. While the branch points would be expected to be trifunctional (two backbone segments and one branch connected to each), creation of effectively tetrafunctional (or higher) branch points could occur due to the presence of two (or more) metallated 4-methystyrene units in close proximity or due to side reactions.

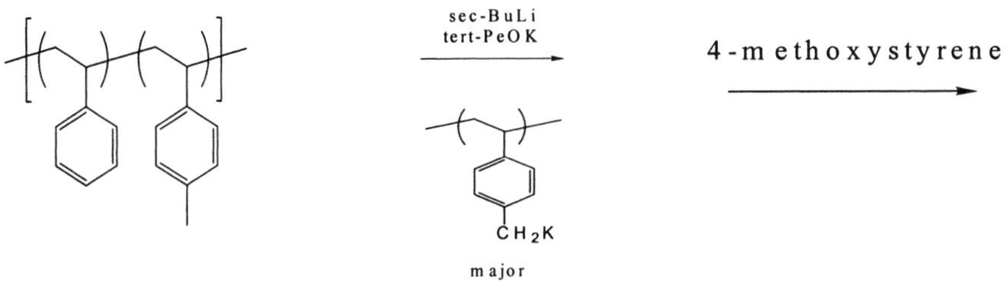

sec-BuLi
tert-PeOK

4-methoxystyrene

CH₂K → CH_2K

major

Figure 1. An example of grafting from. A styrene/4-methylstyrene copolymer is metallated with potassium superbase creating anionic sites that can initiate polymerization of 4-methoxystyrene.

A recent example of the grafting onto strategy is shown in Figure 2.[3] A polybutadiene backbone was prepared by living anionic polymerization in a hydrocarbon solvent resulting in a material containing about 7-8 % 1,2- units randomly distributed along a predominantly 1,4- backbone. These pendant vinyl groups react preferentially with chlorosilanes in hydrosilylation reactions to introduce pendant chlorosilane functionality along the backbone (Fig. 2). PS side chains were synthesized anionically, end capped with a few units of butadiene (to reduce steric hindrance), and grafted onto the backbone by reaction with the chlorosilane groups. Like in the example above, this strategy results in narrow polydispersity backbones and side chains, but the number of branches and their placement are statistically controlled. An advantage of the method depicted in Fig. 2 is the capacity to vary branch point functionality through choice of chlorosilane.

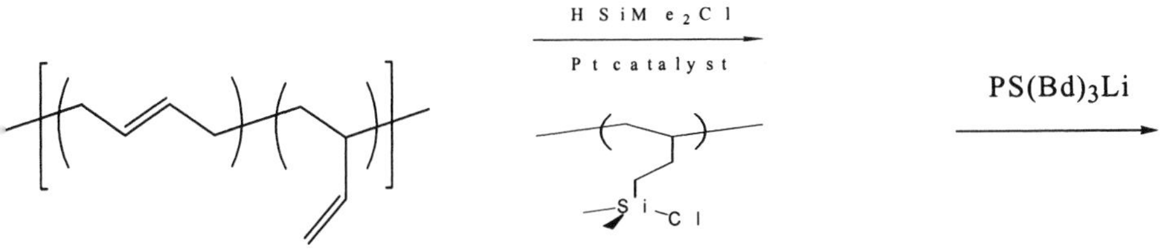

H SiM e₂Cl

Pt catalyst

PS(Bd)₃Li

Figure 2. An example of grafting onto. Hydrosilylation chemistry is used to introduce chlorosilane functionality into 1,2-PBD units. Reaction of living polyanions with these chlorosilane units results in graft copolymer formation.

The macromonomer approach has been widely employed for the synthesis of graft copolymers. An example from work by Hawker and co-workers is shown in Figure 3.[4] These workers utilized TEMPO-mediated living radical copolymerization of a polylactide macromonomer with styrene. Thus, narrow polydispersity backbones and branches were achieved but the reactivity ratios dictated branch point placement. It should be noted that even ideal reactivity ratios of 1 for both monomers does not mean that all chains will contain equal numbers of branches or that their spacing along the backbone will be regular.

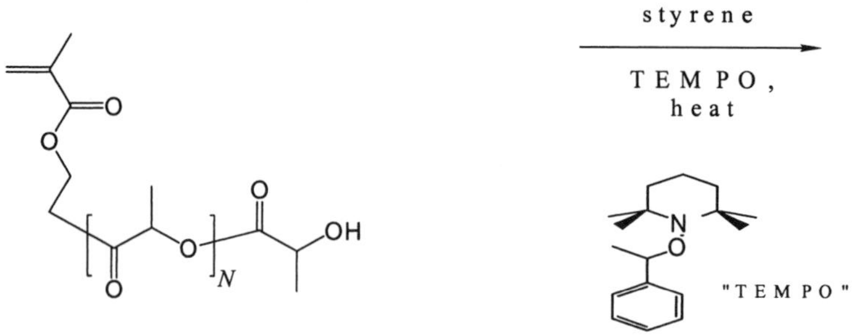

Figure 3. The macromonomer approach. A polylactide macromonomer is copolymerized with styrene in a living radical polymerization to create a graft copolymer.

The obvious strategy for overcoming the problem of control of branch point spacing and extent of branching, as well as for manipulating branch point functionality involves synthesis of segments of the backbone carrying reactive groups at both chain ends, synthesis of branches carrying functionality at one chain end, and controllably linking these segments together. Rempp and co-workers[5] were the first to carry out "polycondensation" of polymers carrying anions at both ends with difunctional electrophiles, such as dibromobutane, to generate a polymer of increased molecular weight. Such a process should obey the kinetics of a step-growth polymerization with polydispersities of about two and the degree of polymerization controlled by the stoichiometry and conversion. Strazielle and Herz[6] reacted triallyloxytriazine (a trifunctional reactant) with dianionic PS (prepared using a dipotassium initiator). Only two of the alylloxy groups reacted with the living dipotassium PS, leaving one allyoxy group at each branch point which was then reacted with living PS made using a monofunctional anionic initiator. This, in principle and after fractionation, should lead to a regular polystyrene comb (branched homopolymer) with uniformity in all five of the molecular parameters outlined above. However, star PS reported in the same paper made using a tetraallyloxy linking agent yielded broader than expected polydispersities, suggesting the presence of side reactions with this chemistry.[6] Recently, a group at Dow Chemical[7] synthesized branched polystyrenes of controlled architecture using the related approach of reacting α,ω-dianionic PS with a mixture of α,α'-dichloro-p-xylene and α,α',α''-trichloromesitylene.

We have recently developed a chlorosilane linking route that is derived from these earlier strategies and applied it to the synthesis of homopolymers and copolymers having multiple, regularly spaced branch points.[8-11] In this paper, we summarize results on the synthesis, molecular characterization, morphology, and tensile properties of copolymers derived from these materials. Described are the "comb", "centipede", and "barbwire" architectures (Figure 4) having polyisoprene (PI) backbones, polystyrene (PS) side chains, and 1, 2, or 4 branches per branch point.

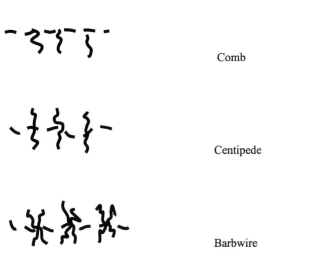

Comb

Centipede

Barbwire

Figure 4. Different multigraft architectures synthesized in this work. In all cases, PI is the backbone and the branches are PS.

Experimental

The synthesis of the copolymers has been previously described in detail.[8,11] Briefly, α,ω-difunctional 1,4-PI is synthesized using a difunctional anionic initiator, (1,3-phenylene)bis(3-methyl-1-phenylpentylidene)dilithium, and reacted with: a. polystyrene carrying a terminal dichlorosilane functionality to create combs; b. PS chains having centrally located dichlorosilane functionality to create centipedes; and c. PS 4-arm stars having centrally located dichlorosilane functionality to create barbwires. The products obtained were polydisperse due to the presence of materials having different numbers of branch points, as expected for a polycondensation. Fractions with high levels of uniformity in terms of number of branch points were isolated by solvent/nonsolvent fractionation as previously described.[8,11] The products were characterized by size exclusion chromatography (SEC), light scattering, and UV and NMR spectroscopy.

Morphological characterization involved films cast from a non-selective solvent (toluene), followed by annealing at 120° C. Osmium tetraoxide staining of the polydiene phase makes the soft phases appear dark in transmission electron micrographs (TEM). X-ray scattering was used to complement and fortify the TEM results. A complete account of the methods used was previously given.[12] Tensile testing was performed on toluene cast annealed specimens using an Instron tensile tester as previously described.[13] Specimens having a 20 mm gauge length were strained at a crosshead speed of 15 mm/min.

Results and Discussion

The synthesis of the three types of branched architectures is summarized in Figure 5. The synthesis of the appropriate dichlorosilane functionalized species is the key step in the synthesis of the regular multigrafts. In the case of combs, the dichlorosilane end capped PS can be synthesized in quantitative yield. Figure 6 shows overlaid, area-normalized SEC chromatograms of a PS branch before and after end capping with methyltrichlorosilane. The chromatograms exactly coincide showing the absence of coupled product, and the use of >100:1 excess of purified chlorosilane assures complete chain end functionalization. With the use of stoichiometric amounts of the two polymeric reagents, we have been able to synthesize comb structures having up to 90 branch points.[10]

Comb

$$PSLi + \text{excess } MeSiCl_3 \longrightarrow (PS)(Me)SiCl_2 + LiCl + MeSiCl_3\uparrow$$

$$LiPILi + (PS)(Me)SiCl_2 \longrightarrow PI\left[(PS)(Me)SiPI\right]_N + LiCl$$

Centipede

$$2\,PSLi + SiCl_4 \longrightarrow (PS)_2SiCl_2 + 2\,LiCl$$

$$LiPILi + (PS)_2SiCl_2 \longrightarrow PI\left[(PS)_2SiPI\right]_N + LiCl$$

Barbwire

$$4\,PSLi + Cl_3Si(CH_2)_6SiCl_3 \longrightarrow Cl(PS)_2Si(CH_2)_6Si(PS)_2Cl + 4\,LiCl$$

$$LiPILi + Cl(PS)_2Si(CH_2)_6Si(PS)_2Cl \longrightarrow PI\left[(PS)_2Si(CH_2)_6Si(PS)_2PI\right]_N + LiCl$$

Figure 5. Synthesis schemes for the various branched architectures.

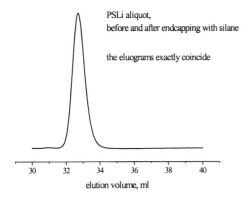

PSLi aliquot,
before and after endcapping with silane
the eluograms exactly coincide
30
32
34
36
38
40
elution volume, ml

Figure 6. SEC chromatograms after end capping with methyltrichlorosilane and after termination with methanol.

Synthesis of the dichlorosilane functionalized polymer intermediates in the synthesis of centipedes and combs is more difficult, due to the inability to achieve a perfect endpoint in the "vacuum titration" [14] of chlorosilanes with poly(styryllithium). Figure 7 shows SEC chromatograms following a vacuum titration to form $(PS)_2SiCl_2$ during the synthesis of a centipede architecture. Notice that at the endpoint it appears that some $PSSiCl_3$ remains. In our experience, it is difficult to push the yield of the coupled product beyond 95-97 % without detecting the presence of 3-arm star species. The imperfect nature of these titrations limits the ability to synthesize centipedes and barbwires with very large numbers of branch points. To date, we have synthesized centipedes with up to 14 branch points and barbwires with up to 5 branch points.[11,12]

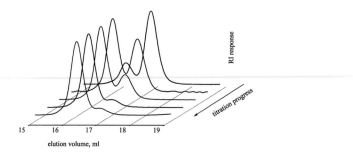

Figure 7. SEC snapshots of the vacuum titration in progress.

The molecular characteristics of the precursors and fractionated products for two barbwire syntheses are given in Table 1. Notice that the polydispersities of the PI backbone segments are 1.13 as opposed to 1.02 to 1.04 for the PS branches. This reflects the slow initiation of isoprene with the difunctional initiator, even in the presence of the polar modifier lithium butoxide. Molecular weights of the repeating units are calculated based on measured molecular weights of the constituents. Functionality of the fractions is calculated by comparing measured molecular weights from light scattering with the unit molecular weight. Three fractions having polydispersities around 1.15 to 1.22 were obtained by fractionating the mother batches of barbwire. Notice that the composition remains almost constant among these fractions while the number of branch points is varied. Thus, these series provide excellent series of model compounds for studying how variations in multigraft architecture and number of branch points affect morphology and mechanical properties.

Table 1. Molecular characteristics of some barbwire copolymers.

	Synthetic Experiment		MG6 1			MG6 2		
PS Graft SEC-MALLS	M_w, kg/mol		8.2			13.0		
SEC-RI	I, M_w/M_n		1.04			1.02		
PI Spacer SEC-MALLS	M_w		86.8			63.3		
SEC-RI	I		1.13			1.13		
Graft Unit			Calculated M_w 120 kg/mol			Calculated M_w 115 kg/mol		
	fractional cut	1	2	3	1	2	3
Multigraft Product ^{1}H NMR	mass% PS	23.5	23.2	23.1	38.3	37.9	38.2
SEC-UV	mass% PS	24.7	24.5	24.3	39.0	38.9	39.1
SEC-MALLS	M_w	705	515	411	409	328	287
SEC-RI	I	1.15	1.16	1.17	1.19	1.22	1.22
Calculated Number of Branch Points		5.2	3.6	2.7	3.0	2.3	1.9
Calculated Composition, mass% PS		24.0	22.8	21.6	38.1	36.4	35.2
Fractionated Final Yield, grams		3.8	4.6	2.8	1.4	2.3	1.8

Milner [15] in 1994 introduced a self-consistent mean field model that predicts the impact of architectural and conformational asymmetry on block copolymer morphology in the strong segregation limit. To a first approximation, the model may be viewed as one that applies to miktoarm stars, star polymers composed of chemically different arms. Milner introduced a parameter, ϵ ($\epsilon \geq 1$), defined as $\epsilon = (n_A/n_B)(l_A/l_B)^{1/2}$ where n_A and n_B are the number of chains of each type connected together in the star and l_A and l_B reflect the differences in chain flexibility of the two different chains. In the case of PS/PI block copolymers, the conformational differences are relatively minor, so the architectural term is dominant. Milner's theory predicts how branched architecture, for polymers having a single branch point, will impact the observed morphology as a function of Φ_B, the volume fraction of the B component is varied for polymers having a single branch point.

To be able to apply Milner's theory to more complex architectures such as the polymers of this work, the concept of the "constituting block copolymer" was introduced empirically and shown to be valid for various architectures.[12,16,17] This concept is demonstrated in Fig. 8. For example, regular combs and centipedes may be imagined as made up of A_2B and A_2B_2 miktoarm stars, respectively. In Figure 9, the morphologies observed for four centipede copolymers are plotted on the Milner phase diagram with the values in the boxes representing the PS content in the sample by volume. All but one of the materials exhibited the predicted morphology. The "67" sample is predicted to lie just inside a bicontinuous morphology window based on the Milner model and the constituting block copolymer hypothesis; instead this material is composed of PI cylinders in a PS continuous phase.

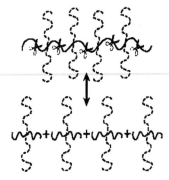

Figure 8. In the multigraft constituting block copolymer concept, the multigraft copolymer is expected to exhibit the same morphology as the simplest constituting miktoarm star copolymer.

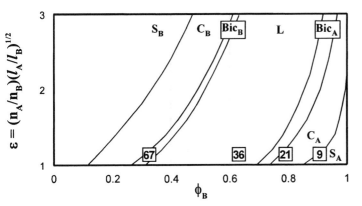

Figure 9. The Milner morphology diagram with the boxes representing four centipede samples having polystyrene volume fractions indicated in the boxes. All but the 67 sample exhibited the morphology predicted by Milner.

Interestingly, when samples made from the same backbone segment and branch but having different numbers of branch points were investigated, they were found to exhibit the same morphology (as expected from the constituting block copolymer hypothesis) but the extent of long range order decreased as the number of branch points was increased (Figure 10).[12] Although every effort was made to anneal these samples to equilibrium morphologies, these differences in long range order could to some extent reflect the increasing hindrances to reorganization imposed by the increasing number of branch points for high molecular weight, highly entangled materials.

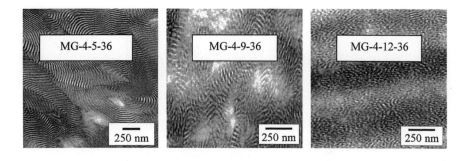

Figure 10. TEM micrographs of three centipede multigrafts having 36 vol% PS and 5, 9, and 12 branch points. Notice the decrease in the extent of long range orientational order on increasing the number of branch points.

The tensile properties of centipede copolymers with various compositions but all with approximately eight branch points are shown in Figure 11. The material having 67 vol % PS is stiff but fails at very low elongation as expected for a material having a PS matrix. The three other materials, exhibiting morphologies of lamellae, PS cylinders, and PS spheres for 37, 22, and 8 % PS, respectively, all exhibit strains at break of >1,000%. A comparison of the tensile properties of two of the centipedes having 22 vol % PS with Styroflex (BASF) and Kraton (Shell) is shown in Figure 12 demonstrating superior mechanical performance of centipedes. The Kraton material used was an SIS triblock of 105,000 in molecular weight and having 21% PS by volume. The Styroflex sample was an SBS material having a statistical SB copolymer as the center block and an overall styrene content of 58 vol % which accounts for its higher stress at break. Clearly, both the stress at break and stress at break of the centipedes increase with increasing number of branch points, and this was confirmed with other specimens.[13] Clearly, the unique architecture of the centipede allows for better stress transfer from the elastomeric backbone to the glassy phases due to the fact that the multiple PS grafts can arrange themselves within multiple PS hard domains.

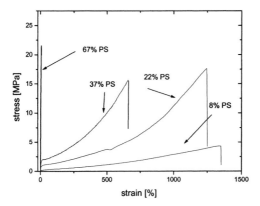

Figure 11. Tensile properties of tetrafunctional multigraft copolymers containing the various volume percents of PS indicated. Strain has been correctly measured using an on-line multisens system.

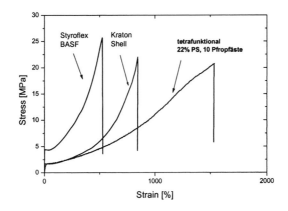

Figure 12. A comparison of the stress-strain properties of two regular centipede graft copolymers with that of two commercial linear triblock TPEs. Strain has been correctly measured using an on-line multisens system.

Finally, it is interesting to examine how an architecturally introduced change in morphology affects the tensile behavior at a constant PS content. In Figure 13, we compare the tensile behavior of centipede and barbwire specimens having essentially the same PS content (21-22 %) and about five branch points. While the elongation at break of the centipede sample, which exhibits a cylindrical morphology, is about same that of the barbwire specimen, the barbwire sample, which exhibits a lamellar morphology, shows twice the tensile strength. Babwires with five branch points are capably of having same tensile strength that of centipedes with about seven to eight branch points.

Clearly the change in morphology from cylindrical to lamellar is due entirely to the change in architecture, and thus this example shows clearly the important impact of block copolymer architecture on their mechanical behavior.

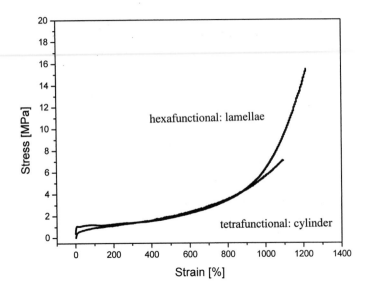

Figure 13. Stress-strain properties of centipede and barbwire specimens having about 24 vol % PS and about 5 branch points. Notice that the shift to a lamellar morphology, which accompanies the change from centipede to barbwire architecture, results in a doubling of the tensile strength of the material.

Conclusions

In summary, well-defined graft copolymers having many precisely positioned branch points of controlled functionality can be synthesized by linking anions (monofunctional branches and difunctional backbone segments) with chlorosilanes. Many branch points may be incorporated into a comb by this strategy but the synthesis of materials with many branch points becomes more difficult for combs and barbwires due to side reactions and steric hindrances to linking. The structure-morphology relationship for multigraft copolymers may be understood by applying Milner's model to the constituting miktoarm star. In multigraft copolymers the observed morphology is independent of the number of branch points but the extent of long range order decreases with increasing number of branch points. Multigraft copolymers show interesting mechanical properties. Strain at break can greatly exceed those of commercial thermoplastic elastomers. Architecturally induced changes in morphology provide a powerful means to manipulate the mechanical properties of graft copolymers.

Acknowledgements

We thank the U. S. Army Research Office for continued support of this research, most recently through DAAGD19-01-1-0544). R. W. acknowledges an Heisenberg-Fellowship from German Science Foundation (DFG). N.H and H.I. acknowledge the financial support of the Greek Ministry of Education (Program: Polymer Science and its Applications) and of the Research Committee of the University of Athens. J.W.M. and K.H. also acknowledges partial support of this research by the Division of Materials Sciences and Engineering, Office of Basic Energy Sciences, U. S. Department of Energy, under contract No. DE-AC05-00OR22725 with Oak Ridge National Laboratory managed and operated by UT-Battelle, LLC.

[1] J. W. Mays, S. P. Gido, in: *"McGraw-Hill Yearbook of Science and Technology 2003"*, McGraw-Hill, New York 2003, p 163.

[2] B. D. Edgecombe, J. M. J. Frechet, Z. Xu, E. J. Kramer, *Macromolecules*, **1998**, *31*, 1292.

[3] M. Xenidou, N. Hadjichristidis, *Macromolecules*, **1998**, *31*, 5690.

[4] C. J. Hawker, D. Mecerreyes, E. Elce, J. Dao, J. L. Hedrick, I. Barakat, P. Dubois, R. Jerome, W. Volksen, *Macromol. Chem. Phys.*, **1997**, *198*, 155.

[5] G. Finaz, Y. Gallot, J. Parrod,, P. Rempp, *J. Polym. Sci.*,**1962**, *58*, 1363.

[6] C. Strazielle and J. Herz, *Eur. Polym. J.*, **1977** *13*, 223.

[7] J. L. Hahnfield, W. C. Pike, D. E. Kirkpatrick, T. G. Bee, *Polymer Preprints*, **1996**, *37(2)*, 733.

[8] H. Iatrou, J. W. Mays, N. Hadjichristidis, *Macromolecules*, **1998**, *31*, 6697.

[9] K. Hong, D. Uhrig, H. Iatrou, Y. Poulos, N. Hadjichristidis, J. W. Mays, *Polymer Preprints*, **1999**, *40(2)*, 104.

[10] Y. Nakamura, Y. Wan, J. W. Mays, H. Iatrou, N. Hadjichristidis, *Macromolecules*, **2000**, *33*, 8323.

[11] D. Uhrig, J. W. Mays, *Macromolecules*, **2002**, *35*, 7182.

[12] F. L. Beyer, S. P. Gido, C. Buschl, H. Iatrou, D. Uhrig, J. W. Mays, M. Y. Chang, B. A. Garetz, N. P. Balsara, N. Beck Tan, N. Hadjichristidis, *Macromolecules*, **2000**, *33*, 2039.

[13] R. Weidisch, S. P. Gido, D. Uhrig, H. Iatrou, J. Mays, and N. Hadjichristidis, *Macromolecules*, **2001**, *34*, 6333.

[14] H. Iatrou , N. Hadjichristidis, *Macromolecules*, **1992**, *25*, 4649.

[15] S. Milner, *Macromolecules*, **1994**, *27*, 2333.

[16] C. Lee, S. P. Gido, Y. Poulos, N. Hadjichristidis, N. Beck Tan, S. F. Trevino,J. W. Mays, *Polymer*, **1998**, *39*, 4631.

[17] C. Lee, S. P. Gido, Y. Poulos, N. Hadjichristidis, N. Beck Tan, S. F. Trevino, J. W. Mays, *J. Chem. Phys.*, **1997**, *107*, 6460.

Macromol. Symp. **2004**, *215*, 127-139

Triblock and Radial Star-Block Copolymers Comprised of Poly(ethoxyethyl glycidyl ether), Polyglycidol, Poly(propylene oxide) and Polystyrene Obtained by Anionic Polymerization Initiated by Cs Initiators

Ph. Dimitrov,[1] *St. Rangelov,*[1] *A. Dworak,*[2] *N. Haraguchi,*[3] *A. Hirao,*[3] *Ch. B. Tsvetanov**[1]

[1] Institute of Polymers, Bulgarian Academy of Sciences, 1113 Sofia, Bulgaria
E-mail:chtsvet@polymer.bas.bg
[2] Institute of Coal Chemistry, Polish Academy of Sciences, 44-102 Gliwice, Poland
[3] Tokyo Institute of Technology, 2-12-1, Ohokayama, Meguro-ku, Tokyo 152-8552, Japan

Summary: The anionic polymerization of ethoxyethyl glycidyl ether (EEGE) initiated by cesium alkoxide was studied. The ring-opening polymerization of EEGE in the presence of cesium alkoxide of 1-methoxy-2-ethanol does not involve any side reactions. The presence of an additional alcohol leads to a significant increase of the initiator efficiency.

Aqueous solutions of poly (ethoxyethyl glycidyl ether) (PEEGE) exhibit lower critical solution temperature (LCST), and the polymer solubility in water is extremely sensitive to its MW. Two novel types of block copolymers based on PEEGE were synthesized: triblock-copolymers of ABA (A'BA') structure, where A is the PEEGE block, A' polyglycidol (PG) and B the polypropylene oxide (PPO) block, and A_2S (A'_2S) and A_4S (A'_4S) heteroarm stars, where S is the polystyrene block.

The synthesis of the ABA block was performed by polymerization of EEGE initiated by bi-functional PPO/Cesium alkoxide macroinitiator. The PEEGE blocks were converted into PG blocks by successful cleavage of the ethoxyethyl group.

Polystyrene/PEEGE and polystyrene/PG three- and five- heteroarm star copolymers were synthesized by a coupling reaction between well-defined chain-end-functionalized polystyrenes carying dendritic benzyl bromide moieties with living anionic polymers of PEEGE with one cesium alkoxide terminal group. The coupling reaction proceeds quantitatively without any side reactions, and thus series of star-branched polymers can be systematically synthesized. Polystyrenes with two or four PG arms have been obtained after the cleavage of the protecting group. The compact structure of these multi-arm star polymers and their amphiphilic character leads to the formation of nanoparticles in aqueous solution with rather uniform size distribution and a mean diameter of 15 nm.

Keywords: block- and heteroarm star copolymers; poly(ethoxyethyl glycidyl ether); polyglycidol; uniform nanoparticles

DOI: 10.1002/masy.200451111

Introduction

Glycidol (2,3-epoxy-1-propanol) is a highly reactive monomer bearing both epoxy and hydroxyl functional groups. Both its composition and structure favor the primary to secondary transitions of the alkoxide active sites, as well as the intermolecular transfers during base-initiated polymerization.[1] The polymerization of glycidol has attracted considerable research interest in the past decade. The propagation may evoke side reactions. Both the anionic[2] and cationic[3,4] polymerizations of this monomer lead to hyper-branched polyethers with low polydispersity. In order to obtain linear polymers of glycidol, its hydroxyl group has to be protected by suitable protecting groups. Spassky et al.[5] have shown that ethoxyethyl glycidyl ether obtained in a reaction of glycidol with ethyl vinyl ether can be polymerized by an anionic mechanism. The protective ethoxyethyl group can then be easily removed thus yielding a linear polyglycidol, which is a highly hydrophilic water-soluble polymer due to the polyether structure with a hydroxyl group in each structural repeat unit. We have recently found that low molecular weight PEEGE of M_n = 2000 g/mol is water-soluble below 11°C.[6] Novel high molecular weight copolymers of ethylene oxide and glycidol or EEGE were obtained by anionic precipitation polymerization by using the calcium amide-alkoxide initiating system.[7]

It is well known that materials made of water-soluble block-copolymers consisting of hydrophobic and hydrophilic blocks or of blocks of varying hydrophobicity are being widely applied in the medical field and in processes requiring the stabilization of emulsions or dispersions. The high interest in these copolymers arises mainly from their unique solution and associative properties, which are a consequence of their molecular structure. An essential requirement for many applications is the precise control of the copolymer structure in terms of its composition, molecular weight, molecular weight distribution, and the hydrophilic/hydrophobic balance.

The purpose of this work is to prepare a family of block-copolyethers with a wide range of hydrophobicity. The investigation is focused on EEGE and propylene oxide (PO) polymerization and copolymerization initiated by cesium alkoxide as well as on the coupling of living poly(ethoxyethyl glycidyl) ether chains with functionalized polystyrene.[8] Thus two novel classes of block-copolymers based on PEEGE were synthesized: triblock-copolymers of ABA (A'BA') structure, where A is the PEEGE block, A' is polyglycidol and B is the polypropylene oxide block, and A₂S (A'₂S) and A₄S (A'₄S) heteroarm stars (palm tree), where S is polystyrene block. In the course of this investigation we also gained insight into the mechanism of polymerization and the structure and properties of the aqueous solutions of the resulting block-copolymers. The micellization of these block-copolymers occurs because of two important factors: hydrophobicity of the polystyrene block and the ability of the PPO and the PEEGE blocks to undergo a temperature-induced phase separation from soluble to insoluble state as their aqueous solutions exhibit a lower critical solution temperature. Thus, depending on the solution temperature regime and the chemical modification via deprotection, almost identical block-copolymers can yield quite different species: double hydrophilic, amphiphilic and hydrophobic block-copolymers.

Experimental

Materials

All solvents were purified by standard methods. Poly(propylene glycol) 2000 purchased from Aldrich was vacuum dried.

Ethoxyethyl glycidyl ether (further referred to as protected glycidol) was synthesized from 2,3-epoxy-1-propanol and ethyl vinyl ether according to Fitton et al [9] and fractionated under reduced pressure. A fraction with purity exceeding 99.8 % (GC) was used. Cesium hydroxide monohydrate 99.5 % (Acros Organics) was used as received. 1-Methoxy-2-ethanol was purified by vacuum distillation.

Chain-end-functionalized polystyrene (PS) with two and four benzylbromide moieties was synthesized according to the procedure described previously.[8] The molecular mass characteristics of the polymer precursors are shown in Table 3.

Polymerization procedure

Preparation of poly(ethoxy ethyl glycidyl ether) with cesium alkoxide terminal group (PEEGEOCs)

Cesium hydroxide monohydrate (0.862 mmol) magnetically stirred in a reaction vessel equipped with argon and vacuum lines, was dissolved with an equimolar amount of 1-methoxy-2-ethanol added at 90°C. After stirring for 1.5 h, the system was switched to the vacuum line for 2 h at the same temperature. Then in some cases, an appropriate quantity of additional alcohol was added at 60°C. The polymerization of EEGE was performed by adding the desired amount of monomer to the fresh ROH/CsOH initiator mixture with the chosen molar amounts for the given experiment. The reaction tube was placed in an oil bath at 60 °C and the formation of PEEGE-OCs proceeded for 48 h. At the end of the reaction time, the growing chains were terminated by adding a large excess of methanol. The monomer was then rapidly stripped under vacuum to avoid further polymerization. The resulting polymer was purified by mixing with cation-binding resin followed by filtration.

Preparation of PEEGE-b-PPO-b-PEEGE block copolymers

The polymerization of EEGE initiated by activated PPO 2000 is described elsewhere.[6]

Coupling reactions

Preparation of PS(PEEGE)n (n = 2 or 4)

0.86 mmol of PEEGE-OCs, dissolved in THF, was added to 5 ml THF solution of 0.043 mmol of bi-functional or tetra-functional polystyrene. The reaction was kept for four weeks in ampoule at 50°C. PS(PEEGE)$_n$ was isolated after precipitation of its acetone solution in hexane.

Cleavage of ethoxyethyl protective groups

The procedure described below is similar to a recent method for deprotection of tetrahydropyranyl ethers.[10] A given amount of star-block-copolymer PS-PEEGE$_n$ or triblock-copolymer PEEGE-PPO-PEEGE was dissolved in MeOH/THF (85:15 v/v). Then AlCl$_3$.6H$_2$O was added and the reaction was kept for one hour at room temperature. The EEGE:AlCl$_3$:MeOH molar ratio was 100:1:800. The reaction product was filtered through Hylfo Super Cel® (diatomaceous earth) and the solvents were evaporated under reduced pressure.

Analyses

Polymer samples were analyzed on a Waters size exclusion chromatography (SEC) system equipped with four Styragel columns with nominal pore sizes of 100, 500, 500, and 1000 Å and a differential refractometric detector. THF was used as the solvent at 40 °C at an elution rate of 1ml/min. Toluene was used as an internal standard for indication of elution volumes. Narrow molecular weight distribution polystyrene samples were used as standards for the molecular weight calibration.

The ^{1}H NMR spectra were recorded at 250 MHz on a Bruker WM 250, using CDCl$_3$, D$_2$O, CD3OD, and THF-d8 as solvents.

Cloud point (CP) transitions of 2% aqueous solutions of the samples were followed on a Specord UV-VIS spectrometer (Carl Zeiss Jena) switched to transmittance regime at $\lambda = 500$ nm using a thermostated cuvette holder. The solutions were initially equilibrated at 0 °C before heating them gradually (0.1 °C).

To determine the critical micellization concentration (cmc) aqueous solutions (2 ml) of a block-copolymer in the concentration range from 0.2 to 20 g/l were prepared at 4°C. 20 µl of a 0.4 mM solution of 1,6-diphenyl-1,3,5-hexatriene (DPH) in methanol were added to each of the copolymer solutions. Solutions were incubated in the dark for 16 h at 20 °C. Absorbance spectra in the range of $\lambda = 300 - 500$ nm were recorded on a Specord UV-VIS spectrometer (Carl Zeiss, Jena) at 30°C. The main absorption peak characteristic of DPH solubilized in a hydrophobic domain was observed at 356 nm.[11]

PS-(PG)$_n$ heteroarm star block copolymer nanoparticles were prepared via the solvent exchange method[12] at room temperature. 5 ml of initial solution (10 g/l) of PS-(PG)$_n$ in MeOH/THF (3:1 v/v) was added drop-wise to 10 ml of vigorously stirred bi-distilled water. Then the organic solvents were evaporated under reduced pressure in order to obtain aqueous PS-(PG)$_n$ micellar solutions. Samples for transmission electron microscopy (TEM) were prepared by removing the solvent under vacuum. Measurements were made using a JEM 200 CX apparatus.

Results and discussion

Homopolymerization of EEGE

It is well known that the anionic polymerization of oxirane proceeds with a fast exchange between free hydroxylic compounds and propagating metal alcoxides. Hence, for a sufficiently long period of reaction time the total number of the growing chains corresponds to the overall amount of hydroxylic reactant introduced into the polymerization medium:

$P_n = \alpha[M]/n_{ROH}$,

where n_{ROH} is the number of moles of hydroxylic reactant introduced into the polymerization mixture, α is the monomer conversion, and P_n is the degree of polymerization. The full conversion of the free alcohol is due to the fact that the initial alcohol is a weaker acid than the chain-end alcohol derived from EEGE, thus its alkoxide is a stronger base and presumably more nucleophilic than the alkoxides at the growing chain ends.[13]

The polymerization reaction proceeds under appropriate experimental conditions (anhydrous reactants, Schlenk tube technique under inert atmosphere at 60°C) and with a very high degree of conversion, thus yielding polymers of comparatively low MW distribution (see Table 1). The in situ formation of the active species from $CH_3OCH_2CH_2OH$, and CsOH was studied by varying the relative amount of reactants. Particular attention was paid to the influence of the ROH/Cs$^+$ ratio on the molecular weight characteristics of the polymer. The efficiency of the MW control implies that the alkoxy end groups introduced in excess with regard to CsOH are all located at the α-end of the resulting macromolecules. Moreover, ^{1}H NMR analyses indicate almost quantitative conversion of the initial alcohol during the polymerization process.

Table 1- Polymerization of EEGE – 1.3 mmol CsOH.H$_2$0, T=60°C, reaction time = 48 h

Sample	$\dfrac{[Cs^+]}{[ROH]}$	Monomer (mmol)	Yield (%)	$\overline{Mn}$ Theor.	$\overline{Mn}$ (NMR)	$\overline{Mn}$ (SEC)	$\overline{Mw}/\overline{Mn}$ (SEC)	Initiator efficiency (f)
1	1/3	27.4	100	1026	980	1000	1.26	1.04
2	1/2	27.4	100	1539	1500	1500	1.23	1.02
3	1/1	13.7	100	1539	2400	2700	1.19	0.64
4	1/1	20.0	100	2246	3100	2800	1.17	0.73
5	1/1	27.4	100	3081	4800	5600	1.17	0.64
6	1/1	75.1	86	7253	8200	8000	1.30	0.88

$f = \dfrac{\alpha[M]}{[ROH].\overline{Pn}}$; α - yield, [M] - monomer quantity (mmol), [ROH] - alcohol quantity (mmol),

$\overline{Pn}$ - average degree of polymerization, determined by ^{1}H NMR

The presence of an excess of alcohol leads to a significant increase of the initiator efficiency, which exceeds the theoretical value of 1. The effective control of the molecular weight as well as the comparatively low molecular weight distribution and the high initiator efficiency of the cesium- alkoxide clearly show that the anionic ring-opening polymerization of EEGE does not involve any side reactions.

Synthesis and aqueous solution properties of PPO/PEEGE and PPO/PG triblock copolymers
Sequential copolymerization reactions were not performed until now, since it is more difficult to control the structure and the composition of the copolymer obtained. The synthesis begins with

preparation of the PPO/Cesium alkoxide macroinitiator, by reacting PPO 2000 with CsOH, as described in Ref. 6.

PEEGE-b-PPO-b-PEEGE block-copolymer was obtained by a direct addition of the second monomer (EEGE) onto the bifunctional PPO macroinitiator. The reaction route is described in Scheme 1.

SEC measurements gave monomodal distribution (Figure 1), which implies that neither PPO nor PEEGE produced any detectable homopolymeric fractions during the copolymer synthesis. The PEEGE blocks were converted into PG blocks by successful cleavage of the ethoxyethyl protective groups (see Scheme 1, NMR spectra). Thus two ABA type tri-block copolymers were obtained with the same length of the blocks, the same central B block of PPO, but with quite different hydrophilic/hydrophobic balance. The results are summarized in Table 2.

The aqueous solution properties of the new block copolymers strongly depend on the role of the central PPO block. When PPO is introduced into water, it develops a hydration shell characterized by an enhanced structuring of the water molecules as evidenced by the negative entropy and enthalpy of mixing.[14] Despite the fact that PPO has the same backbone structure as PEO, an optimal water structure cannot be formed, since the methyl groups of the PO monomer units constitute a steric hindrance. Hence the strain of the water structure results in less hydrogen bond energy as compared to PEO. This is the reason why PPO is partially soluble in water only in the case of low molecular weight polymers. The smaller amount of hydrogen bond energy due to the strain of the water lattice also leads to a LCST property. By increasing the number of methyl groups along the chain it would be expected that the strain in the water structure would increase and thus lead to a less negative value for the heat of mixing and a lower precipitation temperature. Because of the LCST of the PPO block (PPO 2000 has LCST near 19°C) the two new ABA block copolymers are stimuli-responsive. They are water-soluble at temperatures near 0°C, but undergo a temperature induced phase separation from soluble to insoluble state under moderate heating.

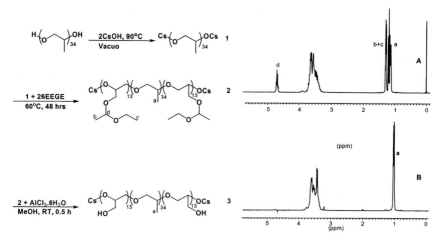

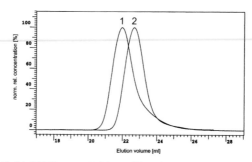

Scheme 1. Synthesis of PEEGE-b-PPO-b-PEEGE and PG-b-PPO-b-PG block copolymers and their ^{1}H NMR spectra: A in CDCl$_3$ and B in D$_2$O

Figure 1. SEC traces of initial PPO macroinitiator (2) and PEEGE-b-PPO-b-PEEGE (1)

PEEGE-b-PPO-b-PEEGE triblock copolymer is built entirely of temperature-sensitive blocks. We have found that PEEGE is more hydrophobic than a PPO block with the same molecular weight.[6] When the central block is more hydrophilic, the tendency for self-assembly of the copolymers is largely weakened as compared to the inverse PPO-b-PEEGE-b-PPO architecture. In contrast, the PG-b-PPO-b-PG copolymer is expected to resemble the association behavior of the most popular nonionic surfactant PEO-b-PPO-b-PEO. At low temperature the aqueous solutions display a double hydrophilic-hydrophilic structure. By raising the temperature the structure becomes amphiphilic, with PPO acting as the hydrophobic block.

The clouding process of both ABA type block copolymers is represented by a sigmoidal curve shown in Figure 2. The thermally induced phase separation leads to a dramatic decrease of the transmittance. As can be seen in Table 2 the CP values and cmc for both structures are quite different. Obviously, the CP and the cmc values of the aqueous solutions of the precursor PEEGE-b-PPO-b-PEEGE are influenced predominantly by the PEEGE blocks. PEEGE, being more hydrophobic can decrease the cmc value. The clouding process in PG-b-PPO-b-PG depends on the PPO block.

Table 2. Characterization of ABA tri-blockcopolymers

Sample	$\overline{M}n$ (NMR)	$\overline{M}n$ (SEC)	$\overline{M}w/\overline{M}n$ (SEC)	CP (20 g/l) (°C)	cmc (g/l)
$EEGE_{13}PO_{34}EEGE_{13}$	5800	5500	1.2	8	0.12
$G_{13}PO_{34}G_{13}$				21	1.90

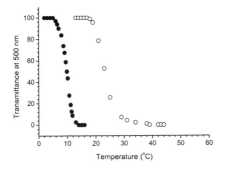

Figure 2. Clouding curves of $EEGE_{13}PO_{34}EEGE_{13}$ precursor (black circles) and $G_{13}PO_{34}G_{13}$ (hollow circles). Concentration =20 g/l

Synthesis of heteroarm star PS(PEEGE)$_n$ and PS(PG)$_n$ copolymers. Formation of nanoparticles.

We attempted to synthesize heterostar-shaped polymers by coupling well defined chain-end-functionalized polystyrenes with 2 and 4 benzyl bromide moieties and living PEEGE. It is well known that the linking reaction in the synthesis of star block-copolymers is usually the slowest, rate-limiting step. Due to the low concentration of alkoxide end-groups on the PEEGE precursor, the reaction was carried out at 50°C with great excess of PEEGE-OCs and for a long period of time (4 weeks). The coupling reactions proceeded quantitatively without any side reactions. This was proved by SEC (Figure 3) and ^{1}H NMR (Figure 4). The reaction route for the synthesis of PS(PEEGE)$_2$ is shown in Scheme 2.

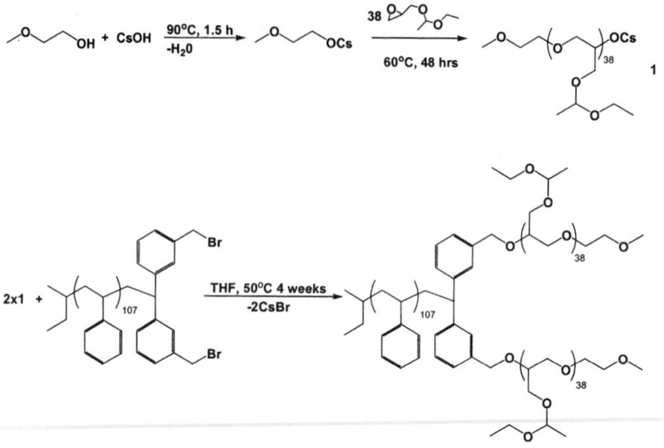

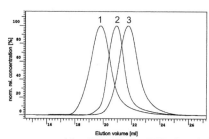

Scheme 2. Synthesis of PS(PEEGE)$_2$ star block-copolymer

Table 3. SEC characterization of PS(PEEGE)$_n$

	PEEGE		PS		PS(PEEGE)n	
	$\overline{Mn}$	$\overline{Mn/Mw}$	$\overline{Mn}$	$\overline{Mw/Mn}$	$\overline{Mn}_{app}$	$\overline{Mw/Mn}$
3 arm	4 500	1.25	11 500	1.02	25 000	1.53
5 arm	4 800	1.36	12 200	1.03	24 200	1.62

Figure 3. SEC traces of arm precursors (3 – PEEGE, 2 – PSBr$_4$) and the product after the linking reaction 1 - PS(PEEGE)$_4$.

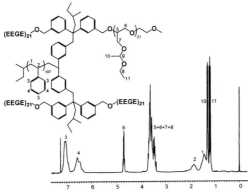

Figure 4. 1H NMR spectrum of heteroarm star-shaped (palm tree) block-copolymer PS(PEEGE)$_4$ in CDCl$_3$

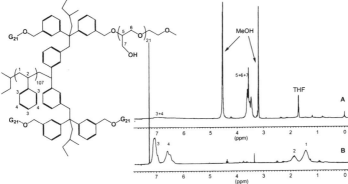

Figure 5. ^{1}H NMR spectra of five arm star-block-copolymers. A – PS(PG)$_4$ in THF-d$_8$/CD$_3$OD; B - PS(PG)$_4$ in CDCl$_3$

The chemical modification of the precursor PEEGE blocks via a deprotection reaction gives access to a new type of heterostar-shaped block copolymers PS(PG)$_n$.

NMR analyses clearly show that CHCl$_3$ is a highly selective solvent for the polystyrene moiety, whereas, the CH$_3$OH/THF mixture is preferred for PG arms (Figure 5).

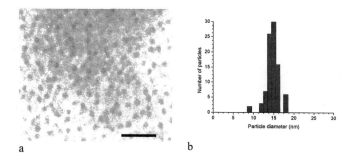

a   b

Figure 6 a) TEM micrograph of PS(PG)$_4$ nano-particles, prepared by the solvent exchange method. The bar scale corresponds to 100 nm. b) Size histogram of particles present in a 5 g/l dispersion of PS(PG)$_4$ in H$_2$O

We used TEM to demonstrate that the PS(PG)$_4$ block copolymers spontaneously form well defined and rather uniform compact spherical nanoparticles in aqueous media. The size distribution (Fig.6b) is monomodal and very narrow. The particle diameters range from 14 to 16 nm. This result supports the fact that in the case of heteroatom star copolymers the association number is considerably lower.[15] The low number of associated unimers could be explained by the fact that each unimer has several stabilizing soluble chains in the micellar shell.

Conclusions

The anionic polymerization of EEGE initiated by cesium alkoxide proceeds as a typical living process without any side reactions. In the PEEGE-b-PPO-b-PEEGE block copolymer each block exhibits a lower critical solution temperature in aqueous media. The PEEG blocks are easily converted into PG blocks. Thus quite different species from an almost identical block polymer backbone can be obtained depending on temperature: double hydrophilic, amphiphilic, and hydrophobic block-copolymers.

Novel polystyrene/polyglycidol heteroarm star block copolymers were synthesized. They form spherical compact and almost uniform micelle-like aggregates in aqueous solution. The polymerization technique used in this study as well as the possibility for the formation of rather short, well defined PEEGE blocks opens the route to the synthesis of a new type of functionalized polymeric micelles.

[1]   A. Sunder, R. Hanselmann, H. Frey, R. Muelhaupt, *Macromolecules*, **1999**, *32*, 4240
[2]   E.J. Vandenberg, *J.Polym.Sci., Polym.Chem.*, **1985**, *23*, 915
[3]   R. Tokar, P. Kubisa, S. Penczek, A. Dworak, *Macromolecules*, **1994**, *2*, 320
[4]   A. Dworak, W. Walach, B. Trzebicka, *Macromol. Chem.Phys.*, **1995**, *196*, 1963
[5]   D. Taton, A.Le Borgne, M. Sepulchre, N. Spassky, *Macromol.Chem.Phys.*, **1994**; *195*, 139
[6]   Ph. Dimitrov, S. Rangelov, A. Dworak, Ch.B. Tsvetanov, *Macromolecules* in print
[7]   Ph. Dimitrov, E. Hasan, S. Rangelov, B. Trzebicka, A. Dworak, Ch.B. Tsvetanov. *Polymer,* **2002**, *43*, 7171
[8]   A.Hirao, N. Haraguchi, *Macromolecules,* **2002**, *35*, 7224
[9]   A. Fitton, J. Hill, D. Jane, R. Miller, *Synthesis,* **1987**, 1140
[10] V.V. Namboodri., R.S. Varma, *Tetrahedron Letters,* **2002**, *43*,1143
[11] P. Alexandridis, J.F. Holzwarth, T.A. Hatton, *Macromolecules,* **1994**, *27*, 2414
[12] C. Allen, D. Maysinger, A. Eisenberg, *Colloids and Surfaces B:Biointerfaces* **1999**,   *16*, 3
[13] J. Hine, M. Hine, *J. Am. Chem. Soc.*, **1952**, *74*, 5267
[14] R. Kjellander, E. Florin, *J.Chem.Soc.,Faraday Trans.*, **1981**, I 77, 2053
[15] M. Stepanek, K. Podhajecka, K. Prochazka, Z. Tuzar, W. Brown, *Langmuir* **2000**, *16*, 6868

Macromol. Symp. **2004**, *215*, 141-150

Mechanism of the Anionic Ring-Opening Oligomerization of Propylene Carbonate Initiated by the tert-Butylphenol/ KHCO₃ System

Sándor Kéki, János Török, György Deák, Miklós Zsuga*

Department of Applied Chemistry, University of Debrecen, 4010 Debrecen, Hungary

Summary: The ring-opening oligomerization reaction of propylene carbonate in the presence of the tert-butylphenol/KHCO₃ initiating system was studied by means of Matrix-Assisted Laser Desorption/Ionization Time of Flight Mass Spectrometry (MALDI-TOF MS) and Electrospray Ionization Time of Flight Mass Spectrometry (ESI-TOF MS). According to the MS spectra obtained, different series of peaks were identified. The MS spectra clearly showed that besides the chain-extension reaction yielding oligomers with all propylene oxide units, the formation of oligomers containing carbonate linkages in the chain, and condensation reaction between the latter two also took place. The structure of the oligomers carrying carbonate linkages was determined by the post-source decay (PSD) MALDI-TOF MS/MS method. Based on the MS results, a mechanism for the oligomerization reaction is proposed.

Keywords: MALDI; oligomers; polyethers; ring-opening polymerization; separation of polymers

Introduction

Cyclic carbonates (e.g. ethylene- and propylene carbonate) undergo ring-opening polymerization at elevated temperatures (> 100 °C) in the presence of appropriate initiators such as Lewis acids, bases, or transesterification catalysts. The oligomerization of EC and PC yields poly(ether carbonates)[1-8] (R= H or CH₃) with the structure shown below:

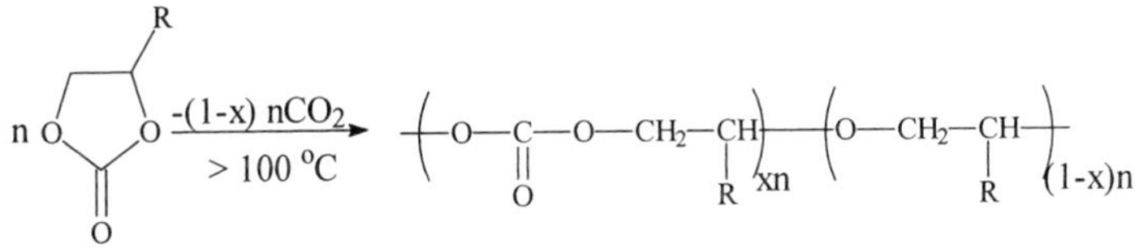

where x is the mol fraction of the carbonate units. It was shown that the carbonate content of the resulting polymer is dependent upon the initiator applied. Initiation with Lewis acids or

    DOI: 10.1002/masy.200451112

transesterification catalysts[1-6] yielded polymers with x=0.4-0.5 carbonate content, and in the presence of base x=0.1-0.3 carbonate was observed[7,8].

The ring opening oligomerization of ethylene- and propylene carbonate in the presence of the bisphenol-A/base initiating system yielding oligomers with all ethylene- and propylene oxide units is of great industrial importance[9]. Previous studies[10,11] have shown that the oligomerization of PC-initiated by the bisphenol-A/base system (in addition to the formation of oligomers with all propylene oxide units), resulted in the formation of oligomers containing carbonate linkages. It was also experienced that the carbonate linkages could be hydrolyzed in an ethanolic solution of potassium hydroxide to yield the desired oligomers (all propylene oxide chain) were obtained[11].

As a continuation of our work, in the present article we have investigated the reaction mechanism of the oligomerization of PC in the presence of a monofunctional phenol derivative using the matrix-assisted laser desorption/ionization time-of-flight (MALDI-TOF)[12,13] electrospray time-of-flight (ESI-TOF)[14,15], and the post-source decay (PSD MALDI-TOF MS/MS)[16,17] mass spectrometric methods.

Experimental

Materials. Methanol, acetonitrile (HPLC grade), ethanol, 2,5-dihydroxybenzoic acid (DHB), sodium trifluoroacetate (NaTFA), potassium trifluoroacetate (KTFA) and tert-butylphenol were received from Aldrich (Germany) and were used without further purification.

Propylene carbonate (analytical grade, Merck, Germany), KOH, $KHCO_3$, and HCl (analytical grade, Reanal, Hungary) were used as received.

Oligomerization. A 50 ml three-necked flask equipped with a magnetic stirrer, thermometer, and nitrogen inlet/outlet in an oil bath was charged with the tert-butylphenol (9.93 g; 0.066 mol), propylene carbonate (27 g; 0.265 mol) and $KHCO_3$ (0.11 g, 0.0011 mol) under nitrogen atmosphere. The reaction mixture was continuously heated to 160 °C in about 1 hour, and it was kept at 160 °C for an additional 23 hours. (Yield: 23.3 g; 91 %).

Hydrolysis of the Oligomers. 2.5 g of the oligomer mixture was added to the solution of KOH in ethanol (1 mol/L) and stirred at room temperature for 2 hours. The reaction mixture was neutralized with concentrated hydrochloric acid, filtered and then the solvent from the filtrate was evaporated using a rotary evaporator. (Yield: 2.13 g; 85 %)

Characterization. *MALDI-TOF MS.* The MALDI MS measurements were performed with a Bruker BIFLEX III™ mass spectrometer (Bruker Daltonik GmbH, Bremen, Germany) equipped with a TOF analyzer. In all cases 19 kV total acceleration voltage was used with pulsed ion extraction (PIE™). The positive ions were detected both in the linear and in the reflectron mode. A nitrogen laser (337 nm, 3 ns pulse width, 10^6-10^7 W/cm^2) operating at 4 Hz was used to produce laser desorption, and 200-250 shots were summed. Samples were prepared with DHB matrix dissolved in methanol (20 mg/ml). Bulk solutions of the oligomer mixture in methanol were made at a concentration of 5 mg/mL. To enhance the cationization, NaTFA or KTFA in methanol (5 mg/mL) was added to the corresponding matrix/analyte solutions. The solutions were mixed in 10:2:1 v/v ratio (matrix:analyte:cationization agent). A volume of 0.5-1.0 μl of these solutions was deposited onto the sample plate (stainless steel), and allowed to air-dry.

MALDI-TOF MS/PSD. All of the PSD spectra were recorded by selection of the ion to be studied, using a pulser allowing an approximately 20 Da window for selection. In each segment the reflectron voltage was decreased. The segments were pasted and calibrated using XMASS 5.0 software from Bruker. The PSD was calibrated using the fragmentation pattern of the Adrenocorticotropic Hormone (ACTH).

The electrospray ionization Time-of-Flight Mass Spectrometry (ESI-TOF MS). ESI-TOF MS measurements were performed on a BioTOF II instrument (Bruker Daltonics, Billerica, MA). Sample solutions were prepared in methanol at 0.01 mg/mL concentration. To enhance the cationization, solutions of NaTFA or KTFA in a concentration of 0.1 mmol/L in methanol were added to the analyte solutions. The solutions were introduced directly into the ESI source by a syringe pump (Cole-Parmer Ins. Co.) at a flow rate of 2 μL/min. The temperature of drying gas (N$_2$) was maintained at 100 °C. The voltages applied on the ESI source were the following: 2000

V capillary, 4000 V cylinder, 4500 V endplate and 120 V capillary exit voltages. The spectra were accumulated and recorded by a digitizer at a sampling rate of 2 GHz.

Liquid Chromatography (HPLC). Samples for HPLC were prepared in acetonitrile (ACN) at a concentration of 1 mg/mL. The measurement were preformed on a Waters 2695 HPLC system equipped with a Waters 2996 photodiode array detector. Separations were made on a C-18 reverse phase column (Waters) thermosted at 35 °C.

30 μL of the analyte solutions were injected and eluted with ACN at a flow rate of 0.7 mL/min.

Results and Discussion

The oligomerization of propylene carbonate was achieved in the presence of the tert-butylphenol/KHCO$_3$ initiating sytem. The reaction mixture was continuously heated to 160 °C within one hour. The evolution of CO$_2$ started at elevated temperature (110 °C) indicating the onset of the reaction. The MALDI-TOF and ESI-TOF MS of the oligomerization reaction mixture obtained after 24 hours reaction time are shown in Fig. 1. The assignments of the peaks appearing in the MS spectra are listed in Table 1.

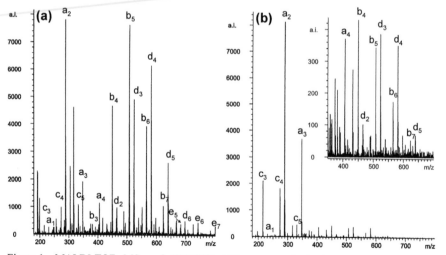

Figure.1. MALDI-TOF MS and ESI-TOF MS spectra obtained on the product of the oligomerization reaction of propylene carbonate initiated by the tert-butylphenol/KHCO$_3$ system. Experimental conditions: 0.25 mol PC, 0.0625 mol tert-butylphenol, 1.8 x 10^{-3} mol KHCO$_3$; Reaction time=24 hours, temperature=160 °C. The subscripts reflect the number of the propylene-oxide units in the given oligomer.

Fig. 1. shows that different series of peaks appeared (**a,b,c,d** and **e** series) in both the MALDI-TOF and the ESI-TOF MS spectra obtained on the resulting reaction mixture. Although there are some differences in the intensity of the oligomer series in the two types of MS spectra, both spectra exhibit practically the same oligomer series. These intensity differences are due to the different sensitivities of the two ionization methods to the oligomer adduct ions.

The mass series of an oligomer mixture in both the MALDI-MS and the ESI-MS (for single charged ions) can be formulated as shown below:

$$M = M_I + M_{endgroups} + nxM_r + M_{Cat} \qquad (1)$$

where M_I, $M_{endgroups}$, M_r and M_{Cat} are the mass of the initiator moiety, the endgroups, repeating unit, and the cation attached to the oligomer, respectively, and n stands for the number of the repeating units (degree of polymerization).

Applying Eq. 1 for the mass series observed in the MS spectra, the types of oligomers formed can be rationalized (Table 1).

Table 1. Assignments of the sodium cationized ($[M+Na]^+$) oligomer series formed in the oligomerization reaction of propylene carbonate initiated by the tert-butylphenol/$KHCO_3$ system.

Series	Structure	Mass of Series
a	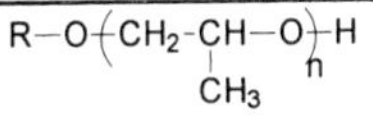 R—O—(CH$_2$-CH—O)$_n$-H, CH$_3$	173+nx58
b	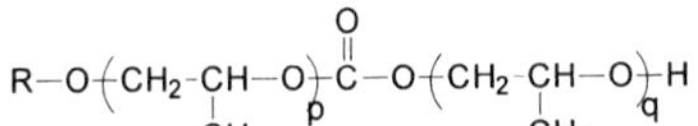 R—O—(CH$_2$-CH—O)$_p$-C(=O)—O—(CH$_2$-CH—O)$_q$-H, CH$_3$ ... CH$_3$ — or — R—O—(CH$_2$-CH—O)$_n$-C(=O)—O—H, CH$_3$ 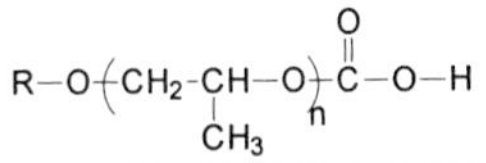	217+(p+q)x58, i.e., 217+nx58
c	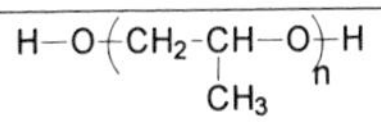 H—O—(CH$_2$-CH—O)$_n$-H, CH$_3$	41+nx58
d	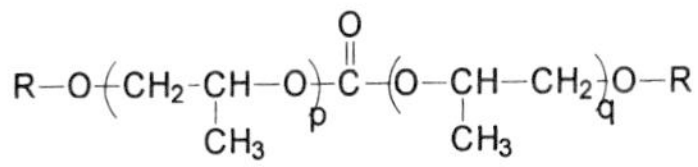 R—O—(CH$_2$-CH—O)$_p$-C(=O)—(O—CH—CH$_2$)$_q$-O—R, CH$_3$ ... CH$_3$	349+(p+q)x58, i.e., 349+nx58
e	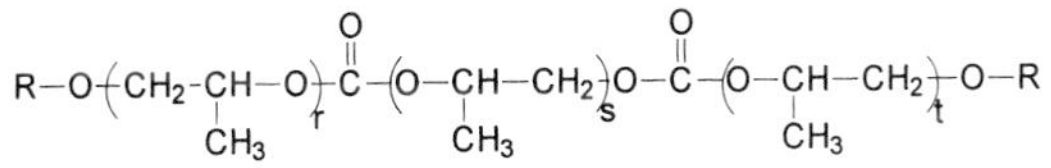 R—O—(CH$_2$-CH—O)$_r$-C(=O)—(O—CH—CH$_2$)$_s$-O—C(=O)—(O—CH—CH$_2$)$_t$-O—R, CH$_3$ ... CH$_3$ ... CH$_3$	393+(r+s+t)x58, i.e., 393+nx58

According to these findings, series **a** corresponds to oligomers containing the initiator moiety with all propylene oxide chains. Series **b** comes from oligomers containing the initiator moiety, propylene oxide repeating units, and one of the propylene carbonate units. The position of the carbonate linkage in series **b** may be in the chain or at the end of the chain. The position of this linkage cannot be judged from simple MS measurements. Series **c** is assigned as propylene glycol oligomers bearing with bearing hydroxyl termini. Series **d** and **e** can be derived from series **a** and **b** as $d = a+b-H_2O$ and $e = 2b-H_2O$.

The presence of the carbonate linkages and their position in the chain in series **b** and **d** were further supported and determined by using the PSD MALDI-TOF MS/MS method. The corresponding [oligomer+Na]$^+$ adduct peaks were selected for PSD and the fragment ions formed under these conditions were recorded. The fragmentations are expected to occur at the carbonates linkages, therefore, by recording the mass of the fragment ions formed from the precursor adduct ions, the position of the carbonate linkages can be determined. Several peaks belonging to the **b** series were selected for PSD, and no fragment ions formed by the loss of a single CO_2 molecule was observed. This implies that carbonate linkages in series **b** are not located at the end of the chain. A representative PSD MALDI-TOF MS/MS spectrum of the precursor ion mass of 507 Da is shown in Fig. 2., which shows that no fragment ions were formed by a loss of a single CO_2 molecule. However, the fragment ions appearing at m/z 197, 215, 259 and 289 derive from structure **1** as depicted in Fig. 2, and peaks at m/z 255, 273 and 317 originate from structure **2** (Fig. 2.). These observations indicate that the precursor ions (at m/z 507) are the mix of three oligomers as shown in Fig. 2; structure **3** is most likely because of the presence of the fragment ions at m/z 347 and 157, and similar fragmentations were observed for the other members of series **b**. It was found in the PSD MALDI-TOF MS/MS spectra of this series that all oligomers carried at least one propylene oxide and one propylene carbonate unit at the chain-end (i.e., q $\geq$2 in Table 1, row 2).

The presence of carbonate linkages in series **d** was also supported by the PSD MALDI-TOF MS/MS method. For example, we have found that the [oligomer+Na]$^+$ adduct ions appearing at m/z 581 are composed of oligomers with p=2; q=2 and p=1; q=3 (see Table 1, row 4).

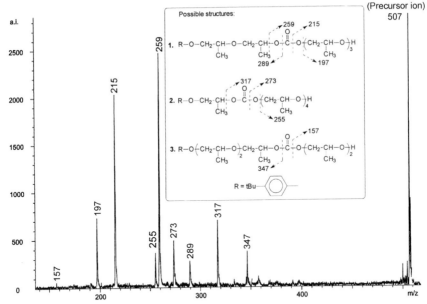

Figure 2. PSD MALDI-TOF MS/MS spectrum obtained on the sodiated oligomer with a precursor ion mass of 507 Da. The inset shows the possible structure and the fragmentation of the oligomer investigated.

To support chemically the presence of the carbonate linkages in the series **b**, **d**, and **e**, the reaction mixtures were treated with an ethanolic solution of KOH. If these oligomers contain carbonate linkages, cleavage at these linkages would occur by hydrolysis according to Eq. 2 and Eq. 3.

$$R_1\text{-}O\text{-}\underset{O}{\overset{\|}{C}}\text{-}O\text{-}R_2 \xrightarrow{\ OH^-\ } R_1\text{-}OH + {}^-O\text{-}\underset{O}{\overset{\|}{C}}\text{-}O\text{-}R_2 \qquad (2)$$

$$R_2\text{-}O\text{-}\underset{O}{\overset{\|}{C}}\text{-}O^- \xrightarrow[-CO_2]{\ H^+\ } R_2\text{-}OH \qquad (3)$$

Indeed, the MALDI-TOF MS and ESI-TOF MS spectra of the reaction mixtures obtained after hydrolysis showed no presence of **b**, **d** and **e** series., i.e., only series **a** and **c** appeared in the MS spectra.

148

Fig. 3 shows the HPLC UV traces recorded at 273 nm of the reaction mixture obtained before and after hydrolysis. Fig. 3 a shows that most of the oligomers are separated by liquid chromatography. The inset in Fig. 3 shows very similar UV spectra indicating that the length of the chain practically does not influence the electronic properties of the aromatic moiety. The assignments of the peaks on the chromatogram presented in Fig. 3 were made by the off-line MALDI-TOF and ESI-TOF MS methods. Fig. 3 b clearly demonstrates that after hydrolysis, the peaks at higher retention time assigned to series b and d completely disappeared.

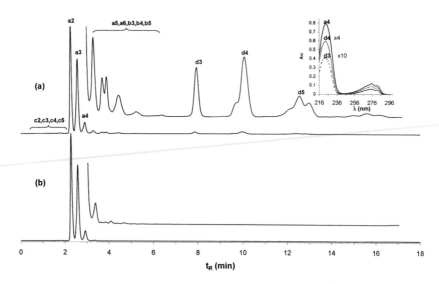

Figure 3. HPLC/UV traces recorded at 273 nm of the oligomer mixture before (**a**) and after hydrolysis (**b**). The inset shows the UV spectra of a_4, d_3 and d_4 fractions. The designation of the fractions is the same as given in Table 1. Experimental conditions: see Fig. 1 caption.

The mechanism of the Oligomerization

Based on the experimental results a mechanism is proposed for the formation of the observed oligomer series (Scheme 1). The phenolate and the alkoxylate chain-ends may react with the propylene carbonate reversibly by carbonyl carbon attack (r_1 and r_1'), or irreversibly by attack on the alkylene carbon (r_2 and r_2'). The carbonyl carbon attack yields alkoxylate chain-ends and carbonate linkage inside the chain, while the alkylene carbon attack produces carbonate chain-ends. However, the nucleophilicity of the carbonate group is much lower than that of the alkoxylate; therefore, oligomerization can no longer proceed from this chain-end. The fate of the carbonate endgroups is to lose CO_2, thereby regenerating alkoxy chain-ends (r_3 and r_3') allowing the oligomerization to continue, or to condense with the oligomers having hydroxy endgroups (r_4, r_5).

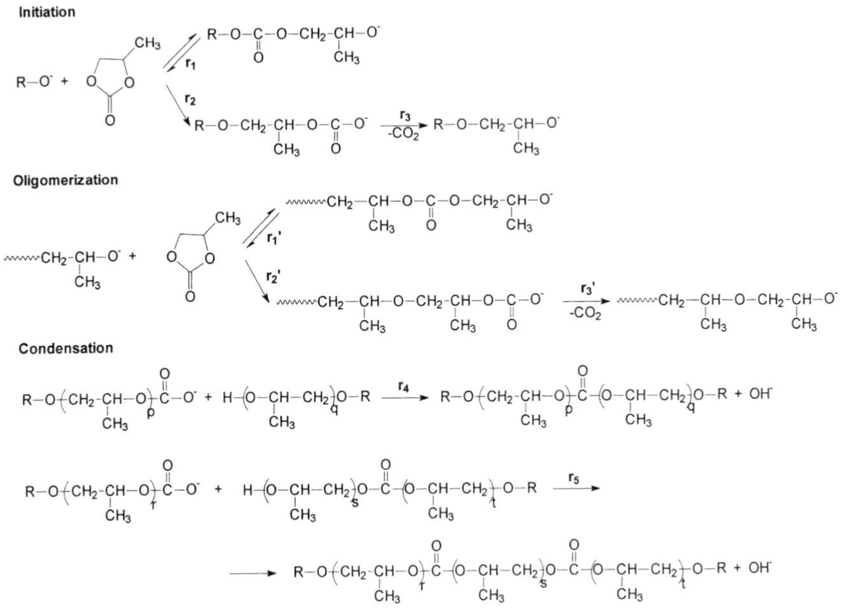

Initiation

$$R-O^- + \text{(propylene carbonate)} \xrightarrow{r_1} R-O-\overset{\overset{\displaystyle O}{\|}}{C}-O-CH_2-\underset{\underset{\displaystyle CH_3}{|}}{CH}-O^-$$

$$\xrightarrow{r_2} R-O-CH_2-\underset{\underset{\displaystyle CH_3}{|}}{CH}-O-\overset{\overset{\displaystyle O}{\|}}{C}-O^- \xrightarrow[-CO_2]{r_3} R-O-CH_2-\underset{\underset{\displaystyle CH_3}{|}}{CH}-O^-$$

Oligomerization

$$\text{wwwCH}_2-\underset{\underset{\displaystyle CH_3}{|}}{CH}-O^- + \text{(propylene carbonate)} \xrightarrow{r_1'} \text{wwwCH}_2-\underset{\underset{\displaystyle CH_3}{|}}{CH}-O-\overset{\overset{\displaystyle O}{\|}}{C}-O-CH_2-\underset{\underset{\displaystyle CH_3}{|}}{CH}-O^-$$

$$\xrightarrow{r_2'} \text{wwwCH}_2-\underset{\underset{\displaystyle CH_3}{|}}{CH}-O-CH_2-\underset{\underset{\displaystyle CH_3}{|}}{CH}-O-\overset{\overset{\displaystyle O}{\|}}{C}-O^- \xrightarrow[-CO_2]{r_3'} \text{wwwCH}_2-\underset{\underset{\displaystyle CH_3}{|}}{CH}-O-CH_2-\underset{\underset{\displaystyle CH_3}{|}}{CH}-O^-$$

Condensation

$$R-O-(CH_2-\underset{\underset{\displaystyle CH_3}{|}}{CH}-O)_p\overset{\overset{\displaystyle O}{\|}}{C}-O^- + H-(O-\underset{\underset{\displaystyle CH_3}{|}}{CH}-CH_2)_q O-R \xrightarrow{r_4} R-O-(CH_2-\underset{\underset{\displaystyle CH_3}{|}}{CH}-O)_p\overset{\overset{\displaystyle O}{\|}}{C}-(O-\underset{\underset{\displaystyle CH_3}{|}}{CH}-CH_2)_q O-R + OH^-$$

$$R-O-(CH_2-\underset{\underset{\displaystyle CH_3}{|}}{CH}-O)_t\overset{\overset{\displaystyle O}{\|}}{C}-O^- + H-(O-\underset{\underset{\displaystyle CH_3}{|}}{CH}-CH_2)_s O-\overset{\overset{\displaystyle O}{\|}}{C}-(O-\underset{\underset{\displaystyle CH_3}{|}}{CH}-CH_2)_t O-R \xrightarrow{r_5}$$

$$\longrightarrow R-O-(CH_2-\underset{\underset{\displaystyle CH_3}{|}}{CH}-O)_t\overset{\overset{\displaystyle O}{\|}}{C}-(O-\underset{\underset{\displaystyle CH_3}{|}}{CH}-CH_2)_s O-\overset{\overset{\displaystyle O}{\|}}{C}-(O-\underset{\underset{\displaystyle CH_3}{|}}{CH}-CH_2)_t O-R + OH^-$$

Scheme 1. Proposed mechanism for the oligomerization of propylene carbonate initiated by the tert-butylphenol/$KHCO_3$ system.

The appearance of series **c** in the MS spectra demonstrates that chain-cleavage at the carbonate links takes place under our conditions according to Eq. 4.

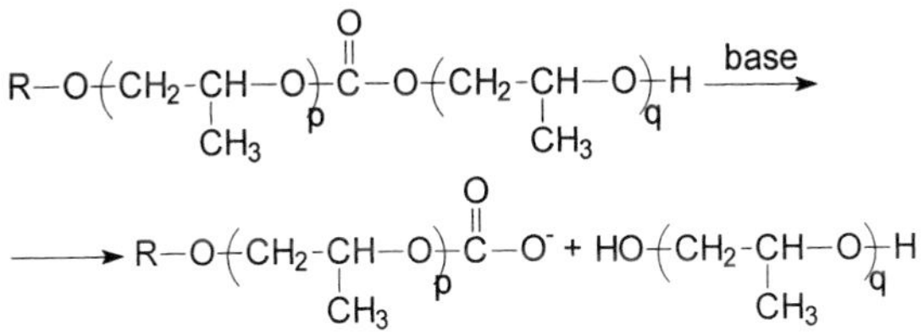

$$R-O-(CH_2-\underset{\underset{\displaystyle CH_3}{|}}{CH}-O)_p\overset{\overset{\displaystyle O}{\|}}{C}-O-(CH_2-\underset{\underset{\displaystyle CH_3}{|}}{CH}-O)_q H \xrightarrow{\text{base}}$$

$$\longrightarrow R-O-(CH_2-\underset{\underset{\displaystyle CH_3}{|}}{CH}-O)_p\overset{\overset{\displaystyle O}{\|}}{C}-O^- + HO-(CH_2-\underset{\underset{\displaystyle CH_3}{|}}{CH}-O)_q H \qquad (4)$$

Conclusions

We investigated the anionic ring-opening oligomerization of propylene carbonate initiated by the tert-butylphenol/KHCO$_3$ system. The analysis of the reaction mixture by MALDI-TOF and ESI-TOF MS methods showed that different oligomer series were formed. The presence of the carbonate linkage in the oligomers was proved by comparison of the MS spectra and HPLC traces of the reaction mixture obtained before and after hydrolysis. It was also demonstrated that liquid chromatography is capable of separating the oligomers. The PSD MALDI-TOF MS/MS method was applied to determine the position of the carbonate link, information which can hardly be obtained by other, non-mass spectrometric methods.

Acknowledgment

This work was financially supported by the grants Nos. T 037448, T 042740, M 28369 and M 36872, MU-00204/2001 given by OTKA (Hungarian Scientific Research Fund), and the grant NKF3A/0036/2002 and the Bolyai János and Békésy György Fellowship. The authors also would like to express their thanks to CELLADAM Ltd., Hungary for providing the BIFLEX III™ MALDI-TOF MS instrument.

[1] Soga, K.; Hosoda, S.; Tazuke, Y.; Ikeda, S. *J. Polym. Sci., Polym. Chem. Ed.* **1976**, *14*, 161.
[2] Soga, K.; Tazuke, Y.; Hosoda, S.; Ikeda, S. *J. Polym. Sci., Polym. Chem. Ed.* **1977**, *15*, 219.
[3] Harris, R.F. *J. Appl. Polym. Sci.* **1989**, *37*, 183.
[4] Harris, R.F.; McDonald, L.A. *J. Appl. Polym. Sci.* **1989**, *37*, 1491.
[5] Vogdanis, L.; Heitz, W. *Makromol. Chem., Rapid Commun.* **1986**, *7*, 543.
[6] Storey, R.F.; Hoffman, D.C. *Macromolecules* **1992**, *25*, 5369.
[7] Vogdanis, L.; Martens, B.; Uchtmann, H.; Hensel, F.; Heitz, W. *Makromol. Chem.,* **1990**, *191*, 465.
[8] Lee, J.C.; Litt, M. H. *Macromolecules* **2000**, *33*, 1618.
[9] Soós, L.; Karácsonyi, B.; Szécsi, I.; Polgári, I. Procedure for the preparation of Di-, Tetra-, Hexa- and Octamer Derivatives of Hydroxyalkyl-Bisphenol-A, Hungarian Patent, No. 209680, 1996.
[10] Soós, L.; Deák, Gy.; Kéki, S.; Zsuga, M. *J. Polym. Sci. Part A. Polym. Chem.* **1999**, *37*, 545.
[11] Kéki, S.; Török, J.; Deák, Gy. Zsuga, M. *Macromolecules*, **2001**, *34*, 6850.
[12] Karas, M.; Hillenkamp, F. *Anal. Chem.* **1988**, *60*, 2299.
[13] Tanaka, K.; Waki, H.; Ido, Y; Akita, S.; Yoshida, Y; Yoshida, T. *Rapid Comm. Mass Spectrom.* **1988**, *2*, 151.
[14] Yamashita, M.; Fenn, J.B. *J. Phys. Chem.* **1984**, *88*, 4451.
[15] Yamashita, M.; Fenn, J.B. *J. Phys. Chem.* **1984**, *88*, 4671.
[16] Spengler, B.; Kirsch, D.; Kaufmann, R. *Rapid Commun. Mass Spectrom.* **1991**, *5*, 198.
[17] Spengler, B.; Kirsch, D.; Kaufmann, R.; Jaeger, E. *Rapid Commun. Mass Spectrom.* **1992**, *6*, 105.

Macromol. Symp. **2004**, *215*, 151-163 151

Synthesis and Self-Association of Stimuli-Responsive Diblock Copolymers by Living Cationic Polymerization

Sadahito Aoshima,[1] *Shinji Sugihara,*[1] *Mitsuhiro Shibayama,*[2] *Shokyoku Kanaoka*[1]

[1] Department of Macromolecular Science, Graduate School of Science, Osaka University, Toyonaka, Osaka 560-0043, Japan
E-mail: aoshima@chem.sci.osaka-u.ac.jp
[2] The Institute for Solid State Physics, The University of Tokyo, Kashiwa, Chiba 277-8581, Japan

Summary: The living cationic polymerization of several functional monomers in the presence of an added base is investigated as a possible preparation of a new series of water-soluble or stimuli-responsive copolymers. Under appropriate conditions, the polymerization allows the selective preparation of polymers with various shapes and different sequence distributions of monomer units, including stimuli-responsive block copolymers, gradient copolymers, poly(vinyl alcohol) graft copolymers, and star-shaped polymers. The stimuli-induced self-association of the diblock copolymers is also examined. An aqueous solution of the diblock copolymer with a thermo-sensitive segment undergoes rapid physical gelation upon warming to the critical temperature to give a transparent gel, and returns sensitively to the solution state upon cooling. The sharp transition of stimuli-responsive segments with highly controlled primary structure turns out to play an important role in the self-association. Small-angle neutron scattering, dynamic light scattering, and electron microscopy studies reveal that the physical gelation involves a thermosensitive micellization of diblock copolymers (core size: 18–20 nm) and subsequent micelle macrolattice formation (bcc symmetry). Based on the gelation mechanism, several stimuli-responsive gelation systems are achieved using other stimuli such as the addition of a selective solvent or compound, cooling, pH change, and irradiation with ultraviolet light.

Keywords: living cationic polymerization using an added base; macrolattice formation of the micelles; star-shaped polymers; stimuli-induced self-association; stimuli-responsive block copolymer

Introduction

Recently, nano-organized self-assemblies using amphiphilic block copolymers or hydrophobically modified water-soluble polymers have attracted attention both academically and for possible practical applications.[1-3] A key to successful synthesis of more intelligent polymers would be the utilization of living polymerization, as the chemical structure, molecular weight, and hydrophobicity of the amphiphilic polymers are known to control the morphology of the self-aggregates. However, the development of such well-controlled amphiphilic block copolymers has been limited because of difficulties in the formation of a polar hydrophilic moiety by living polymerization. Recent progress in living polymerization, particularly in controlled/living radical polymerization, has led to the design and synthesis of novel well-defined amphiphilic block copolymers.

In the field of cationic polymerization, living polymerization has been achieved through the use of nucleophilic counteranions, externally added bases, or added salts. In the first two cases, the key point is the stabilization of propagating carbocation by the nucleophilic interaction. We are currently investigating the living cationic polymerization of polar monomers in the presence of an added base[4] toward preparation of living polymers with various properties. In particular, the production of stimuli-responsive polymers with a variety of functional groups would make it possible to prepare novel intelligent polymers.[5-10]

In this work, toward designing a new strategy for preparing such stimuli-responsive polymers (mainly diblock copolymers) by living cationic polymerization, the polymerization of monomers with polar functional groups was carried out, and the synthesis of polymers of various shapes was investigated. A new stimuli-induced self-association of the diblock copolymers was also examined.

Results and Discussion

1. *Synthesis and Properties of Various Functional Polymers by Living Cationic Polymerization*
Synthesis. As a typical example of living cationic polymerization of a monomer with a polar side group, homo- and block-copolymerizations of BMSiVE [2-(*tert*-butyldimethylsilyloxy)ethyl vinyl

ether] were carried out using $Et_{1.5}AlCl_{1.5}$ in the presence of ethyl acetate in toluene at 0 °C.[9,11] As shown in Fig. 1(A), the polymer molecular weight increased in direct proportion to conversion. The polymers exhibited very narrow, nearly monodisperse molecular weight distributions (MWDs), as shown in Fig. 1(B). ^1H NMR structural analysis confirmed that the silyloxy pendants remained entirely intact. The MWD of the polymers by sequential block copolymerization is shown in Fig. 1(C). EOVE was polymerized first, and BMSiVE was added neat to the polymerization mixture when EOVE has been consumed almost quantitatively. The MWD clearly shifted toward higher molecular weighs, but still it remained very narrow ($M_w/M_n \leq 1.1$). The absence of tailing to lower molecular weights reflects the quantitative formation of block copolymers. These results demonstrate that the presence of an added base induces the living cationic polymerization of BMSiVE despite the presence of acid-sensitive silyloxy groups.

Scheme 1 summarizes the properties of the living polymers obtained (water-soluble polyalcohols and polycarboxylic acid in the frame were obtained using a corresponding protected monomer). In these systems, the propagating carbocations were stabilized by nucleophilic interaction with the added base to prevent not only the conventional chain transfer and termination, but also acid-catalyzed side reactions.

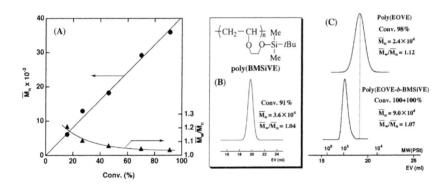

Figure 1. Relationships between conversion and M_n or M_w/M_n (A), MWDs of poly(BMSiVE) (B), and poly(EOVE-b-BMSiVE) (C) obtained with $1/Et_{1.5}AlCl_{1.5}$ in the presence of ethyl acetate in toluene at 0°C: [BMSiVE]$_0$=0.8M, [**1**]$_0$=4.0mM, [$Et_{1.5}AlCl_{1.5}$]$_0$=20mM, [ethyl acetate]=1.0 M; **1**: $CH_3CH(OiBu)OCOCH_3$, poly(EOVE): -[$CH_2CH(OCH_2CH_2OC_2H_5)$]$_n$-.

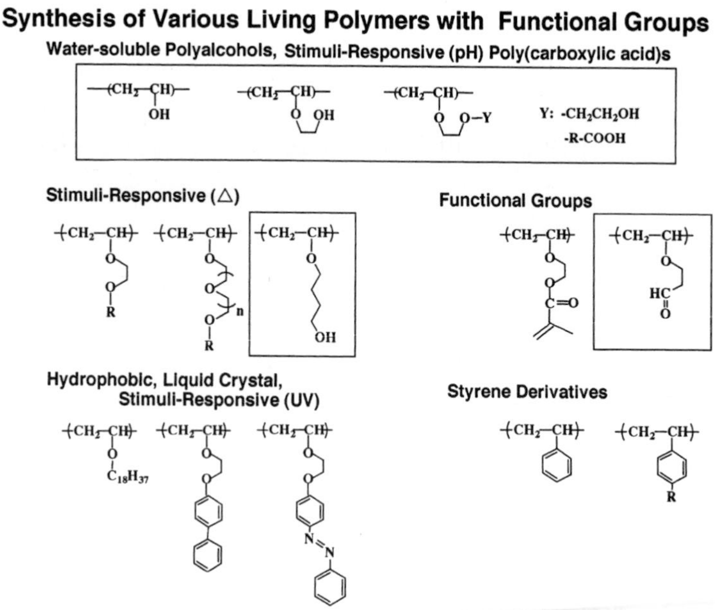

154

Synthesis of Various Living Polymers with Functional Groups
Water-soluble Polyalcohols, Stimuli-Responsive (pH) Poly(carboxylic acid)s

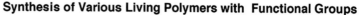

Scheme 1. Synthesis and properties of various living polymers obtained with an added base.

Properties. Heating a transparent aqueous solution of the polymer with oxyethylene chains and a ω-alkyl group (POEVE) caused the solution to become cloudy at a certain critical temperature.[5-7] As shown in Fig. 2, the phase separation was quite sensitive and reversible on heating and cooling. The suggested mechanism of phase separation is as follows. At low temperature, the polymers are soluble in water by hydration. When the temperature reaches the critical point, dehydration occurs, giving rise to aggregation (phase separation) by hydrophobic interaction. POEVEs are the only living polymers to date that exhibit such a sharp phase separation, and the effects of molecular weight, MWD, and sequence distribution on the critical temperature T_{ps} have been investigated extensively. The sequence distributions have been shown to affect the phase separation behavior remarkably, and represents a means of preparing novel thermo-sensitive gelation systems.[5,7]

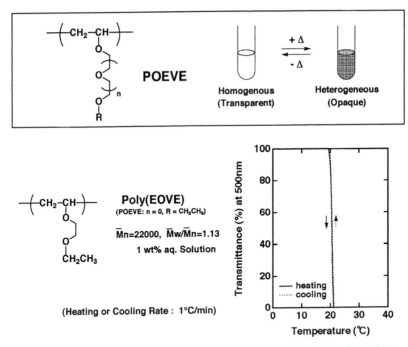

Figure 2. A typical phase diagram, traced by the transmittance at 500nm, of 1 wt% aqueous poly(EOVE) solution.

As another possibility for thermosensitive phase separation, more generalized methods have recently been examined. That is, instead of thermo-sensitive homopolymers with an appropriate hydrophilic/hydrophobic balance (the upper two polymers in Scheme 2), [5-11] thermo-sensitive random copolymers of hydrophilic and hydrophobic monomers were synthesized via living cationic copolymerization (Scheme 2).[12] On heating, an aqueous solution of a random copolymer undergoes thermally induced phase separation at a critical temperature (Fig. 3). This phase separation is quite sensitive and reversible on heating and cooling. Interestingly, the randomness of the sequence distribution is indispensable to realizing such highly sensitive phase separation.

Random Copolymers as Thermo-Sensitive Polymers

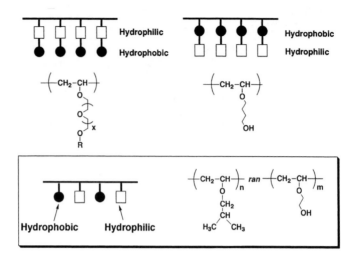

Scheme 2. Thermo-sensitive homopolymers with an appropriate hydrophilic/hydrophobic balance and random copolymers of hydrophilic and hydrophobic monomers.

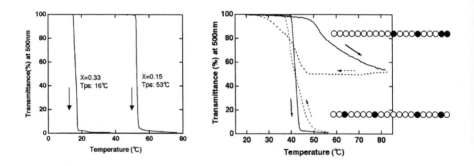

Figure 3. Temperature dependence of the transmittance at 500 nm of 1.0 wt% aqueous solutions of random copolymers (IBVE$_x$-ran-HOVE$_{1-x}$)$_{200}$ (x = 0.33 and 0.15; $M_w/M_n \leq 1.05$) and the effects of sequence distribution on phase separation behavior (x = 0.20): heating rate 1.0 °C/min; poly(HOVE): -[CH$_2$CH(OCH$_2$CH$_2$OH)]$_n$-.

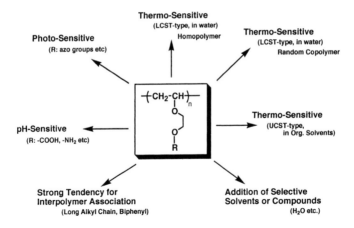

Scheme 3. Synthesis of sensitive stimuli-responsive polymers with various stimuli.

More recently, we have investigated the synthesis of stimuli-responsive polymers, sensitive to other stimuli, as shown in Scheme 3.

2. *Synthesis of Various Shapes of Polymers*

Scheme 4 summarizes the selective preparation of polymers with various shapes and different sequence distributions of monomer units by living cationic polymerization with an added base: (i) star-shaped polymers, (ii) PVA graft copolymers, and (iii) gradient copolymers.

Star-shaped polymers of vinyl ethers were prepared through linking reactions of linear living polymers with a small amount of a bifunctional vinyl ether by living cationic polymerization. Under optimum conditions, as shown in Fig. 4, quantitative conversion of living poly(IBVE)s (arms) was achieved to give star-shaped polymers with a very narrow MWD ($M_w/M_n = 1.11$). Multiangle laser light scattering (MALLS) and dynamic light scattering (DLS) analyses revealed that the M_w and size of the star-shaped polymer obtained with living poly(IBVE) ($M_n = 1.64 \times 10^4$, $M_w/M_n = 1.07$) were 19.5×10^4 (f (the number of arms per molecule) = 11) and 9.5 nm (diameter), respectively.

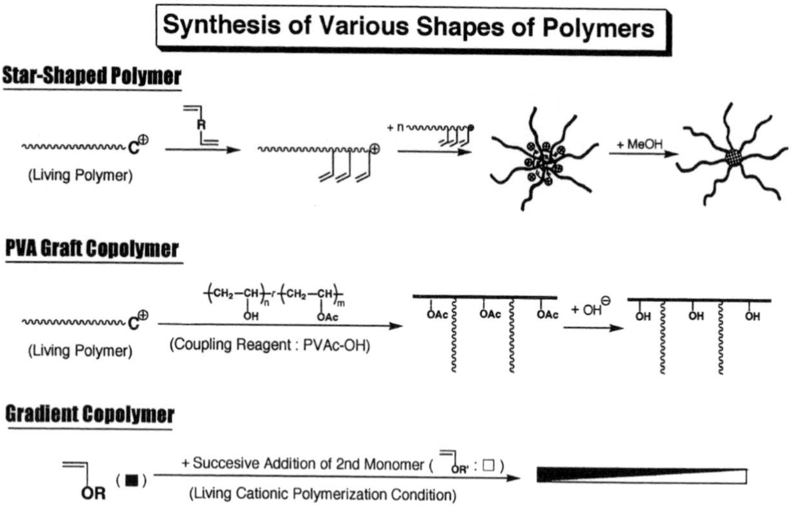

Synthesis of Various Shapes of Polymers

Star-Shaped Polymer

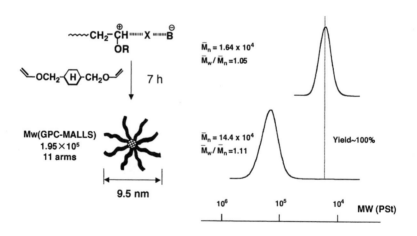

(Living Polymer)

+ MeOH

PVA Graft Copolymer

(Living Polymer)   (Coupling Reagent : PVAc-OH)

Gradient Copolymer

+ Successive Addition of 2nd Monomer ($\square$)
(Living Cationic Polymerization Condition)

Scheme 4. Selective syntheses of polymers with various shapes and different sequence distributions by living cationic polymerization.

$$\sim\sim CH_2-\overset{\oplus}{\underset{OR}{CH}}\cdots\cdots X \cdots\cdots \overset{\ominus}{B}$$

$\bar{M}_n = 1.64 \times 10^4$
$\bar{M}_w / \bar{M}_n = 1.05$

$-OCH_2-\langle H\rangle-CH_2O-$ 7 h

Mw(GPC-MALLS)
1.95×10^5
11 arms

$\bar{M}_n = 14.4 \times 10^4$
$\bar{M}_w / \bar{M}_n = 1.11$

Yield~100%

9.5 nm

10^6 10^5 10^4 MW (PSt)

Figure 4. Synthesis of a star-shaped polymer of IBVE with 1/EtAlCl$_2$ in the presence of ethyl acetate in hexane at 0°C; DP_n(arm)~160; [divinyl compd.]$_0$/[**1**]$_0$=10.0; Living poly(IBVE), [IBVE]$_0$=1.5M , [CH$_3$COOC$_2$H$_5$]=1.0M, [**1**]$_0$/[EtAlCl$_2$]$_0$=10.0/20.0, mM.

Poly(vinyl alcohol) (PVA) graft copolymers were also prepared, consisting of a PVA main-chain and poly(vinyl ether) branches.[13,14] The synthetic route involves the quantitative coupling reaction of living poly(vinyl ether) cationic chains with partially saponified PVAc (PVAc-OH). The latter (PVAc-OH) is soluble in several organic solvents such as toluene, and no side reactions occurred in the course of coupling. The polymer main-chain could be readily converted to PVA by conventional saponification after the coupling reactions. Using this method, various types of grafts, such as two different grafts and diblock grafts, were prepared and the difference in phase separation behavior was examined.

Gradient copolymers have a very characteristic sequence distribution, in which the instantaneous composition changes continuously along the main chain.[15] During living polymerization of the first monomer in the presence of an added base, a second monomer was successively added, and the polymerization proceeded in a living fashion to give narrow-MWD polymers with compositional change along the polymer chain.[16]

3. *Stimuli-Induced Self-Association of Diblock Copolymers*

The rapid formation of a physical gel occurred on warming an aqueous solution of a diblock copolymer having a water-soluble polyalcohol segment and a thermoresponsive segment, -[$CH_2CH(OCH_2CH_2OH)$]$_m$-*block*-[$CH_2CH(OCH_2CH_2OC_2H_5)$]$_n$-, to the lower critical solution temperature (LCST) of phase separation for the thermoresponsive segment.[9,17-19] The viscosity change was quite abrupt and reversible on heating and cooling. Figure 5 shows a temperature scan of G' and G''. Around 20 °C, G' changes by more than 4 orders of magnitude within the temperature range of less than 1 °C.

On the basis of dynamic viscoelasticity, small-angle neutron scattering (SANS), transmission electron microscopy (TEM), and DLS measurements, the mechanism of physical gelation was deduced to begin with the formation of micelles with well-controlled size and structure, followed by close packing of the micelles (bcc macrolattice formation) or weak interaction between the micelles. [17-19]

160

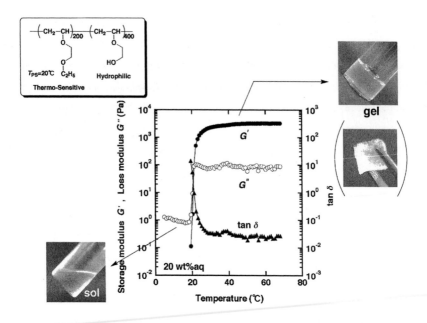

Figure 5. Temperature dependence of storage modulus G', loss modulus G'', and tan δ for 20wt% EOVE$_{200}$-b-HOVE$_{400}$ aqueous solution (frequency=1Hz).

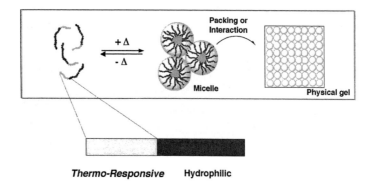

Scheme 5. Schematic representation of thermally-induced physical gelation of diblock copolymers.

Based on the gelation mechanism, several systems with various patterns for physical gelation behavior (sol-gel, gel-sol, or sol-gel-sol) have been developed, which are responsive to other stimuli such as the addition of a selective solvent[20] or compound, cooling, pH change, and irradiation with ultraviolet light. Scheme 6 illustrates the physical gelation behavior for a series of block copolymers.

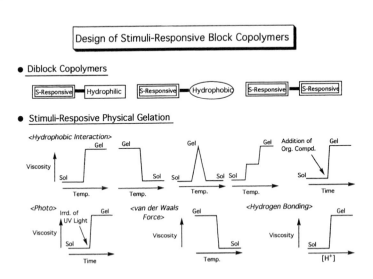

Scheme 6. Various stimuli-responsive physical gelation behavior with diblock copolymers.

Experimental

Materials. Si-containing vinyl ethers and OEVEs were prepared from a corresponding hydroxyalkyl vinyl ether or 2-chloroethyl vinyl ether as reported previously. The cationogen $CH_3CH(OiBu)OCOCH_3$ (**1**) was prepared from IBVE (isobutyl vinyl ether) and acetic acid.

Polymerization Procedure. Polymerization was carried out at 0 °C under a dry nitrogen atmosphere in a glass tube equipped with a three-way stopcock. The reaction was initiated by the addition of $Et_{1.5}AlCl_{1.5}$ toluene solution to a mixture of a monomer, an added base, and the cationogen in toluene or hexane at 0 °C. After a certain period, the polymerization was quenched with prechilled methanol containing a small amount of aqueous ammonia solution. The product polymer was recovered from the organic layer by evaporation and vacuum-dried overnight. The conversion of a monomer was determined by gravimetry. Desilylation of the product polymers was carried out with the addition of 3.0 N aqueous HCl-EtOH to a purified polymer in THF-EtOH at 0 °C.

Characterization. The MWD of the polymers was measured by size exclusion chromatography (SEC) in chloroform at 40 °C on three polystyrene gel columns. The number-average molecular weight (M_n) and M_w/M_n were calculated from SEC curves in reference to polystyrene calibration. The M_w of the star-shaped polymers was also determined by MALLS in chloroform at 40 °C on a Dawn E instrument (Ga-As laser, $\lambda = 690$ nm). The size of the star-shaped polymers was determined from the R_h by DLS in chloroform at 30 °C.

Aqueous solutions of the polymers were prepared by dissolving the polymer in Mill-Q water (18 MΩ cm). The phase-separation temperatures of the polymer solutions were determined from the transmittance at 500 nm, monitored at a rate of 1 °C /min on heating and cooling scans. A high-sensitivity differential scanning calorimeter was used to study the endothermic enthalpy. The (apparent) viscosity based on flow properties and the dynamic viscoelasticity of the polymer solutions were measured using a stress-controlled rheometer with a cone-plate of 4 cm diameter and angle of 2°. Dynamic viscoelastic measurement was carried out at various temperatures at an angular frequency of 6.283 rad/s (1.0 Hz). SANS, small-angle x-ray scattering (SAXS), and freeze-fracture TEM measurements were performed as described in detail elsewhere. [17,19]

[1] C. L. McCormick, *"Stimuli-Responsive Water Soluble and Amphiphilic Polymers"*, ACS Symposium Series 780, Oxford, New York, 2000.
[2] A. S. Hoffman, *Macromol. Symp.* **1995**, *98*, 645.
[3] P. Alexandridis, B. Lindman, *"Amphiphilic Block Copolymers: Self-Assembly and Applications"*, Elsevier, Amsterdam, 1997.
[4] S. Aoshima, T. Higashimura, *Macromolecules* **1989**, *22*, 1009.
[5] S. Aoshima, H. Oda, E. Kobayashi, *J. Polym. Sci. Part A: Polym. Chem.* **1992**, *30*, 2407.
[6] S. Aoshima, H. Oda, E. Kobayashi, *Kobunshi Ronbunshu* **1992**, *49*, 937.
[7] S. Aoshima, E. Kobayashi, *Macromol. Symp.* **1995**, *95*, 91.
[8] S. Aoshima, S. Sugihara, *J. Polym. Sci. Part A: Polym. Chem.* **2000**, *38*, 3962.
[9] S. Aoshima, K. Hashimoto, *J. Polym. Sci. Part A: Polym. Chem.* **2001**, *39*, 746.

[10] S. Sugihara, S. Aoshima, *Kobunshi Ronbunshu* **2001**, *58*, 304.

[11] S. Sugihara, K. Hashimoto, Y. Matsumoto, S. Kanaoka, S. Aoshima, *J. Polym. Sci. Part A: Polym. Chem.* **2003**, *41*, 3300.

[12] S. Sugihara, S. Kanaoka, S. Aoshima, submitted.

[13] S. Aoshima, M. Ikeda, K. Nakayama, E. Kobayashi, H. Ohgi, T. Sato, *Polym. J.* **2001**, *33*, 610.

[14] S. Aoshima, Y. Segawa, Y. Okada, *J. Polym. Sci. Part A: Polym. Chem.* **2001**, *39*, 751.

[15] K. Matyjaszewski, M. J. Ziegler, S. V. Arehart, D. Greszta, T. Pakula, *J. Phys. Org. Chem.* **2000**, *13*, 775.

[16] S. Aoshima, T. Kikuchi, S. Matsuzono, H. Miyazawa, *Polym. Prepr., Jpn.* **2000**, *49*, 389.

[17] S. Okabe, S. Sugihara, S. Aoshima, M. Shibayama, *Macromolecules* **2002**, *35*, 8139.

[18] S. Okabe, S. Sugihara, S. Aoshima, M. Shibayama, *Macromolecules* **2003**, *36*, 4099.

[19] S. Sugihara, K. Hashimoto, S. Okabe, M. Shibayama, S. Kanaoka, S. Aoshima, *Macromolecules* in press.

[20] S. Sugihara, S. Matsuzono, H. Sakai, M. Abe, S. Aoshima, *J. Polym. Sci. Part A: Polym. Chem.* **2001**, *39*, 3190.

Advances in the Design of Photoinitiators, Photo-Sensitizers and Monomers for Photoinitiated Cationic Polymerization

J. V. Crivello,* J. Ma, F. Jiang, H. Hua, J. Ahn, R. Acosta Ortiz

Department of Chemistry, New York Center for Polymer Synthesis, Rensselaer Polytechnic Institute, Troy, New York 12180-3590, USA
E-mail: crivej@rpi.edu

Summary: The present article reports on recent work from this laboratory that has led to advances in several areas of the field of photoinitiated cationic polymerization. Two different classes of novel photoinitiators 5-arylthianthrenium salts and 4-hydroxyphenyl dialkylsulfonium salts have been prepared and their use in the polymerization of various monomer systems studied. Also described is the development of both monomeric and polymeric photosensitizers that may be employed to broaden the spectral sensitivity of various onium salt photoinitiators. Acceleration of rate of the ring-opening polymerization of epoxide monomers has been achieved through the use of benzyl alcohols. Further, benzyl alcohols bearing sensitizing groups were found to be especially interesting as combination accelerators-photosensitizers. Lastly, epoxy monomers having groups with easily abstractable protons display exceptional reactivity in photoinitiated ring-opening cationic polymerization. A mechanism has been proposed to explain this observation.

Keywords: cationic photoinitiators; cationic photopolymerization; epoxide monomers; photosensitizers; ring-opening polymerization

Introduction

Both academic and industrial interest in cationic photopolymerizations has been increasing as the number of commercial applications of this chemistry has multiplied and as the economic and environmental advantages of this technology have become apparent.[1] Today, photoinitiated cationic polymerizations are not only being employed for coating and printing ink applications where the high speed and solvent-free use had been the main driving force in their development, but also in such applications as stereo- and photolithography where the high photosensitivity and excellent mechanical properties are the crucial factors in their adoption. As the

DOI: 10.1002/masy.200451114

interest in these photosensitive systems moves to ever-higher performance applications, there is an increasing interest in the development of advanced systems with better performance characteristics. Accordingly, in this laboratory, we have pursued efforts to prepare novel, more photosensitive photoinitiators, improve their spectral absorption characteristics through photosensitization, increase the rates of polymerization of monomers through the use of accelerators and attempt to prepare new monomers through rational design. In this article, we will report on the results we have recently obtained in each of these fields.

Results and Discussion

Photoinitiators

Among the best cationic photoinitiators yet developed are triarylsulfonium salts. These compounds combine a unique set of almost paradoxical properties. Triarylsulfonium salts display excellent photosensitivity with quantum yields estimated in the range of 0.6-0.9.[2] At the same time, these photoinitiators are extraordinarily thermally stable. This allows even highly reactive monomers to be processed at temperatures approaching 150 °C without inducing spontaneous thermal polymerization. It has, however, been noted that the major region of photosensitivity of simple triarylsulfonium salts such as I ($MtX_n^- = BF_4^-$, PF_6^-, AsF_6^-, SbF_6^-, etc.) lies in the short wavelength region of the UV spectrum.[3]

I

Increasingly, there is need to carry out photoinitiated cationic polymerizations using long wavelength UV and visible radiation. For example, many new imaging systems are being developed that employ either lasers or light emitting diodes as monochromatic or narrow band irradiation sources. Typically, these efficient light sources emit light in the long wavelength UV and visible regions. Lastly, there is considerable present interest in developing visible wavelength responsive photoinitiator systems that can be used to conduct photopolymerizations with solar irradiation. To address such applications, it is necessary to design photoinitiators that have photoactive absorption bands at specific, predetermined wavelengths. With this rationale in mind, work in this laboratory has been exploring the development of synthetic methods for the preparation of novel photoinitiators.

5-Arylthianthrenium salts with absorption in the mid (300-400 nm) range of the UV spectrum have been prepared as shown in equation 1 by the direct condensation of thianthrene-5-oxide in the presence of methanesulfonic acid (MSA) and phosphorous pentoxide.[4]   This method is particularly attractive for the synthesis of 5-arylthianthrenium salts bearing electron-donating substitutents, R.   A large number of these compounds were prepared with different counterions of low nucleophilicity and then examined as photoinitiators for a variety of vinyl and ring-opening cationic polymerizations.

$$\text{(thianthrene-5-oxide)} + \text{(arene-R)} \xrightarrow[\text{P}_2\text{O}_5]{\text{MSA}} \xrightarrow{\text{NaMtX}_n} \text{II} \quad \text{MtX}_n^-$$

II   eq. 1

Figures 1 and 2 respectively, show the photopolymerizations of vinyl ether and epoxy monomers.   These studies demonstrate that 5-arylthianthrenium salts are excellent photoinitiators for these two types of cationically polymerizable monomers.

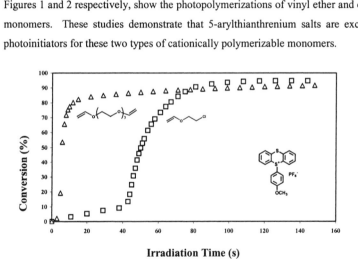

Irradiation Time (s)

Figure 1.   Comparison of the photopolymerizations of 2-chloroethyl vinyl ether and triethyleneglycol divinyl ether (DVE-3) carried out in the presence of 0.5 mol% of 5(4-methoxyphenyl)thianthrenium SbF_6^-. (light intensity 206 mJ/cm^2)

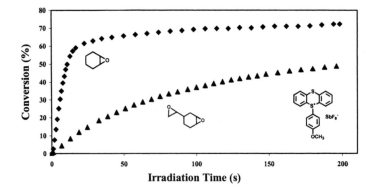

Figure 2. Study of the photopolymerization of cyclohexene oxide and 4-vinylcyclohexene dioxide carried out in the presence of 0.5 mol% of 5(4-methoxyphenyl)thianthrenium SbF$_6^-$. (light intensity 128 mJ/cm^2)

The synthetic method described above in equation 1 has also been effectively applied to the synthesis of another class of sulfonium salt cationic photoinitiators. 4-Hydroxyphenyl dialkylsulfonium salts, **III**, can be prepared by the reaction of appropriately substituted phenols with dialkylsulfoxides as shown in equation 2.[5]

$$\text{III} \qquad \text{eq. 2}$$

In a similar fashion, a series of 2-hydroxyphenyl dialkylsulfonium salts can be prepared starting with 2,4-disubstituted phenols as substrates. Using this method, it is possible to prepare sulfonium salts with alkyl groups, R$_1$ and R$_2$ of any desired chain length. Such photoinitiators can be designed to be soluble even in quite nonpolar, lipophilic monomers. Whereas the photolysis of 5-arylthianthrenium and other triarylsulfonium salts proceeds by a mechanism in which the substrate undergoes the irreversible cleavage of a carbon-sulfur bond, the photolysis of 4- and 2-hydroxyphenyl dialkylsulfonium salts takes place by a reversible mechanism.[6] This mechanism is depicted in equation 3 for a typical member of this class of photoinitiators.

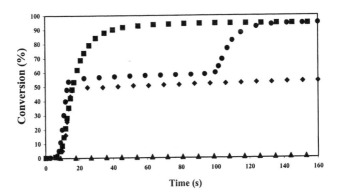

IV   eq. 3

During photolysis, the photoexcited sulfonium salt releases the acid, $HMtX_n$, and simultaneously generates the corresponding ylide, **IV**. Upon cessation of the irradiation, ylide and acid recombine to regenerate the starting sulfonium salt. This reversibility can be readily observed when hydroxyphenyl dialkylsulfonium salts such as **III** are used as photoinitiators.

Figure 3. Study of the photopolymerization of 2-chloroethyl vinyl ether carried out in the presence of 1.0 mol% of 4-hydroxy-3,5-dimethylphenyl 1-n-decyl methyl sulfonium hexafluorophosphate. (▲) irradiation for 5 seconds; (■) continuous irradiation; (◆) 15 seconds irradiation and (●)15 seconds irradiation followed by 85 seconds dark and then continued irradiation (light intensity 850 mJ/cm^2)

Figure 3 shows an investigation of the photopolymerization of 2-chloroethyl vinyl ether under various irradiation conditions. As can be noted in Fig. 3, continuous irradiation with UV light results in the rapid consumption of the monomer. The photopolymerization exhibits a 10 second inhibition period during which trace impurities in the system react with the acid that is photochemically generated. If irradiation is conducted for a time shorter than the induction period, essentially no polymerization is observed. On the other hand, irradiation for fifteen seconds results in approximately a 50% consumption of the monomer. Once the shutter is closed, the polymerization essentially stops. Repeating this latter experiment and allowing the polymerization to stand in the dark for 85 seconds followed by continuing the irradiation results in immediate resumption of the polymerization. These results strongly support the reversibility of the photolysis mechanism.

Photosensitizers

The development of novel synthetic methods for the preparation of onium salt cationic photoinitiators has provided photoinitiators with a high degree of photosensitivity in the mid- to short wavelength UV. Progress towards the development of photosensitive cationic systems that are responsive at long UV wavelengths has been achieved through the use of photosensitization. Previously it has been demonstrated that electron-transfer photosensitization is a very effective means of broadening the spectral sensitivity of sulfonium and iodonium salts into the long wavelength UV and visible regions of the spectrum.[7,8] Briefly, electron-transfer photosensitization of onium salts takes place as shown in Scheme 1.

$$\text{Scheme 1}$$

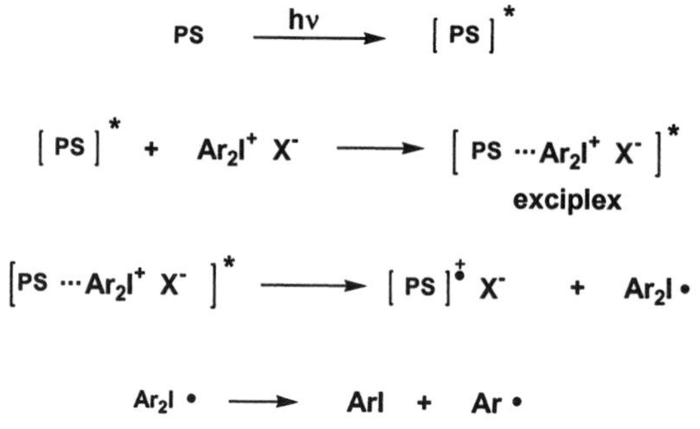

$$PS \xrightarrow{\ h\nu\ } [\,PS\,]^{*}$$

$$[\,PS\,]^{*} + Ar_2I^{+}\ X^{-} \longrightarrow [\,PS\cdots Ar_2I^{+}\ X^{-}\,]^{*}$$
$$\text{exciplex}$$

$$[\,PS\cdots Ar_2I^{+}\ X^{-}\,]^{*} \longrightarrow [\,PS\,]^{\ddagger}\ X^{-} + Ar_2I\,\bullet$$

$$Ar_2I\,\bullet \longrightarrow ArI + Ar\,\bullet$$

Light is first absorbed by the photosensitizer (PS) and the excited photosensitizer interacts directly with the onium salt or by the initial formation of an excited state complex. Subsequently, these species undergo a formal electron transfer in which the photosensitizer is oxidized and the onium salt is reduced. In a subsequent step, irreversible fragmentation of the diaryliodine free radical takes place to yield additional radical products. The photosensitizer cation-radical ($PS \overset{+}{\bullet}$) generated by the above photoredox reaction initiates the polymerization of a cationically polymerizable monomer by any of several complex processes shown in Scheme 2.

Scheme 2

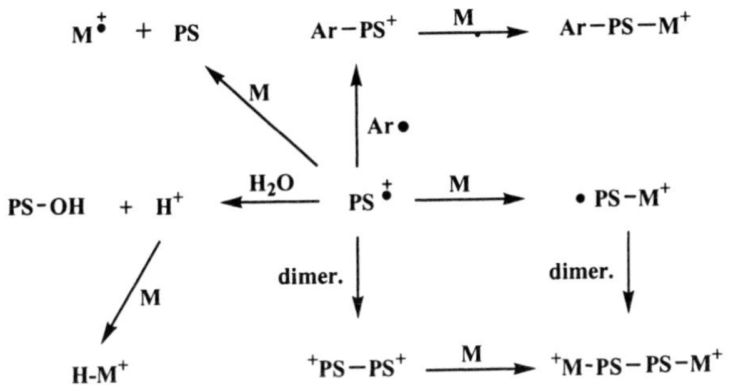

Most of the early work in this laboratory focused on the use of polynuclear aromatic hydrocarbons that display excellent long wavelength absorption characteristics and efficient electron-transfer photosensitization for all types of onium salt cationic photoinitiators. However, the disadvantage in using photosensitizers such as pyrene, anthracene and perylene is their purported toxicity. Accordingly, alternative photosensitizers have been sought.

The design of photosensitizers bearing cationically polymerizable functional groups was seen as a solution to the above problem.[9,10,11] Such photosensitizers as **V-VII** shown below were prepared and shown to undergo copolymerization with either vinyl or heterocyclic compounds that polymerize by cationic ring-opening reactions as well as provide a photosensitization function.

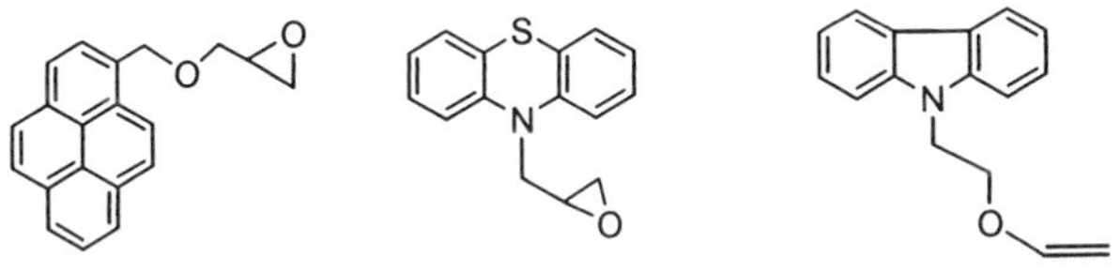

| V | VI | VII |

Figure 4 depicts one such study. The inclusion of VI together with the monomer, limonene dioxide, results in the marked overall acceleration of the photopolymerization.

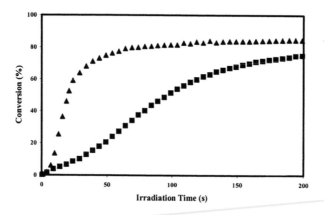

Figure 4.   Comparison of the photopolymerization of limonene dioxide in the presence (▲) and absence (■) of 0.5 % of 10-(2,3 epoxypropyl)-phenothiazine), **VI**, using 1.0 % (4-n-decyloxyphenyl) diphenylsulfonium SbF_6^- as photoinitiator. (Light intensity 250 mJ/cm^2·min)

An interesting study of the photosensitized polymerization of limonene dioxide is shown in Figure 5. Both N-vinyl carbazole (NVK) and its polymer (PVK) display excellent photosensitization activity and to the same degree. Thus, PVK and similar types of polymers bearing pendant carbazole, anthracene, pyrene, perylene and phenothiazine groups are excellent photosensitizers for onium salt induced cationic photopolymerizations. These polymeric photosensitizers have the further benefits of being nonvolatile, nontoxic and easily handled.

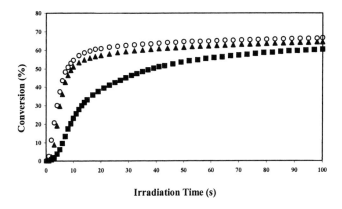

Irradiation Time (s)

Figure 5.   FT-RTIR comparison of the polymerization of limonene dioxide in the presence of 2.0% NVK (**O**) and PVK (**▲**) no photosensitizer (**■**). (light intensity 145 mJ/cm^2·min; 0.05% (4-n-decyloxyphenyl) diphenylsulfonium SbF$_6^-$)

Accelerators

During the course of exploring alternative means for increasing the speed of cationic ring-opening polymerizations of epoxide monomers other than photosensitization, it was observed that the addition of small amounts of benzyl alcohols markedly accelerates the rates of those polymerizations.[12] For example, in Figure 6 is depicted a study of the photoinduced polymerization of 4-vinylcyclohexene dioxide (VCHDO) in the presence of an iodonium salt and various benzylic alcohols.

Acceleration of the photopolymerization of this difunctional epoxide monomer can be attributed to two effects related to the benzyl alcohol. First, as has been demonstrated by Penczek, Kubisa and their coworkers, the addition of alcohols to the ring-opening polymerization of epoxides results in rapid transfer. Higher and more rapid conversions are generally observed for the polymerizations of such monomers since gelation is retarded. The alcohol is incorporated into the growing polymer chain as a benzyl ether end group. A second mechanism that takes place is illustrated in Scheme 3.

174

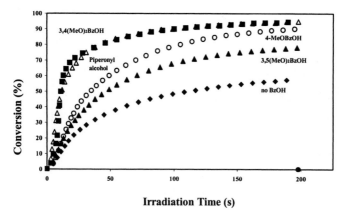

Irradiation Time (s)

Figure 6. Comparison of the accelerating effects of various benzyl alcohols (20%) bearing electron donating groups on the cationic ring-opening photopolymerization of VCHDO. (photoinitiator, 1.0 % IOC10; light intensity 190 mJ/cm^{2}min).

Scheme 3

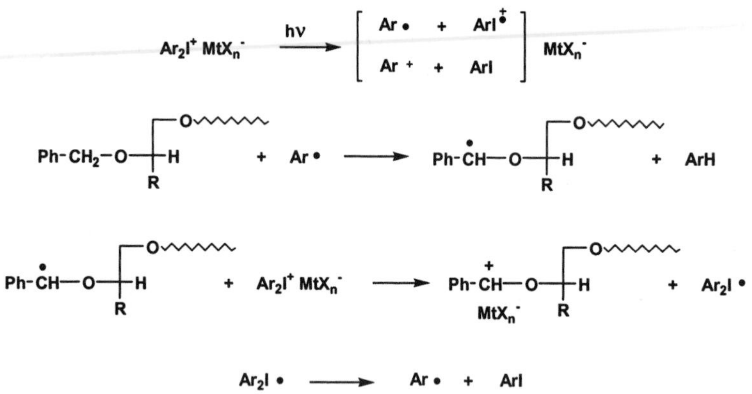

The photolysis of diaryliodonium salts generates a number of species among which are aryl radicals, radical cations and cations. Photopolymerization is generally attributed to the cationic species or their byproducts (i.e. protonic acids). Generally, the radical products are neglected in this process. However, when easily abstractable benzylic ether hydrogen atoms are present, they may be abstracted to form benzylic radicals that are attached to the ends of the growing polymer chains. Such radicals are readily oxidized to the corresponding benzyl cations in the presence of diaryliodonium salts. Simultaneously, the diaryliodine radical that is formed decomposes to yield an aryl radical and an aryl iodide. The aryl radical participates in a chain reaction in which the iodonium salt is consumed by a photoinitiated process that produces benzyl cations. In such a process, a large number of cationic initiating species are produced by the absorption of a few protons. This amplification effect is observed as acceleration in the rate of the polymerization of the monomer. Figure 6 also reveals that benzyl alcohols bearing electron-donating groups are most effective as accelerators. Particularly effective in this regard are 3,4-dimethoxybenzyl alcohol and piperonyl alcohol (3,4-methylenedioxybenzyl alcohol). This observation can be explained by noting that such groups stabilize both radical and cationic intermediates proposed in the mechanism of Scheme 4.

A further evolution of this concept is shown in Figure 7 in which the benzylic alcohol contains a group that can also perform a photosensitizing function. Exceedingly rapid polymerization of cyclohexene oxide is observed in the presence of 1-pyrenemethanol due to its ability to function as a photosensitizer, chain transfer agent and free radical accelerator.[13]

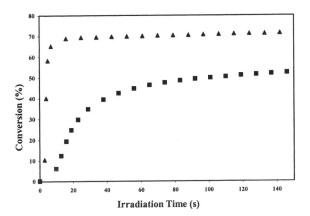

Figure 7. Kinetic study of the cationic photopolymerization of cyclohexene oxide using 0.01% IOC10 in the absence of a photosensitizer (■) and with 0.2% (▲) 3-perylenemethanol. (light intensity 178 mJ/cm^2 min)

Novel Monomers

Several new monomers have been designed to take advantage of the radical induced decomposition process described in Scheme 3.[14,15] The structures of some of these monomers are depicted below.

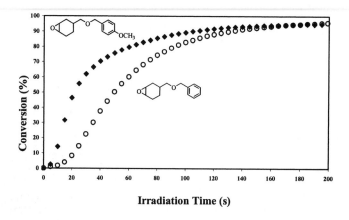

VIII   **IX**   **X**

Each of these monomers contains a benzylic ether group to provide easily abstractable hydrogen atoms that can interact with the photoinitiator in a manner similar to that proposed in Scheme 4. In Figure 8 the photopolymerization of monomers **VIII** and **IX** is shown to be extremely rapid even under the low light intensity illumination conditions (200 mJ/cm^{2}min) used.

Figure 8.   Comparison of the cationic photopolymerizations of **VIII**, (**O**) and **IX**, (♦);(photoinitiator, 1.0 % (4-n-decyloxyphenyl)phenyliodonium SbF$_6$$^-$; light intensity 200 mJ/cm^{2}min)

Conclusions

Based on the results presented in this article, it can be concluded that many aspects of photoinitiated cationic polymerizations can be manipulated to achieve desired results. The photosensitivity and spectral response of these systems can be adjusted through the synthesis of novel photoinitiators and photosensitizers. The rates of cationic photopolymerizations can be accelerated through the use of benzyl alcohol accelerators and by the design of monomers that incorporate abstractable hydrogen atoms.

Acknowledgements

The authors are grateful to the National Science Foundation and to the Petroleum Research Fund administered by the American Chemical Society for financial support of this research.

[1]   J.V. Crivello, in: *"Ring-Opening Polymerization"*, D.J. Brunelle, Ed., Hanser, Munich, **1993**, p.157/196ff.
[2]   C. Selvaraju, A. Sivakumar, P. Ramamurthy, *J. Photochem. Photobiol. A: Chem.* **2001**, *138*, 213.
[3]   J.V. Crivello, J.H.W. Lam, *J. Polym. Sci., Polym. Chem. Ed.* **1979**, *17(4)*, 977.
[4]   J.V. Crivello, J. Ma, F. Jiang, *J. Polym. Sci., Part A: Polym. Chem. Ed.* **2002**, *40,(20)*, 3465.
[5]   J. Ahn and J.V. Crivello, *Polymer Preprints* **2002**, *43(2)*.
[6]   J.V. Crivello, J.L. Lee, *Macromolecules* **1981**, *14(5)*, 1141.
[7]   J.V. Crivello, J.H.W. Lam, *J. Polym. Sci., Polym. Chem. Ed.* **1979**, *17(4)*, 1059.
[8]   J.V. Crivello, in: *"UV Curing Science and Technology"* S.P. Pappas, Ed., Technology Marketing Corp., Stamford, Conn., **1978**, p.24/77ff.
[9]   Y. Hua, J.V. Crivello, *Macromolecules* **2001**, *34(8)*, 2488.
[10]   J.V. Crivello, Y. Hua, Z. Gomurashvili, *Macromol. Chem. Phys.* **2001**, *202*, 2133.
[11]   Z. Gomurashvili, J.V. Crivello, *Macromolecules* **2002**, *35(8)*, 2962.
[12]   J.V. Crivello, R. Acosta Ortiz, *J. Polym. Sci., Part A: Polym. Chem.* **2002**, *40*, 2298.
[13]   Y. Hua, F. Jiang, J.V. Crivello, Chem. Mater. **2002**, *14(5)*, 2369.
[14]   J.V. Crivello, R. Acosta Ortiz, *J. Polym. Sci.: Part A: Polym. Chem. Ed.* **2001**, *39(14)*, 2385.
[15]   J.V. Crivello, R. Acosta Ortiz, *J. Polym. Sci., Part A: Polym. Chem. Ed.* **2001**, *39(20)*, 3578.

On the Preparation and Polymerization of *p*-Methoxystyrene Miniemulsions in the Presence of Excess Ytterbium Triflate

Séverine Cauvin,[1] *François Ganachaud**[1,2]

[1]Laboratoire de Chimie des Polymères, UMR 7610 CNRS/Université Pierre et Marie Curie T44 E1, 4 Place Jussieu 75252 Paris Cedex, France
[2]New address: Laboratoire de Chimie Macromoléculaire, UMR 5076 CNRS/ENSCM, 8 Rue de l'Ecole Normale 34296 Montpellier Cedex 5, France
Fax: 33 4 67 14 72 20; E-mail: ganachau@enscm.fr

Summary: The cationic polymerization of p-methoxystyrene using an acid initiator and ytterbium triflate as a cocatalyst was studied in miniemulsion. Conductometry measurements revealed that ytterbium triflate dissociates in water. The high ionic strength implemented by ytterbium cation releasing requires the use of an electrosteric surfactant, i.e. sodium dodecylpolyoxyethylene(8) sulfate, as an efficient stabilizer against particle coalescence. The catalyst increases significantly the polymerization rate but only moderately affects the molar masses. A tentative polymerization scheme is proposed based on these results and several other "blank" experiments.

Keywords: cationic polymerization; miniemulsion; p-methoxystyrene; rare earth catalyst

Introduction

Ionic polymerization in emulsion is a subject developed in our team over a decade, both on ring opening polymerization (cyclosiloxanes,[1,2] phenyl glycidyl ether[3]) and polymerization of vinyl monomers (n-butyl cyanoacrylate[4], p-methoxystyrene[5]). We reported in this last study on the cationic polymerization of p-methoxystyrene (*p*MOS) catalyzed by an acid surfactant, namely dodecylbenzene sulfonic acid.

Scheme 1

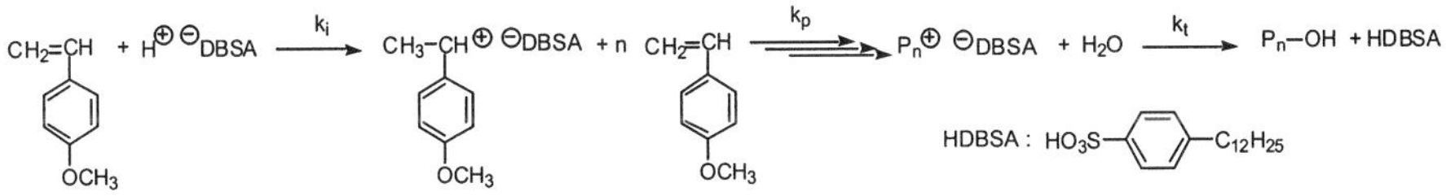

A simple polymerization scheme applies (Scheme 1). All reactions are interfacial. Initiation is the limiting step, whereas propagation is fast because the active centers pair with the surfactant head, which acts as a bulky counter-ion. Termination with water limits strongly the final molar masses.

Recently, Sawamoto and coworkers described the polymerization of p-methoxystyrene with the aid of rare earth triflates,[6-8] a water-resistant Lewis acid extensively used in organic synthesis (see ref. [9] for a recent review). The HCl adduct of the monomer was first used as an initiator, and ytterbium triflate as a cocatalyst.[6,7] It was shown that chains could be controlled up to 30% conversion, after which transfer reactions with water were too high to maintain control of the polymerization. Final molar masses were in the order of 3000 g/mol, with low polymolecularities (less than 1.4). Sulfonic acids were also used as initiators,[8] with a neat increase in the rate of polymerization, presumably by bringing protons at the interface. In this latter case, however, molar mass distributions were higher (1.7). Other metal triflate were also found to be efficient Lewis acids for catalyzing pMOS polymerization.[10] Another team reported on the polymerization of pMOS in water using again ytterbium triflate together with phosphonic acids as initiator, which allowed longer chains to be produced (typically 10,000 g/mol).[11] In all these studies, no mention of the actual colloidal nature of their dispersions were given, though it is revealed to be a striking parameter for the process.

We present here some results on polymerization of pMOS in miniemulsion using ytterbium triflate as a catalyst. After considering the molecular structure of ytterbium triflate in water, we focus our discussion on colloidal stability aspects and show how the presence of the catalyst affects the polymerization scheme proposed before (Scheme 1).

Experimental part

All reagents used in this study were characterized at least by ^{1}H NMR before use. p-methoxystyrene (pMOS) (ACROS, 96%) was used as supplied. Disponil surfactant Fes32is, polyoxyethylene(8) lauryl sulfate sodium (1) was kindly donated by Cognis and characterized by MALDI-TOF mass spectrometry. Alkyl chains are predominantly composed of 12 carbons, with less than 30% with 14 and 16 carbon atoms. The ethoxylated part is widely distributed and centered at 8 ethylene oxide units. All chains bear a sulfate moiety.

$$C_{12}H_{25}\text{---}\left(O\text{--}CH_2\text{--}CH_2\right)_8\text{---}OSO_3^- \; Na^+$$

1

HCl 35% (Prolabo) and H_2SO_4 98% (BDH) were diluted in distilled water prior to their addition in the emulsion. Ytterbium triflate was donated by Rhodia Terres Rares (La Rochelle, France). The pH of ytterbium triflate solution is unchanged compared to pure water, which is an indication that no triflic acid formed in the sample during storage.

All analysis methods, e.g. ^1H NMR, particle size measurements by QELS, SEC in THF were described previously.[5] Conductometric measurements were performed on a Consort K120 digital conductometer, equipped with a platinum electrode and a temperature probe. KCl and acetic acid were used as, respectively, strong and weak electrolytes to validate the conductometry measurements. TEM analysis was performed on a JEM100CXII UHR microscope from JEOL, with a 100 kV acceleration tension. One drop of the sample was deposited on a carbon coated grid and allowed to evaporate before analysis.

Regular miniemulsions were obtained by sonication using a 450 Branson Ultrasonics Corporation sonifier at power 7 (25 W). pMOS was added to the surfactant solution maintained by means of an ice bath at a temperature of 25°C during sonication (1 min 30 sec). Small volumes of HCl or H_2SO_4 solutions were then added to the reactor to start the polymerization (typically 3 to 5 equivalents compared to surfactant content). The emulsion was then transferred into the reactor where polymerization proceeded under thermal regulation (60°C) and mechanical stirring (350 rpm).

At regular time intervals, 0.5 mL aliquots were withdrawn. One drop from the sample was poured in a diluted surfactant solution ($c=10^{-4}$ M< CMC) prior to QELS particle diameter measurement. 1 mL of methanol and 1 mL of water were poured into the sample to stop the reaction and break the emulsion (adding NaOH better inhibits the reaction but also catalyzes the formation of ytterbium oxides, which precipitate and disturb the emulsion treatment). CH_2Cl_2 was also systematically added in the tube to ensure that no organic product remained in water. The organic phase was efficiently separated from the aqueous one by centrifugation (30 mins at 5300 rpm on a bench centrifugator), dried with $MgSO_4$, evaporated, redissolve in THF and filtered (0.45 µm filters from MILIPORE) prior to SEC analyses.

Results and discussion

1. Yb(OTf)$_3$ molecular structure in water

Among the numerous reports that use ytterbium triflate as a catalyst in water, none really described so far the actual configuration of this salt in water. Triflate is, similar to chloride or perchlorate, a non-coordinating ligand linked to ytterbium through its outer-sphere, as revealed by various analytical methods.[12] Since ytterbium triflate has been shown to dissociate totally in polar solvents such as acetonitrile,[13] it is suspected that dilution in water would also produce the 1:3 electrolytes.

To confirm this point, conductometric analyses were carried out on Yb(OTf)$_3$ and YbCl$_3$ solutions of various concentrations. Only one paper reports values for conductometry of Yb(OTf)$_3$ in water at two concentrations.[12] YbCl$_3$ is known to totally dissociate in water.[14] Figure 1 shows ytterbium equivalent molar conductance ($\Lambda_{mol,Yb}$) as a function of YbCl$_3$ and Yb(OTf)$_3$ water concentrations expressed as the square root of their normality, for the purpose of comparison with the literature.[14] Equivalent molar conductances were calculated by dividing the molar conductance by 3 (the valence of the ytterbium cation) and substracting the molar conductance for Cl$^-$ (76.3 S/m^2/mol) and OTf (44.5 S/m^2/mol).

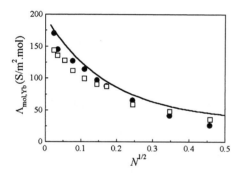

Figure 1: Equivalent molar conductance for ytterbium cation as revealed by conductometric measurements. ($\square$) YbCl$_3$; ($\bullet$) Yb(OTf)$_3$; (—) YbCl$_3$ from the literature.[14]

$\Lambda_{eq,Yb}$ slowly decreases with ytterbium triflate concentration. Similar behavior is observed for YbCl$_3$, in agreement with previous reports from the literature (line in Figure 1). Such a decrease is not due to a weak electrolyte behavior. Instead, rare earth cations can complex with 8 or 9 molecules of water, depending on the type of lanthanide and its concentration in water, which readily affect their ionic sphere and thus their mobility in water. Note that our results compare well with those from the literature (Figure 1), although we used a typical conductometer.

Dissolution of Lewis acids in water is known to involve two steps: first, dissociation of the salt and water complexation, and second, hydrolysis of the metal cation (scheme 2).[15] Rare earth atoms are considered "water-resistant" due to their stability towards hydrolysis at neutral pH;[16] where other Lewis Acids, such as AlCl$_3$ or TiCl$_4$, form hydroxides and quickly deactivate through aggregation.[15] By evaporation, the crystal salts reform and the catalyst can thus be reused.

Scheme 2

$$Yb(OTf)_3 \xrightarrow{\text{dissociation}} 3\ OTf^- + Yb(H_2O)_m^{3+} \xrightarrow{\text{hydrolysis}} Yb(H_2O)_{m-n}(OH)_n^{(3-n)+} + nH^+$$

There are three main implications for the polymerization process studied here: i) a high ionic strength will develop in the water medium, and thus extremely robust surfactants are sought to produce stable emulsions; ii) the catalyst dissociated in water will not be capable of catalyzing a polymerization reaction under this state; however, by saturating the water, a few ytterbium triflate molecules may reach the interface and catalyze the polymerization; iii) some ligands are known to penetrate the inner sphere of the ytterbium atom and thus form stable complexes even in water. Sulfates were for instance demonstrated by conductometry to inner-coordinate rare earth atoms.[17] Recent studies from organic chemistry showed that this property was actually used to prepare Lewis Acid Surfactant Complex (LASC) by mixing sodium dodecyl sulfate and ytterbium chloride in water.[16]

2. Colloidal stability during the preparation and polymerization of *p*MOS

Table 1 summarizes the various experiments carried out in the present study. Increasing loads of ytterbium triflate were added and different blank experiments, e.g. without catalyst or adding only sodium triflate and ytterbium chloride in excess, were also performed. HCl and H$_2$SO$_4$ were added as a proton source for starting the polymerization.

As stated above, the dissociation of ytterbium triflate to release trivalent cation and monovalent anions has profound impact on the ionic strength of the dispersion. Usual surfactants, e.g. non-ionic, cationic or anionic ones, could not stabilize the particles even at low catalyst content. One family of surfactants, i.e. Disponil™ from Cognis, is used in conventional radical emulsion polymerization in industry, when efficient stability against particle coalescence is sought. These surfactants indeed bear a polyethylene oxide moiety between the alkyl chain and the sulfate end-group, which provides rather efficient electrosteric stabilization. Among these, polyoxyethylene(8) lauryl sulfate sodium was chosen since it provides satisfactory resistance against a trivalent cation such as ytterbium.

Table 1: Recipes for all experiments reported in this article.[a]

Run	m_{pMOS}	m_1	m_{water}	m_{acid}	$m_{additive}$	Additive[d]	Particle	σ[b]	R_p	$\overline{M_n}$	$\overline{M_w}/\overline{M_n}$
	(g)	(g).	(g)	(g)	(g)		size (nm)		(%/h)	(g/mol)	
1	2.40	0.56	4.14	0.14	-	-	240	0.04	8.5	990	1.17
2	2.43	0.57	4.07	0.14	0.22	Yb(OTf)₃	240	0.19	13.0	950	1.22
3	2.41	0.56	4.00	0.14	0.58	Yb(OTf)₃	-	-	16.0	1180	1.22
4	2.41	0.56	4.00	0.15	1.11	Yb(OTf)₃	325	0.90	24.0	1230	1.30
5[c]	1.83	0.44	3.02	0.09	0.83	Yb(OTf)₃	740	1	12.5	1260	1.31
6[c]	2.41	0.56	4.07	0.12	1.11	Yb(OTf)₃	620	1	9.0	1270	1.29
7	2.41	0.56	4.09	0.16	0.31	YbCl₃	-	-	14.5	850	1.15
8	1.83	0.43	3.02	0.05	0.89	Alun salt	-	-	16.0	800	1.18
9	2.40	0.56	4.02	0.16	0.69	NaOTf	-	-	16.5	970	1.21

[a] Other parameters: temp. 60°C, stirring rate: 350 rpm; [b] Particle size distribution (0<σ<1); [c] Catalyst: H₂SO₄; [d] Alun salt: AlKO₈S₂,12H₂O: NaOTf: sodium triflate.

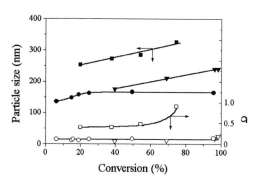

Figure 2: Particle size (diameter d_z, in nm, plain symbols) and polydispersity (σ, without units, open symbols) as a function of conversion for experiments carried out in presence of ytterbium triflate (in wt.%, see Table 1): (●) 0% (run 1). (▼) 2.7% (run 2). (■) 13.5% (run 4).

Figure 2 shows the particle size and polydispersity evolutions with conversion for two contents of catalyst Yb(OTf)₃ and one control reaction in the absence of catalyst (other results are provided in Table 1). In all cases, a stable dispersion was obtained with the final particle sizes increasing slightly with ytterbium content (typically between 150 and 500 nm). Emulsion with low content of catalyst gives a fine particle distribution, whereas for the experiment containing 13.5wt.% of ytterbium triflate, a polydispersity coefficient of 1 is reached before the end of polymerization. For experiments carried out with H_2SO_4 as the acid, much larger particles were obtained (Table 1, runs 5 and 6). Such results may be due to a further increase of the ionic strength by sulfate divalent salt.

Figure 3 shows a micrograph of a dispersion obtained for 13.5wt.% $Yb(OTf)_3$. Even if a multi-population of particles is generated in the presence of ytterbium triflate, it still gives a dispersion of monomer droplets in water. The poor contrast and average diameter superior to that obtained by QELS is due to the smoothness of the particles, that spread slightly on the microscope grid.

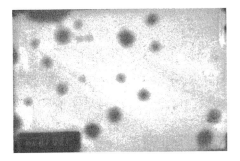

Figure 3: Transmission electron micrograph of the emulsion containing the highest content of ytterbium triflate (13.5wt.%, run 4 in Table 1). Scale: 1µm/cm.

3. Kinetics aspects

Figure 4 plots conversion versus time for experiments conducted at 60°C and various contents of ytterbium triflate. Ytterbium triflate clearly increases the polymerization rate (Table 1), and conversion goes up to a maximum of 85% for the largest ytterbium triflate content. Since particle sizes are slightly higher when the ytterbium triflate is introduced, and thus their surface area is smaller, an increase of the overall rate of polymerization is evidence for co-catalyst participation in the polymerization process.

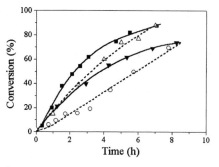

Figure 4: Influence of the content of ytterbium triflate on the conversion to polymer. Ytterbium content (in wt.%, see Table 1): (○) 0% (run 1); (▼) 2.7% (run 2); (△) 6.7% (run 3); (■) 13.5% (run 4).

Figure 5 presents the average molar masses for the above experiments. The molar mass increase more sharply at higher ytterbium triflate content, but to only about two or three monomer unit on average. A pseudo-plateau is observed in all cases, though the molar masses level off at larger conversions at higher ytterbium triflate contents (50%, compared to 20% conversions without ytterbium triflate, see Figure 5).

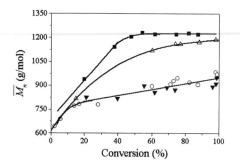

Figure 5: Influence of ytterbium triflate content on the average molar masses with conversion. Same symbols as Figure 4.

The type of acid introduced as a proton source is quite important. We saw before that adding H_2SO_4 dramatically affects the particle sizes, but even more sharply the polymerization rates (Table 1, runs 5 and 6), surely due to the strong complexation of sulfate with ytterbium. However, neither the acid content nor the type of acid modify the average molar mass evolution with conversion (Figure 6). This result may indicate that chloride and sulfate ions, in the presence of excess ytterbium triflate, are only "spectators" of the polymerization.

4. Understanding the mechanism of polymerization

Since some obvious variations were found in terms of polymerization rates and average molar masses while adding ytterbium triflate to the recipe, designed experiments were conducted to show the roles of each actor in the polymerization scheme. YbCl$_3$ or triflate sodium in respective molar contents to Yb(OTf)$_3$ did not show striking differences in kinetics and produced molar masses similar to the polymerization carried out without ytterbium catalyst (Table 1, runs 7,8). Increasing consequently the ionic strength resulted in lower final average molar masses (Table 1, run 9). The presence of ytterbium *and* triflate is important for the process.

Molar mass/conversion plots presented here (Figures 5 and 6) are close to those obtained in our previous study using DBSA as an INISURF.[5] Two main conclusions can be drawn from our previous observations: i) the fact that the molar masses reach a plateau shows that, in the present system, polymerization is interfacial and the critical DP limits the average molar masses;[5] ii) the polymolecularity always evolve with conversion (Figure 6), showing that chains are stopped by water and not reactivated. Indeed, a controlled/living system would show a decrease of polymolecularity with conversion.[18] Another indicator against controlled polymerization is given by the evolutions of molar mass with conversion, which are independent of the type and content of acid introduced in the recipe (Figure 6).

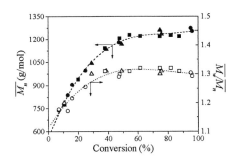

Figure 6: Influence of type and content of acid on the average molar masses (plain symbols) and polymolecularity index (open symbols) with conversion. (■) HCl, 3.5 eq; (●) H$_2$SO$_4$, 1.9 eq; (▲) H$_2$SO$_4$, 2.1 eq.

Scheme 3

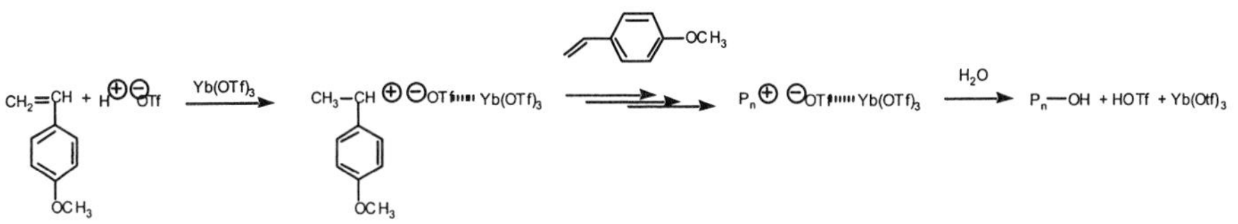

A tentative polymerization scheme is proposed in Scheme 3. A triflate anion, released from ytterbium triflate dissociation in water, captures a proton and reaches the interface to start pMOS polymerization. Triflic acid is believed to be the initiator, since triflate ions are in large excess compared to chloride ions and HCl partitions preferentially in water compared to triflic acid or other superacids.[19] This is relevant to experiments with H_2SO_4, where sulfates are inactivated by complexing ytterbium. One of the few ytterbium triflate molecules present in the dispersion then associates with the triflate to form a bulky counter-anion. Such Lewis Acid/Brønsted Acid complexes have been proposed previously by Sawamoto and coworkers as the main catalyst of pMOS polymerization.[6-8] Besides, polymerization is obviously interfacial and chain-stopping events with water are predominant. The fact that longer chains are produced when ytterbium triflate is used, may be due either to the lower content of water at the particle interface due to high ionic strength, or the faster propagation (compared to termination) rates, or both.

Conclusion

This short communication has disclosed some of the constraints imposed by the addition of a catalyst salt in an emulsion process. Ytterbium triflate dissociation increases drastically the ionic strength; only surfactants that provide electrosteric stability avoid particle coalescence. It seems clear that very few ytterbium triflate molecules participate in the reaction,[6-8] which requires concentrated catalyst conditions. In this case, the co-catalyst helps at generating, with the triflate anion, a reactive bulky ion pair at the interface, which increase the polymerization rates and produce slightly higher final average molar masses.

Obviously, average molar masses in the present *direct emulsion* process are much lower than those obtained in other studies (typically 3000 g/mol).[6-8] We will present later on results obtained using DBSA as surfactant and excess ytterbium triflate, for which an *inverse (water-in-monomer) emulsion* process explains such discrepancies.[20]

Acknowledgments

This work was supported by a grant from ADEME and Rhodia Recherches. FG thanks Joël Belleney for analyzing the surfactant by MALDI-TOF and Franck D'Agosto for his comments on the manuscript. SC acknowledges Rosalina Dos Santos for assisting her on some of the experiments. TEM micrograph was kindly provided by Patricia Beaunier.

[1] De Gunzbourg, A.; Favier, J.-C.; Hémery, P. *Polym. Int.* **1994**, *35*, 179.

[2] Barrère, M.; Ganachaud, F.; Bendejacq, D.; Dourges, M.-A.; Maitre, C.; Hémery, P. *Polymer* **2001**, *42*, 7239.

[3] Maitre, C.; Ganachaud, F.; Ferreira, O.; Lutz, J.-F.; Paintoux, Y.; Hémery, P. *Macromolecules* **2000**, *33*, 7730.

[4] Limouzin, C.; Caviggia, A.; Ganachaud, F.; Hémery, P. *Macromolecules* **2003**, *36*, 667-674.

[5] Cauvin, S.; Sadoun, A.; Dos Santos, R.; Belleney, J.; Ganachaud, F.; Hémery, P. *Macromolecules* **2002**, *35*, 7919.

[6] Satoh, K.; Kamigaito, M.; Sawamoto, M. *Macromolecules* **1999**, *32*, 3827.

[7] Satoh, K.; Kamigaito, M.; Sawamoto, M. *Macromolecules* **2000**, *33*, 4660.

[8] Satoh, K.; Kamigaito, M.; Sawamoto, M. *J. Polym. Sci. Part A* **2000**, *38*, 2728.

[9] Kobayashi, S.; Sugiura, M.; Kitagawa, H.; Lam, W. W.-L. *Chem. Rev.* **2002**, *102*, 2227.

[10] Satoh, K.; Kamigaito, M.; Sawamoto, M. *Macromolecules* **2000**, *33*, 5836.

[11] Scheuer, A. D.; Storey, R. F. *Polym. Prep.* **2002**, *43 (2)*, 932.

[12] Seminara, A.; Rizzarelli, E. *Inorg. Chim. Acta* **1980**, *40*, 249.

[13] Di Bernardo, P.; Choppin, G. R.; Portanova, R.; Zanonato, P. L. *Inorg. Chim. Acta* **1993**, *207*, 85.

[14] Spedding, F. H.; Porter, P. E.; Wright, J. M. *J. Am. Chem. Soc.* **1952**, *74*, 2055.

[15] Fringuelli, F.; Pizzo, F.; Vaccaro, L. *J. Org. Chem.* **2001**, *66*, 4719.

[16] Kobayashi, S.; Nagayama, S.; Busujima, T. *J. Am. Chem. Soc.* **1998**, *120*, 8287.

[17] Spedding, F. H.; Jaffe, S. *J. Am. Chem. Soc.* **1954**, *76*, 882.

[18] Litvinenko, G.; Müller, A. H. E. *Macromolecules* **1997**, *30*, 1253.

[19] Walker, F. H.; Dickenson, J. B.; Hegedus, C. R.; Pepe, F. R. *Prog. Org. Coat.* **2002**, *45*, 291.

[20] Cauvin, S.; Ganachaud, F.; Hémery, P. *Macromolecules* **2004**, submitted.

Carbocationic Polymerizations for Profit and Fun

*Ralf M. Peetz, Joseph P. Kennedy**

Institute of Polymer Science, The University of Akron, OH 44325-3909, USA
E-mail: kennedy@polymer.uakron.edu

Summary: This presentation consists of two largely independent parts: The first "for Profit" part concerns a bird's eye view of recently commercialized carbocationic processes and materials created by these processes in the author's laboratories whose marketing started during the past ~5 years by various companies. These materials/processes include liquid telechelic polyisobutylene (PIB) for architectural sealants, poly(styrene-*b*-isobutylene-*b*-styrene) (PSt-*b*-PIB-*b*-PSt) triblocks for thermoplastic elastomers, PIB/PSt-based blocks for coating of medical devices, and PIB-based microemulsions for surface protection of painted metal surfaces. It is concluded that in order to enhance and solidify research in polymer synthesis it would behoove the scientific community to pay increased attention to intellectual property protection. Appropriately managed patenting and publishing activities are self-reinforcing and may be quite profitable.

The second "for Fun" part concerns a brief review of the design, synthesis and characterization of two novel fully aliphatic star-block copolymers: $\varnothing$(PIB-*b*-PNBD)$_3$ and $\varnothing$(PNBD-*b*-PIB)$_3$ (where PNBD = polynorbornadiene). The constituent moieties of these star-blocks are identical except their block sequences are reversed. Motivation for the synthesis of $\varnothing$(PIB-*b*-PNBD)$_3$, consisting of a low Tg (~-73°C) PIB inner-corona attached to a high Tg (~320°C) PNBD outer corona, was the expectation that this star-block would exhibit thermoplastic elastomer characteristics, and that it could be used in applications where similar polyaromatic-based TPEs cannot be employed (e.g., magnetic signal storage). The other star-block, $\varnothing$(PNBD-*b*-PIB)$_3$, comprises the same building blocks with the PIB and PNBD sequences reversed. We found that the secondary chlorine at the PNBD chain end, in conjunction with TiCl$_4$, is able to initiate the polymerization of isobutylene. Details of the carbocationic polymerization of NBD, together with the microstructure of PNBD, will be discussed.

Keywords: block copolymers; living polymerization; polyisobutylene; poly-norbornadiene; thermoplastic elastomers

I. Introduction

This presentation concerns two main themes: Carbocationic polymerization research for fun and

---

* Parts of this presentation have been published in references 1 and 2.

for profit. The first part concerns the **profit** motive and illustrates with a few examples recently commercialized products made by carbocationic processes originally invented and investigated at the University of Akron; subsequently, in the second part we summarize recent **fun** research we carried out to produce novel fully aliphatic star-block polymers.

II.   The Lead Question and the Profit Motive

The lead question (you may call it the guiding principle) of our research is this: _How can we translate data into knowledge and knowledge into useful things?_ To generate data is relatively easy (any chemist can do a rate study of a methyl, ethyl, isopropyl, t-butyl series). To translate data into knowledge, however, is more difficult; and, ultimately, to upgrade newly generated knowledge to create something useful is extremely difficult. Some research/researchers may never reach this final stage.

Let us start by taking a bird's-eye view of products, recently commercialized by various companies around the world that are based on fundamental knowledge generated in our laboratories. First of all, there is EPION®, introduced in 1997 by Kaneka Inc., of Japan.

$$
\begin{array}{c}
\underset{\underset{\displaystyle OCH_3}{|}}{\overset{\overset{\displaystyle OCH_3}{|}}{CH_3-Si-CH_2CH_2CH_2}}\!\!\sim\!\!\sim\!\!\sim\!\!\sim\!\!\sim\!\!PIB\!\!\sim\!\!\sim\!\!\sim\!\!\sim\!\! CH_2CH_2CH_2\underset{\underset{\displaystyle OCH_3}{|}}{\overset{\overset{\displaystyle OCH_3}{|}}{-Si-CH_3}}
\end{array}
$$

EPION®

Epion is a low molecular weight liquid methoxysilane ditelechelic polyisobutylene (PIB) used as architectural sealant. This PIB derivative is obtained by hydrosilation of allyl ditelechelic PIB with methyl dimethoxysilane first described by Wilczek and Kennedy.[3, 4] The combination of properties of hermetic seals of large glass panels used in skyscrapers made with EPION® are superior to other existing sealants, including butyl rubber-based sealants.

Next, we have developed thermoplastic triblock elastomers of poly(styrene-_b_-isobutylene-_b_-styrene) [5, 6] as the cationic chemist's response to anionically prepared poly(styrene-_b_-butadiene-_b_-styrene) (Shell's Kratons). The polyisobutylene-based TPEs exhibit unparalleled oxidative and chemical resistance, combined with outstanding softness and barrier properties, and good processing characteristics. Subsequently, Kaneka Inc. developed a large-scale manufacturing process and proceeded to introduce to the market SIBS (for styrene/isobutylene/styrene) triblocks for various TPE applications.

| Polystyrene | ~~~~~~~~~~~~~~~~~~~PIB~~~~~~~~~~~~~~~~~~ | Polystyrene |

SIBS

Last but not least, Capital Chemicals Co. recently developed and commercialized POLYPHANE®, a polyisobutylene-based hydrophobizing microemulsion that provides outstanding temporary protection of painted metal surfaces, for example for automobiles, railroad rolling stock, etc.

The major conclusions of the first part of our presentation are that academic research can be quite profitable, and that it would behoove academic researchers to patentably protect their inventions. Inventing useful things can be quite profitable both to the academic inventor and to the industrial developer. Finally, we emphasize that, when properly managed, inventing/patenting for profit, and publishing for fun and toward the advancement of science are highly rewarding self-reinforcing activities. And this thought smoothly leads us to the second part of this presentation:

III.   Carbocationic Polymerization of Bicyclic Olefins and the Synthesis of Star-Block Copolymers [1, 2]

III.1. Objectives and Background

Our overall objective was the synthesis of well-defined star-block copolymers that combine soft rubbery PIB segments with hard, high T_g segments of cycloaliphatic polyolefins. Block copolymers comprising soft and hard segments are of great current interest for gaining insight into the structure/property relationship of segmented polymers in general and thermoplastic elastomers (TPEs) in particular. In addition, the envisioned systems are of special interest since both block segments are aliphatic hydrocarbons. Scheme 1 shows the formulae of the monomers considered for these investigations together with the characteristic repeat units obtainable by carbocationic polymerization, and the T_g's of the polymers.

194

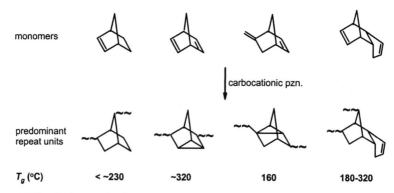

monomers				
		carbocationic pzn.		
predominant repeat units				
T_g (°C)	< ~230	~320	160	180-320

Scheme 1. T_g's of select cycloaliphatic polyolefins

Thus we set out to synthesize two architecturally closely related three-arm star-block copolymers: One consisting of a tricumyl (tCum) core out of which radiate three poly(isobutylene-*b*-cycloolefin) arms, tCum[poly(isobutylene-*b*-cycloolefin)]₃, and the other with the block sequence reversed, tCum[poly(cycloolefin-*b*-isobutylene)]₃.

Because of their rigid repeat structures, polycycloolefins exhibit a combination of desirable properties, e.g., outstanding chemical resistance, high heat distortion temperature, stiffness and strength, optical transparency, and low dielectric constants.[7, 8] These properties render polycycloflefins useful for lenses,[9] compact discs,[10] waveguides,[11] photoresists,[12] electronic packaging,[10] medical applications,[9, 10, 13] potential solar energy storage devices,[14] and integrated circuits.[15, 16, 17]

We concentrated our efforts on polynorbornene (PNB) and polynorbornadiene (PNBD), and particularly on the latter which shows the highest T_g reported for hydrocarbon polymers.

Our synthetic strategy included the preparation of a three arm star prepolymer, tCum(PIB-Cl')₃, followed by the blocking of the second monomer NBD from the active -Cl' chain ends. Scheme 2 illustrates the synthetic route to tCum(PIB-*b*-PNBD)₃. To ensure the envisioned blocking of PNBD, the conditions for the living/controlled polymerization of this cycloolefin were worked out by model experiments (see III.3.).

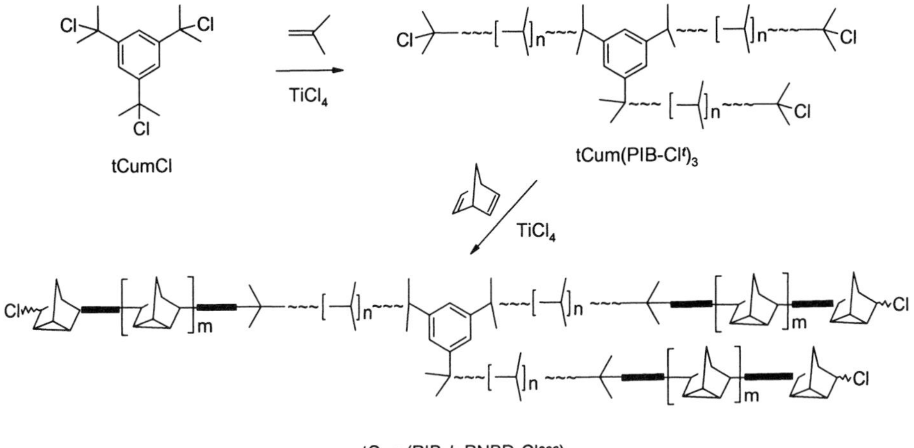

tCum(PIB-b-PNBD-Cl^sec)₃

Scheme 2. Synthetic Strategy for tCum(PIB-b-PNBD)₃

III.2. Experimental

The materials, equipment, and procedures (both synthetic and characterization) have been described.[1, 2]

III.3.1.   Orienting Experiments with Norbornadiene

Preparatory to systematic data gathering, orienting experiments were carried out. The carbocationic polymerization of NBD initiated most likely by adventitious protic impurities in the presence of AlCl₃, AlBr₃, MoCl₅, MoCl₅/EtAlCl₂, and Et₃Al, as well as by *tert*-BuCl/EtAlCl₂ has been documented.[14, 18, 19, 20, 21, 22] Based on this information and experience with other olefins and diolefins,[23] we decided to examine polymerizations initiated by 2-chloro-2,4,4-trimethylpentane (TMPCl)/TiCl₄ combinations. TMPCl is an excellent model for the tert-Cl terminus of PIB produced by living carbocationic polymerization.[24] Controlled initiation with TMPCl would yield trimethylpentyl headgroups and thus leads to the possibility of blocking the cyclic olefin from such PIBs.

Based on orientating polymerizations by the use of CH₂Cl₂ diluent, NBD was chosen for further detailed experimentation. Under similar conditions, 2-norbornene (NB) gave substantially lower yields and molecular weights than NBD so that investigations with this monomer were discontinued. In addition, the higher T_g of PNBD prompted us to focus experimentation on this polymer.

III.3.2. Living/Controlled Polymerization of NBD

Efforts were made to develop conditions for the living/controlled carbocationic polymerization of NBD. The effect of monomer concentration, nature of solvents, reagent concentrations and temperature on the rate and molecular weights were explored. Figure 1 (molecular weights vs. conversion and $\ln([M]_0/[M])$ vs. time) summarizes the results.

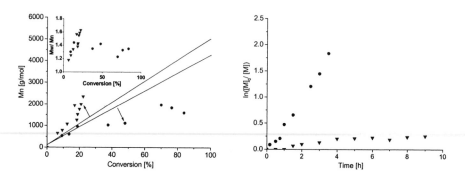

Figure 1. NBD polymerization: M_n vs. conversion (left), theoretical lines assume livingness, inset M_w/M_n vs. conversion; $\ln([M]_0/[M])$ vs. time (right); solvent CH_3Cl (●) and $CH_3Cl/CHCl_3$ 30/70 (v/v) (▼)

The series with CH_3Cl (●) was carried out by the use of eight test tubes with: $[NBD]_0 = 843$ mM, $[TMPCl] = 18.15$ mM, $[DtBMP] = 8.9$ mM, $[DMA] = 32.7$ mM, 33 mL CH_3Cl at −60 °C. The precooled coinitiator $[TiCl_4] = 415$ mM was added last. After given time intervals, the reactions were quenched with precooled methanol.

The M_n vs. conversion plot shows M_n's lower than theoretical above ~ 20% conversion, which indicates chain transfer in this region. The corresponding $\ln([M]_0/[M])$ vs. time plot is linear up to ~ 45 min, after which the rate increases significantly. M_w/M_n ~ 1.35 over the entire conversion range (< 84%). According to this evidence the system may be considered living up to ~ 20% conversion.

The series with $CH_3Cl/CHCl_3$ 40/60 (v/v) (▼) was effected with $[NBD]_0 = 650$ mM, [TMPCl] = 12.0 mM, [DtBMP] = 4.9 mM, [DMA] = 21.5 mM, 500 mL $CH_3Cl/CHCl_3$ 40/60 (v/v), −35 °C. The precooled coinitiator $[TiCl_4] = 474$ mM was added last. After given time intervals, samples were taken and quenched with precooled methanol. These conditions mimic the blocking of PNBD from tCum(PIB-Cl^t)$_3$ (see below). The use of this solvent mixture is needed on the one hand to dissolve the high molecular weight monodisperse prepolymer ($CHCl_3$), and on the other hand to provide sufficient polarity (CH_3Cl) for the polymerization of NBD.

The M_n vs. conversion plot shows M_n's above the theoretical line. At low conversions (<23%) this is a signature for slow initiation.[25,26] The M_w/M_n values increase with conversion which is another indication for slow initiation. The induction period in the $\ln([M]_0/[M])$ vs. time plot further confirms slow initiation. Evidently, the low polarity solvent mixture reduces the rate. It is known that even a minor polarity reduction retards living polymerization.[27]

III.3.3. PNBD Microstructure

Scheme 3 helps to visualize the mechanism of initiation and propagation, and shows the repeat structure of PNBD, together with the head- and tail-groups, and branching units.

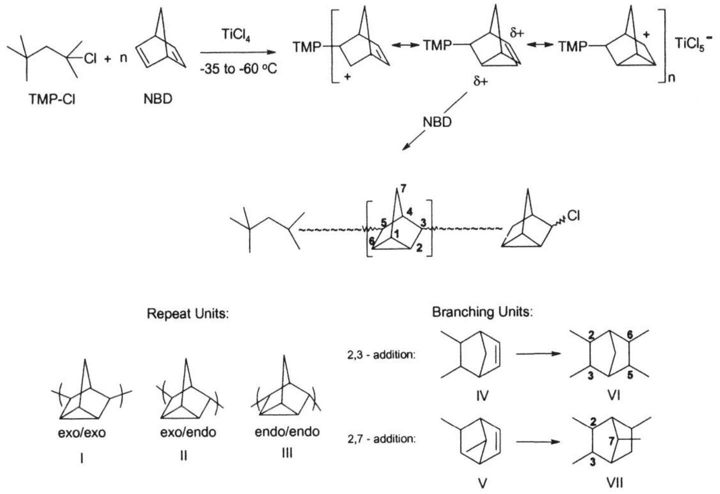

Scheme 3.   Synthesis and Structure of PNBD

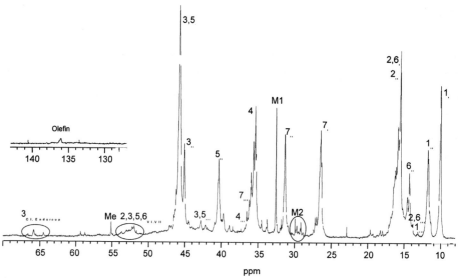

Figure 2. ^{13}C-NMR-Spectrum of PNBD (750 MHz, CDCl$_3$, prepared by TMPCl/TiCl$_4$/CH$_3$Cl/-60 °C/2h)

The microstructure of PNBD was investigated by ^{1}H and ^{13}C-NMR spectroscopy. Figure 2 shows the ^{13}C-NMR-spectrum of a representative PNBD (M_n = 2250, M_w/M_n = 1.28). The initiator fragment (head-group), tail-group, repeat unit connectivity and branching units were analyzed and identified.

The initiator residue (CH$_3$)$_3$CCH$_2$C(CH$_3$)$_2$- and the -CH$_2$C(CH$_3$)$_2$CH$_2$C(CH$_3$)$_2$Cl tail group in PIB prepared with TMPCl/TiCl$_4$ have been identified.[28, 29] PNBD obtained by the TMPCl/TiCl$_4$ system shows the three terminal methyl carbons of the head-group (M1) at δ = 32.5, the methylene carbon (Me) at δ = 55.1, and the two adjacent methyl carbons (M2) that produce multiple signals at δ ~ 29.5 ppm. The multiplicity is due to the endo- and exo-positions of the first NBD repeat unit, and to the magnetic nonequivalence of the methyl groups caused by steric congestion.

The tail-group gives distinctive signals at δ = 64.5 and 65.8 ppm, due to the C^3 bearing the terminal chlorine (exo- and endo-Cl's). According to 2d C-H correlation spectroscopy (750 MHz, CDCl$_3$, not shown) the tail proton bonded to the terminal C^3 resonates at δ ~ 3.9 ppm.

For further details see references 1 and 2.

We obtained M_n = 2,540 g/mol by integrating the ^{1}H NMR spectrum of a representative PNBD (not shown), and M_n = 2,900 g/mol by GPC (RI trace, PIB standards). Integration was effected by relating the intensity of the terminal (C^3) proton signal at ~ 3.9 ppm against the sum of all other protons in the 0.5 – 3.0 ppm range, including the 17 protons in the initiator fragment.

Thus structure analysis by high resolution NMR spectroscopy led to insight regarding repeat unit connectivity, branching, and to a definition of the head- and tail-groups of PNBD prepared by the TMPCl/TiCl$_4$ initiating system.

III.3.4. T_g Versus Molecular Weight

One of the important objectives of our investigations was to prepare PNBDs with T_g's in the 250 – 300 °C range for potential thermoplastic applications. We found conditions under which the molecular weights increased to ~ 2,500 g/mol with increasing conversions (see above). Thus we turned to determine the effect of molecular weight on T_g. Figure 3 shows the T_g as function of PNBD molecular weights.

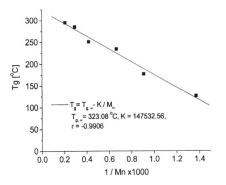

Figure 3. T_g of PNBD as a function of $1/M_n$

Fox and Flory suggested an empirical equation to express the dependence of the T_g on the M_n of different linear polymers: [30, 31, 32]

$$T_g = T_{g\infty} - K/M_n$$

where $T_{g\infty}$ is the T_g of the infinite molecular weight polymer and K is a characteristic material constant. The dependence is due to the decrease in excess free volume, due to chain ends, with increasing chain length.[33] Our data yields $T_{g\infty}$ = 323 °C and K = 14.75 x 10^4 (similar to the K values for PVC or PVAc).[34] This $T_{g\infty}$ value is in good agreement with T_g~320 °C published for carbocationically polymerized NBD (M_n 14,970 g/mol M_w / M_n = 3.73, polystyrene standards).[35] The agreement between our (extrapolated) and previous literature data suggests very similar microstructures of the PNBDs. This corroborates our conclusions reached by NMR spectroscopy that the microstructures of PNBDs produced by TMPCl/TiCl$_4$ and AlCl$_3$ are very similar.

III.4. Block Copolymerization of Isobutylene and Norbornadiene: Ter- and Star Blocks

III.4.1 Proof of Concept: Novel Linear Triblocks

The initial results by the use of CH$_2$Cl$_2$ diluent in model polymerizations were promising. Thus these scouting experiments were extended to blocking NB and NBD from a –Cl^t ditelechelic PIB (dCum(PIB-Cl^t)$_2$: M_n = 5,800 g/mol, M_w/M_n = 1.07 by GPC-LLS; dCum = dicumyl) as initiator under established conditions (Table 1). These trials lead to two novel linear triblocks, PNB-b-PIB-b-PNB and PNBD-b-PIB-b-PNBD. DSC-analysis indicated the presence of the T_g's of the hard segments, indicating phase separation.

Table 1. Blocking NB and NBD from dCum(PIB-Cl^t)$_2$

Olefin	Conditions		Results			
	[dCum(PIB-Cl^t)$_2$]/[Olefin]/[TiCl$_4$]/ [DtBP](mmol/L); Solvent, T(°C), t(h)		Star-Block (g)	M_n (g/mol)	M_w/M_n	$T_{g,\text{hard segment}}$ (°C)
(structure)	20.5/ 1380/ 400/ 112; CH$_2$Cl$_2$, -60, 3		3.4g	7.1	1.09	118
(structure)	20.5/ 813/ 400/ 112; CH$_2$Cl$_2$, -60, 3		3.7g	9.1	1.07	218

The synthesis of PNB-b-PIB-b-PNB and PNBD-b-PIB-b-PNBD demonstrates that our synthetic strategy conceptualized for the preparation of aliphatic TPEs is indeed operational. These results encouraged us to pursue the controlled synthesis of star blocks.

III.4.2.   Synthesis of tCum(poly(isobutylene-*b*-norbornadiene))₃

Scheme 2 outlines the strategy for the synthesis of tCum(PIB-*b*-PNBD)₃ and shows the structure of the three-arm star-block copolymer. We have demonstrated that the TMPCl/TiCl₄ combination efficiently initiates NBD polymerization (see above). Thus we prepared the tCum(PIB-Cl')₃ macroinitiator by our well-documented technique [36,37] and used it in conjunction with TiCl₄ to induce the polymerization of NBD. Preliminary experiments showed that mixtures of polar/ nonpolar solvents (i.e., ~ 50/50 (v/v) CH₃Cl/nC₆H₁₄, CH₃Cl/methylcyclohexane (MeCH), CH₂Cl₂/nC₆H₁₄, CH₂Cl₂/MeCH) conventionally used for the preparation of PIB blocks covalently linked to glassy blocks [38] are unsuitable for the envisioned synthesis because NBD polymerizations do not proceed in the presence of even small amounts (~ 20%) of hexanes or MeCH, and because PNBD is insoluble in these aliphatic hydrocarbons. Extended orienting experiments with various solvents and solvent mixtures indicated that under well-defined conditions 30/70 (v/v) CH₃Cl/ CHCl₃ mixtures at −35 °C gave satisfactory blocking. CHCl₃ is a solvent for high molecular weight narrow polydispersity PIB and keeps the tCum(PIB-*b*-PNBD)₃ in solution at −35 °C even in the presence of 30% CH₃Cl. At higher CH₃Cl concentrations (i.e., 40/60 (v/v) CH₃Cl/CHCl₃) or at lower temperatures (i.e., -45 to -55 °C) the PIB prepolymer precipitated from solution. Thus, satisfactory star-block syntheses could be effected with 30/70 (v/v) CH₃Cl/ CHCl₃ at −35 °C; under these conditions all the species of a charge (i.e., prepolymer, star-block, and PNBD byproduct (see below)) are in solution, and the 30% CH₃Cl provides sufficient polarity for cationic blocking.

In a representative synthesis 25 mL precooled TiCl₄ were added to a mixture of 5 g prepolymer ($M_n \sim 102,000$ g/mol, $M_w / M_n = 1.03$), 20 mL NBD, 0.5 g DtBMP, 1 mL DMA in 700 mL 30/70 (v/v) CH₃Cl/ CHCl₃ at −35 °C. After 1 h additional 25 mL precooled TiCl₄ were added. The reaction was quenched after 10 h by precooled methanol. Under similar conditions, blocking of PNBD did not occur using 50/50 (v/v) mixtures of hexanes/CH₃Cl, methylcyclohexane/CH₃Cl, and CH₂Cl₂/CH₃Cl. The solubility of the macroinitiator was much lower in CH₂Cl₂ than in hexanes, MeCH or CHCl₃.

The molecular weight of the prepolymer increased to $M_n = 107,300 - 109,200$ g/mol with exceptionally low values for M_w/M_n (~ 1.01).

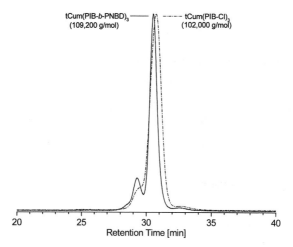

Figure 4.   GPC traces of tCum(PIB-Cl^t)$_3$ and tCum(PIB-b-PNBD)$_3$ (LLS detector)

Figure 4 shows the GPC trace of a prepolymer together with the corresponding star-block. The small but distinct shift of the peak associated with the prepolymer toward lower elution counts after blocking with NBD (i.e., from ~ 30.8 to 30.5 min) indicates blocking of a small amount of PNBD.

In addition to the main peak at ~ 30.8 min associated with the star-block, non-purified polymerization products showed a peak at ~ 39 min, indicating the presence of PNBD homopolymer (M_n ~ 1,600 g/mol (M_w/M_n = 1.28; RI – detector, linear PIB standards) most likely formed by chain transfer. The homopolymer can be removed by precipitation (from THF into ice-cold 1,2-dichloroethane), and essentially uncontaminated, pure star-block can be obtained after two precipitations.

The position of the peak due to the prepolymer suggests M_n =102,000 g/mol, M_w/M_n ≈ 1.03 (LLS – detector). In line with this GPC analysis the arm molecular weight of the prepolymer is ~ 34,000 g/mol. The position of the peak due to the star-block suggests M_n =109,200 g/mol, M_w/M_n ≈ 1.01. The shoulder at ~ 29 min indicates the presence of a small amount of a species of M_n ~ 206,000 g/mol, i.e., about double that of the prepolymer. This species most likely arose by star/star coupling. In this process a growing arm intramolecularly cationates an arm of another star with a -CH$_2$-C(CH$_3$)=CH$_2$ terminus formed by chain transfer (i.e., dehydrochlorination) of a

$CH_2C(CH_3)_2Cl$ terminus. Such star/star coupling reactions were observed previously.[39,40] The very small hump centered at ~ 33 minutes (~ 43,000 g/mol) may be due to a very small amount of product formed by premature chain breaking during star formation.

The molecular weight increase from the prepolymer to the star-block is ~ 7,200 g/mol which is equivalent to an increase of ~ 2,400 g/mol per arm in the star-block. The latter value is comparable to the molecular weight of the PNBD homopolymer contaminant. Thus GPC analysis suggests a star-block consisting of $M_{n,\,PIB}$ = 34,000 and $M_{n,\,PNBD}$ = 2,400 g/mol per arm.

III.4.3.   Microstructure of tCum(poly(isobutylene-*b*-norbornadiene))₃

The microstructure of tCum(PIB-*b*-PNBD-Cl^{sec})$_3$ was investigated by NMR spectroscopy. Figure 5 shows the ^{1}H-NMR spectrum of a representative star-block and of the prepolymer, together with assignments.

The high resolution allowed the detection and quantitative analysis of the aromatic initiator fragment and chlorine-containing tail-groups in both the prepolymer and the star-block, despite their relatively high M_n's. On hand of this information it was possible to calculate molecular weights from ^{1}H-NMR data.

In the prepolymer, the resonance Ar at δ = 7.14 ppm is due to the three aromatic protons of the cumyl initiator fragment. The six (per arm) methyl protons (a) resonate at δ = 0.81 ppm. The terminal methyl and methylene protons (e and d) resonate at δ = 1.70 and δ = 1.98 ppm. It is possible to assign signals for the methyl protons of the tricumyl core (tCu at δ ~ 1.83) but not the adjacent methylene units, because the resonances of the latter are too near the large resonances from the methylene and methyl protons (b and c) of the IB repeat units.

In the star-block, the PNBD-Cl^{sec} terminus is a tricyclic structure with a chlorine in the 3 position. The proton adjacent to this Cl may be in exo or endo position with signals appearing in the δ = 3.8 – 4.0 ppm range. While the signals are, of course, weak and do not allow exo/endo resolution, they are clearly discernible. The assignment of this terminal proton is discussed above.[1]

According to quantitative NMR spectroscopy [1,2] the depicted three-arm star-block consists of a tricumyl core out of which emanate three PIB-*b*-PNBD-Cl^{sec} branches composed of $M_{n,PIB}$ = 34,500 and $M_{n,\,PNBD}$ = 2,100 g/mol, abbreviated by tCum[PIB(34.5K)-*b*-PNBD(2.1K)-Cl^{sec}]$_3$.

The molecular weights determined by proton NMR spectroscopy are in excellent agreement with those determined by GPC, i.e., $M_{n,\ prepolymer} = 103,900$ and $102,000$ g/mol, and $M_{n,\ star\ block} = 110,100$ and $109,200$ g/mol, respectively.

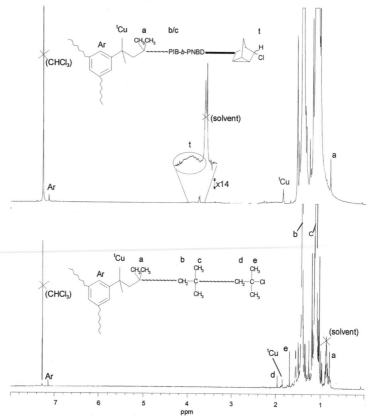

Figure 5. ^{1}H NMR spectra of tCum(PIB-Cl^t)$_3$ and tCum(PIB-b-PNBD-Cl^{sec})$_3$ with assignments (750 MHz, CDCl$_3$)

III.4.4. Synthesis of tCum[poly(norbornadiene-*b*-isobutylene)]$_3$

The synthetic strategy for the synthesis of tCum(PNBD-*b*-PIB-Cl^t)$_3$ was essentially the same as in the case of tCum(PIB-*b*-PNBD-Cl^{sec})$_3$, except that the sequence of blocking was reversed: the polymerization of NBD was effected by the tCumCl/TiCl$_4$ combination and, subsequently, the tCum(PNBD-Cl^{sec})$_3$ prepolymer was used, in conjunction with TiCl$_4$ to induce the polymerization of IB. The first step of the synthesis was the preparation of the three-arm star tCum(PNBD)$_3$ prepolymer. On the basis of our experience with the TMPCl/TiCl$_4$ system (see above), we expected and indeed found that tCumCl readily induces NBD polymerization in the presence of TiCl$_4$. The resulting tCum(PNBD-Cl^{sec})$_3$ showed $M_n \sim 2{,}900$ g/mol (RI-detector, against PIB standards).

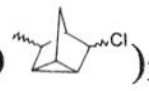

The second step, the blocking of IB from the  terminus, also proceeded with ease. It is of interest that this *sec*-chlorine initiates IB polymerization because the structurally similar isopropyl chloride is an inefficient initiator under similar conditions.[41] The ease of initiation with  is most likely due to the strained tricyclic structure.

The molecular weight of the star-block was $\sim 14{,}200$ g/mol ($M_w/M_n \sim 1.99$). That suggests ~ 900 g/mol per inner PNBD arm and 3,800 g/mol per outer PIB arm: tCum(PNBD(0.9K)-*b*-PIB(3.8K))$_3$.

III.4.5.   Microstructure of tCum[poly(norbornadiene-*b*-isobutylene)]$_3$

Figure 6 depicts the ^{1}H-NMR spectrum of the star-block, together with relevant assignments. The resonances at $\delta = 1.98$ and 1.68 ppm (c and d) indicate the presence of the terminal methylene and methyl protons, respectively. The PNBD segments are not resolved under the dominating methyl and methylene signals (a and b) of the PIB segments but base broadening of these signals (in the 0.8 – 1.8 ppm range) clearly suggests the presence of PNBD. The signal for the aromatic protons (Ar) is not resolved either, despite the high magnetic field applied. The rigid PNBD segments may cause signal broadening by strain on the core as suggested by signal intensity in the 7.1 – 7.25 ppm range. The presence of a weak broad signal (t) at $\delta \sim 3.9$ ppm is due to unreacted PNBD chain ends (for a detailed analysis see above) and suggests inefficient initiation. The resolution of the signals is insufficient for a quantitative evaluation of the repeat structure.

206

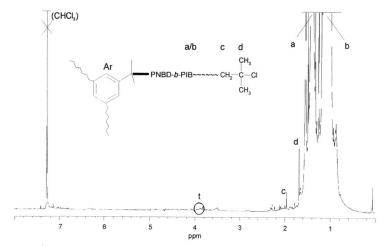

Figure 6. ¹H NMR spectrum of the star block tCum[PNBD(0.9K)-*b*-PIB(3.8K)]₃ (750 MHz, CDCl₃)

III.5. Thermal Characterization of the Star Blocks

Both tCum(PNBD-*b*-PIB)₃ and tCum(PIB-*b*-PNBD)₃ showed two T_g's, indicating the separation of rubbery and glassy phases (Figure 7).

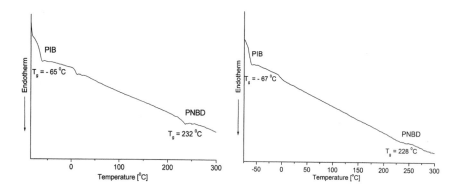

Figure 7. DSC thermograms of tCum(PIB-*b*-PNBD)₃ (left) and tCum(PNBD-*b*-PIB)₃ (second heating, M_n's = 109,200 (GPC-LLS) and 14,200 g/mol (GPC-RI) respectively)

The DSC thermogram of tCum[PIB(34.5K)-b-PNBD(2.1K)]$_3$ showed one transition at ~ -65 °C due to the PIB phase and another at ~ 232 °C which is associated with the glassy PNBD domains. The transition at ~ 0 °C is most probably due to moisture. The presence of the two transitions indicates phase separation, which is remarkable in view of the relatively low molecular weight of the PNBD segments. With the Fox-Flory relationship (see above), the T_g ~ 232 °C for the PNBD in the star-block corresponds to M_n ~ 1,620 g/mol. This value is somewhat lower than those calculated from NMR and GPC data (i.e., NMR gives ~ 253 °C for M_n ~ 2,100 g/mol; GPC gives ~ 263 °C for M_n ~ 2,400 g/mol).

The DSC thermogram of tCum[PNBD(0.9K)-b-PIB(3,8K)]$_3$ also showed two T_g's, one at - 67 °C, marking the rubbery PIB phase, and another at 228 °C, indicating the presence of the glassy PNBD phase. It is surprising to observe this phase separation, considering that the relatively small rigid PNBD inner core would be expected to be masked by the voluminous PIB outer core. With the Fox-Flory relationship (see above), the T_g ~ 228 °C for the PNBD phase would correspond to M_n ~ 1,560 g/mol. This value is similar to the PNBD molecular weight between two PIB blocks: I.e., each arm of the star-block bears a PNBD segment af ~900 g/mol. Two PNBD segments are connected via the cumyl core of the star-block.

Acknowledgements

The authors are grateful to Ahmed F. Moustafa for carrying out numerous experiments. The work was supported by the National Science Foundation (grants DMR-0243314 and DMR-9988808).

208

[1]   R. M. Peetz, A. F. Moustafa, J. P. Kennedy, *J. Polym. Sci., Part A, Polym. Chem.* **2003**, *41*, 732-739
[2]   R. M. Peetz, A. F. Moustafa, J. P. Kennedy, *J. Polym. Sci., Part A, Polym. Chem.* **2003**, *41*, 740-751
[3]   L. Wilczek, J. P. Kennedy, *J. Polym. Sci., Part A, Polym. Chem.* **1987**, *25*, 3255-3265
[4]   J. P. Kennedy, D. R. Weyenberg, L. Wilczek, A. P. Wright, US Patent 4,758,631
[5]   G. Kaszas, J.E. Puskas, J. P. Kennedy, W.G. Hager, *J. Polym. Sci., Part A, Polym. Chem.* **1991**, *29*, 427-436
[6]   G. Kaszas, J. E. Puskas, J. P. Kennedy, W.G. Hager, US Patent 4,946,899
[7]   H. Cherdron, M.-J. Brekner, F. Osan, *Angew. Makromol. Chem.* **1994**, *223*, 121-33
[8]   A. Toyota, M.Yamaguchi, *Polym. Mat. Sci. and Eng.* **1997**, 76, 24
[9]   T. Kohara, *Macromolecular Symposia* **1996**, *101* (5th International Conference "Challenges in Polymer Science and Technology", 1994), 571-579
[10]   H. T. Land, F. Osan, T. Wehrmeister, *Polym. Mat. Sci. and Eng.* **1997**, *76*, 22-23
[11]   X.-M. Zhao, R. A. Shick, R. Ravikiran, P. S. Neal, L. F. Rhodes, A. Bell, PCT Int. Appl. **2002**, WO 2002010231
[12]   S. Jayaraman, B. L. Goodall, L. F. Rhodes, R. A. Shick, R. Vicari, R. D. Allen, J. Opitz, R. Sooriyakumaran, T. Wallow, PCT Int. Appl. **1999**, WO 9942502
[13]   D. Rueppel, B. Hahn, A. Schneller, K. Berger, F. Osan, Ger. Offen. **2000**, DE 19834025
[14]   N. Tonuglu, N. Balcioglu, *Macromol. Rapid Commun.* **1999**, *20(10)*, 546
[15]   *The International Technology Roadmap for Semiconductors* (Semiconductor Industry Association, San Jose, CA), 2000
[16]   S. J. Martin, J. P. Godschalx, M. E. Mills, E. O. Shaffer II, P. H. Townsend, *Adv. Mater.* **2000**, *12(23)*, 1769
[17]   P. G. Apen, PCT Int. Appl. **2001**, WO 2001004954 or 2000-US18830
[18]   J. P. Kennedy, J. A. Hinlicky, *Polymer* **1965**, *6*, 133
[19]   D. W. Grattan, P. H. Plesch, *Makromol. Chem.* **1980**, *181*, 751
[20]   A. Mizuno, M. Onda, T. Sagane, *Polymer* **1991**, *32(16)*, 2953
[21]   J. T. G. Huckfeldt, W. Risse, *Macromol. Chem. Phys.* **1998**, *199*, 861
[22]   Balcioglu, N.; Tunoglu, N. J Polym Sci Part A Polym Chem 1996, 34, 2311
[23]   J. P. Kennedy, in: *"Cationic Polymerization of Olefins: A Critical Inventory"*, Wiley: New York 1975,Ch. I
[24]   G. Kaszas, J. E. Puskas, C. C. Chen, J. P. Kennedy, *Macromolecules* **1990**, *23(17)*, 3909
[25]   M. Zsuga, T. Kelen, *Polym. Bull.* **1986**, *16*, 285
[26]   M. Zsuga, J. P. Kennedy, T. Kelen, *J. Macromol. Sci. – Chem.* **1989**, *A26*, 1305
[27]   J. P. Kennedy, B. Ivan, in: *"Designed Polymers by Carbocationic Macromolecular Engineering: Theory and Practice"*, Hanser: Munich 1992, pp 72-73
[28]   A. V. Lubnin, J. P. Kennedy, *J. Macromol. Sci. Pure Appl. Chem.* **1994**, *A31(60)*, 655
[29]   J. Si, J. P. Kennedy, *J. Polym. Sci., Part A, Polym. Chem.* **1994**, *32*, 2011
[30]   T. G. Fox, P. J. Flory, *J. Polym. Sci.* **1954**, *14*, 315
[31]   T. G. Fox, P. J. Flory, *J. Am. Chem. Soc.* **1948**, *70*, 2384
[32]   T. G. Fox, P. J. Flory, *J. Phys. Chem.* **1951**, *55*, 221
[33]   G. B. McKenna, in: *"Comprehensive Polymer Science, Vol.2, Polymer Properties"*, Booth, C.; Price, C., Eds.; Pergamon: Oxford, 1989; pp 311-362
[34]   R. F. Boyer, *Macromolecules* **1974**, *7*, 142
[35]   M. B. Roller, J. K. Gillham, J. P. Kennedy, *J. Appl. Polym. Sci.* **1973**, *17*, 2223
[36]   R. Faust, J. P. Kennedy, *J. Polym. Sci., Part A, Polym. Chem.* **1987**, *25*, 1847
[37]   A. Nagy, R. Faust, J. P. Kennedy, *Polym. Bull.* **1985**, *13*, 97
[38]   J. P. Kennedy, B. Iván, in: *"Designed Polymers by Carbocationic Macromolecular Engineering: Theory and Practice"*, Hanser: Munich 1992, Ch. III.3.
[39]   S. Jacob, I. Majoros, J. P. Kennedy, *Polym. Bull.* **1998**, *40*, 127
[40]   S. Jacob, J. P. Kennedy, *Adv. Polym. Sci.* **1999**, *146*, 1
[41]   J. P. Kennedy, E. Maréchal, in: *"Carbocationic Polymerization"*, Wiley: New York 1982; p. 106

Telechelic Polyisobutenes with Asymmetrical Reactivity

Arno Lange,* Hans-Peter Rath, Gabriele Lang

BASF AG, Polymer Research, GKS/M – B1, D 67 056 Ludwigshafen, Germany
E-mail: arno.lange@basf-ag.de

Summary: 3-Chlorocyclopentene-(1) has been used as initiator in the quasiliving cationic polymerization of isobutene. Polyisobutenes in the molecular weight range Mw = 300 – 60000 Dalton with a molecular weight distribution MWD between 1.17 and 1.34 have been synthesized. The linear polymer was selectively hydrosilated on one or both ends. With a furan coupling agent, the PIB (polyisobutene) was dimerized, yielding polymers with a cyclopentene head- and tail group.

Keywords: cationic polymerization; functionalization of polymers; initiators; NMR; polyolefins

Introduction

In the field of the living cationic polymerization of isobutene, a discrepancy between academical advancement and industrial realization has been noted: The issue is versatility versus cost. Numerous authors[1] have shown the ample opportunities for a "molecular engineer". For many industrial applications, a bifunctional polyisobutene in the molecular weight range Mw = 500 – 5000 Dalton is the educt of choice. Whereas widely used starters like 1,4-bis(1-chloro-1-methylethyl)benzene are well suited from a technical / chemical point of view, their inherent cost structure is a major obstacle for many industrial applications.

Polymerizations with 3-Chlorocyclopentene (CCP)

In the living cationic polymerization of isobutene, well-defined telechelic polymers with reactive end groups may be produced[1]. The reactivity of these functional PIBs together with the stability and elasticity of the polymer backbone makes them versatile and hence attractive macromers for sealants, adhesives or industrial rubbers. On the other hand, the need for an absolutely proton free environment, low temperatures and special initiators in the polymerization puts a burden on the industrial synthesis which the markets often are not willing to bear. In the 1000 – 2000 Dalton

© 2004 International Union of Pure and Applied Chemistry

DOI: 10.1002/masy.200451117

molecular weight range, an economic initiator is the cornerstone of a successful industrial process.

3-Chlorocyclopentene seemed to be a prospective candidate for such an initiator: Allylic activation of the chlorine atom ensures the proper reactivity. It is generated by simple addition of HCl gas to cyclopentadiene according to a procedure in Organic Synthesis[2]. (Reaction of dichlorocarbenes with dicyclopentadiene[3] is an alternative to this route.)

We applied standard polymerization conditions to a CCP / IB (isobutene) system. The polymers **Ia** have been prepared with molecular weights from 300 Dalton to 60,000 Dalton and molecular weight distribution (MWD) between 1.17 and 1.34.

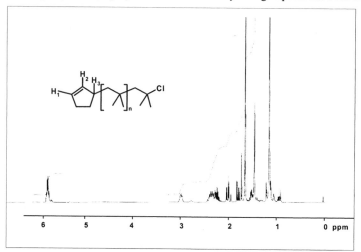

Low concentrations of the catalyst TiCl$_4$ together with elevated temperatures during the work-up procedure[4] directly lead to a 3:1 mixtures of the olefins **Ib** and **Ic**. By end-capping with trimethylallylsilane, a polymer with reactive vinyl end groups **Id** has been obtained.

Figure 1: ^{1}H-NMR of Polymer **Ia**, M$_n$ = 350, solvent CDCl$_3$, 16 scans at 500 MHz

The structure of the polymers was determined by ^{1}H-NMR. As model component, a 1:1 adduct of IB and CCP (**Ia**, n=0) was prepared according to Mayr[5]. Figure 1 shows the ^{1}H-NMR spectrum of a low molecular weight (M_n = 350) PIB **Ia**. The signals in the olefinic region are attributed to protons H_1 and H_2 , the signal at δ = 2.95 ppm to the allylic proton H_3. Especially this allylic signal is a clear indication for the proposed structure. In the initiator CCP, H_3 is attached to a C-Cl group and it's ^{1}H-NMR signal is significantly shifted to δ = 5.05 ppm.

In Figure 2 the CCP / IB ratio is plotted against the number molecular weight average M_n. As expected from a quasiliving system, the M_n is a linear function of the CCP / IB relation.

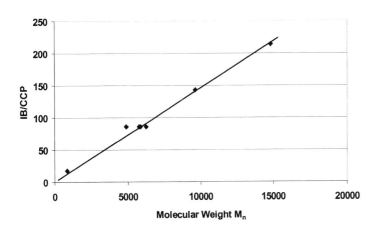

Figure 2: Plot of the IB/CCP relation versus M_n. Typical reaction condition: 4.28 mol IB, 0.05 mol CCP, 300 ml Hexane, 300 ml CH2Cl2, 0.026 mol   TiCl4, 0.002 mol DBP (2,6-Di-t.butylpyridine), 2 h -50° – -70°C

Functionalization

The different reactivity of the terminal olefins in the dienes **Ib-Id** has been used to tailor-make new telechelic PIBs. Depending on reaction conditions, both ends can be functionalized. Addition of H$_3$C-SiHCl$_2$ to PIB **Ib** with H$_2$PtCl$_6$ * 6H$_2$O ("Speiers reagent"[7]) as catalyst yielded the bifunctional structure **IIb**[a].

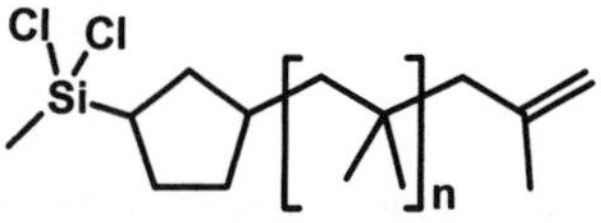

IIa 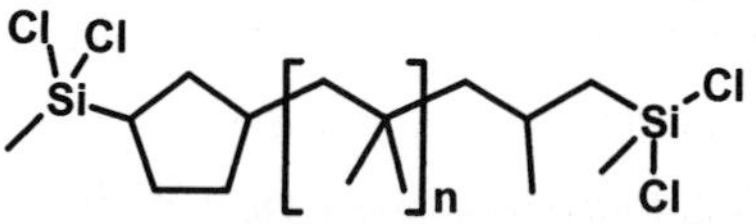   **IIb**

Catalyzed by a 1,3-divinyl-1,1,3,3-tetramethyl-disiloxan-Pt(0) complex, the reaction proceeded to the mono addition product **IIa**[a].

A living cationic polymerization of IB may be terminated by a coupling reagent, yielding a polymer with twice the molecular weight.[6] Reaction with 2,5-bis(furanyl(2)-methyl)furan led to the dimerized PIB **III**.

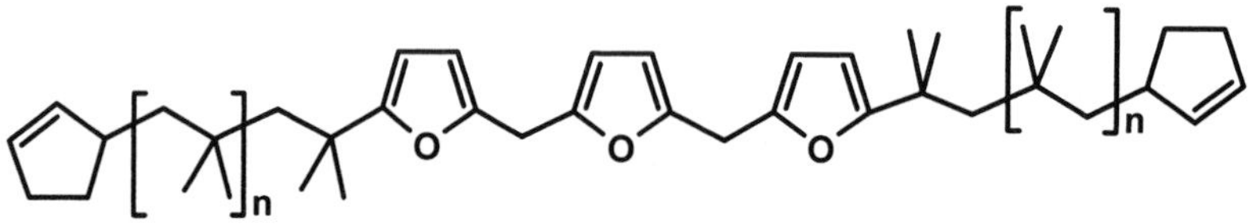

III

The terminal cyclopentene double bonds in structure **III** are more reactive than a $-CH_2-(CH_3)=CH_2$ or $-CH=C(CH_3)_2$ group and offer an ample synthetic potential. In many cases, the use of an expensive end-capping reagent like allyltrimethylsilane may be avoided.

Conclusion

The use of CCP as initiator in the living cationic polymerization of IB leads to a cost efficient synthesis of the new polyisobutenes I. They are very versatile intermediates for symmetrical and asymmetrical (R-PIB-R') telechelics. The (functionalized) polyisobutenes are useful as macromers e.g. for sealants, adhesives or industrial rubbers.

[a] The cyclopentane ring is shown to be substituted in the 1,3- position. It should be noted however that the polymer is actually a mixture of 1,2- and 1,3-isomers.

Experimental

1. Polymerization (typical experiment)

A 2l 4-necked flask was equipped with stirrer, dry ice condenser, septum inlet and 2 dropping funnels with cooling jackets. Both dropping funnels contained 3 Å molecular sieves over glass wool. A mixture of 300 ml CH_2Cl_2 and 300 ml n-hexane was conditioned at $-78°C$ for 20 min in a funnel, than added to the reaction vessel. IB (400 ml) was treated in the same way. With vigorous stirring, 6 mmol triethoxyphenylsilane, 50 mmol CCP and 26 mmol $TiCl_4$ were injected through the septum at $-70°C$. The temperature increased to $-55°C$ and was kept between $-50°C$ and $-70°C$ for 2 h. The polymerization was terminated by the addition of 50 ml iso-propanol. The solution was washed three times with water and the solvent was driven off with a rotary evaporator at 180°C until a final pressure of 5 mbar was reached.

The product was 300g of clear polymer, Mn = 4,890; MWD = 1.18

2. Bis-Hydrosilylation

A 2l 4-necked flask was equipped with stirrer, condenser, thermometer and a dropping funnel. Polymer from experiment 1 (200g, 41 mmol) was dissolved in 1200ml toluene. A solution (1ml) of 0.1 M H_2PtCl_6 * $6H_2O$ in isopropanol and then 11.5 g (100 mmol) dichloromethylsilane were added. The reaction mixture was kept at 90°C for 12h. The solvent was driven off with a rotary evaporator at 180°C until a final pressure of 5 mbar was reached.

The product was 195g of clear polymer. 1H-NMR (500MHz, 16 scans, $CDCl_3$) indicated the absence of olefinic protons. A new signal at $\delta = 0.80$ ppm corresponded to $-Si(CH_3)Cl_2$ groups.

3. Mono-Hydrosilylation

A 2l 4-necked flask was equipped with stirrer, condenser, thermometer and a dropping funnel. polymer from experiment 1 (200g, 41 mmol) was dissolved in 1000ml THF (tetrahydrofuran). A solution (4ml) of 0.1 M 1,3-divinyl-1,1,3,3-tetramethyl-disiloxan-Pt(0) complex in polydimethyl-siloxane and then 5.7 g (50 mmol) dichloromethylsilane were added. The reaction mixture was kept at 60°C for 3h. The solvent was driven off with a rotary evaporator at 180°C until a final pressure of 5 mbar was reached.

The product was 200g of clear polymer. ^{1}H-NMR (500MHz, 16 scans, CDCl$_3$) indicated the absence of cyclo-olefinic protons. A new signal at $\delta = 0.80$ ppm corresponded to a $-Si(CH_3)Cl_2$ group.

4. Dimerization

The polymerization was run as in experiment 1, but before termination, 50 mmol 2,5-bis (furanyl(2)-methyl)furan was added to the polymerization mixture. After an additional period of 120 min at $-70°C$, the mixture was terminated and worked up as described above.

[1] J.P.Kennedy, B.Iván Carbocationic Macromolecular Engineering, , Hanser Publishers 1992
[2] R.B.Moffett, Org. Synth. *32*, 41
[3] A.B.Kostitsyn et al. Synlett (**1990**) 713
[4] A.G.Lange and H.P.Rath DOS 100 61 727
[5] H.Mayr, H.Klein and G.Kolberg, Chem. Ber. *117* (8), 2555-79 (**1984**)
[6] R.Faust and Y.C.Bae, Macrom. *30*, 198 (**1997**)
[7] J.L.Speier, J.A.Webster and G.H.Barnes J.Am.Chem.Soc. *79*, 974 (**1957**)

Macromol. Symp. **2004**, *215*, 215-229

Amphiphilic Polymers Based on Poly(2-oxazoline)s - From ABC-Triblock Copolymers to Micellar Catalysis

Oskar Nuyken,* Ralf Weberskirch, Martin Bortenschlager, Daniel Schönfelder

Lehrstuhl für Makromolekulare Stoffe, TU München, Lichtenbergstr. 4, D-85747 Garching, Germany
Fax: 0049-89-28913562; E-mail: oskar.nuyken@ch.tum.de

Summary: In this presentation we give an overview on our results in the field of poly(2-oxazolines). By means of living cationic polymerization in combination with the initiation and termination method, functionalized poly(2-oxazoline)s have been prepared, that were used as (i) macromonomers to form graft copolymers, (ii) lipopolymers to prepare stealth liposomes, (iii) ABC like polymers to form two compartment micellar networks and (iv) macroligand for micellar catalysis application. Within this report, we will discuss in detail the synthesis and characterization of various poly(2-oxazoline)s for the above mentioned research areas.

Keywords: amphiphiles; block copolymers; cationic polymerization; micellar catalysis; poly(oxazoline)s

Introduction

2-Substituted-2-oxazolines represent a very versatile monomer system, that can be polymerized or copolymerized via a living, ring-opening cationic polymerization mechanism.[1] During the past two decades there has been a tremendous increase in the application of these monomers in the field of block and graft copolymer synthesis and hyperbranched and star-like polymers with a broad field of different applications.[2] In this report we aim to give an overview on our activities in the area of poly(2-oxazoline)s, (1) studying the mechanism of termination and its usage for macromonomer synthesis, (2) investigating the preparation of lipopolymers and their interaction with bilayer model membranes, (3) the synthesis of amphiphilic telechelic poly(2-methyl-2-oxazoline)s with fluorocarbon and hydrocarbon endgroups and their aggregation behavior in aqueous solution and (4) studying the potential use of functionalized diblock copolymers as macroligands in the field of micellar catalysis.

                    DOI: 10.1002/masy.200451118

Termination Studies and Macromonomer Synthesis

In our initial studies, we were interested in the preparation of new polymers with different architecture and composition. A suitable approach is the so called macromonomer route, where a polymer precursor is functionalized with a monomer moiety, that can be polymerized in a second step. The living polymerization mechanism of 2-oxazolines made them a very interesting system to incorporate the monomer via a functionalized termination reagent. Thus, different nucleophiles were studied in the termination reaction of 2-alkyl-2-oxazoline polymerization with respect to efficiency. Based on our results (see Table 1) secondary amines were most efficient termination agents.[3]

Table 1. Termination of a living polymerization of 2-alkyl-2-oxazoline, studied via ^{1}H NMR[a)]

Nu	conversion in % after 10 min	time needed for 100 % conversion
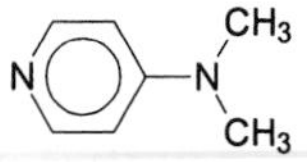	67	60 min
	5	22 h
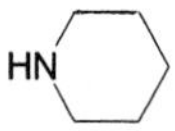	100	< 10 min
H_2O	0	> 8 d

[a)] model studies with **1**:

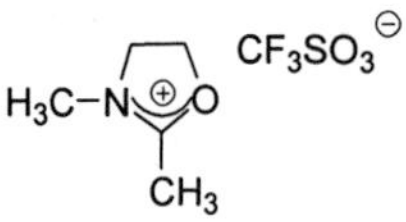

Based on these results, a mono-functionalized piperazine derivative **3** as termination reagent was prepared bearing a styrene unit for subsequent polymerization reactions. Macromonomer synthesis via termination is described in the following scheme:[4]

Table 2. Results of the macromonomer synthesis via termination route

R$_2$	solvent	T	t	n$_{th}$	n$_{NMR}$	M$_n$	M$_w$
		°C	h				M$_n$
Ph	acetonitrile	70	72	10.5	10.1	1390	1.13
Ph	acetonitrile	70	72	9.7	9.3	1260	1.15
C$_9$H$_{19}$	1,2-dichlorobenzene / nitrobenzene (1 : 1)	70	48	10.0	10.4	2550	1.05

The good agreement between nth and nNMR (table 2) and the low PDI's < 1.15 show clearly, that this synthetic route is applicable for the preparation of macromonomers with controlled molar masses and perfect functionality. These macromonomers were afterwards successfully applied for the synthesis of graft copolymers with methyl methacrylate as monomer.

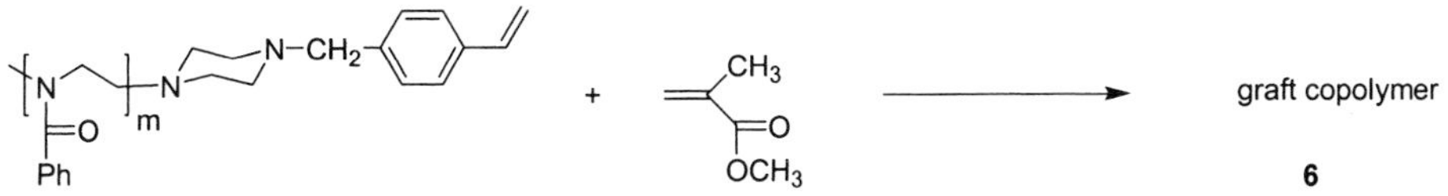

Lipopolymers with Segmented Deuterated Blocks

A second area of interest was concerned with the preparation of lipopolymers.[5] Such polymeric structures are interesting candidates to prepare so called *stealth liposome*, that can be potentially used as controlled release systems for therapeutics. A question still under debate within this research is concerned with the limited understanding how polymer structure and chain conformation affect the long term stability of such liposomes *in vivo* and its accumulation in specific organs whereas other tissues are not affected at all. A possibility to get better insights in chain conformations is neutron scattering of partly deuterated lipopolymers. For that purpose block copolymers with different positions of the deuterated block became available via sequential polymerization of 2-methyl-2-oxazoline (H) with 2-trideuteromethyl-2-oxazoline (D) using 2,3-distearylglyceryltriflate **7** (L) as a initiator.[5,6] Piperidine was used for termination. By using this approach polymers with the following composition L-DHHH, L-HDHH, L-HHDH and L-HHHD were prepared.

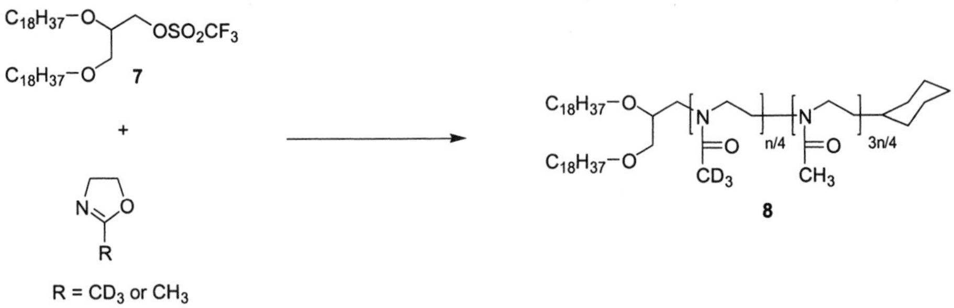

The increase of molecular weight was followed via GPC for L-HDHH and is shown in figure 1.

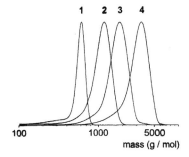

Figure 1. GPC of L (1), L-H (2), L-HD (3) and L-HDHH (4) with L = distearylglyceryltriflate, H = poly(2-methyl-oxazoline) block and D = poly(2-trideuteromethyl-2-oxazoline) block (see **8**).

These polymers were applied for neutron scattering of bilayers composed of lipopolymer **8** (25 mol%) and distearoylphosphatidylcholine (DSPC, 75 mol%). These measurements indicated clearly that brush conformation is favored over pancake and mushroom (Fig. 2), since the periodical length, which depends on the distance between D and the phosphorous atom of the DSCP, becomes significantly smaller with decreasing distance between D and P, which would make no difference in case of pancake and mushroom conformation.

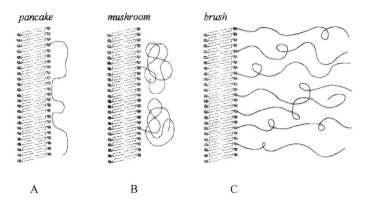

Figure 2. Idealized polymer conformation at the interface of a lipid bilayer and water: (A) pancake, (B) mushroom and (C) brush conformation.

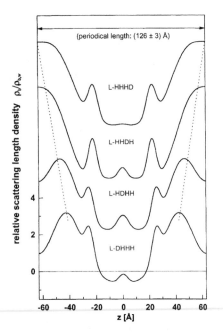

Figure 3. Scattering length density profiles of bilayer from a pospholipid (75 %) and **8** (25 %).

ABC-type Block Copolymers and Micellar Networks

Another area of interest was concerned with the preparation of multifunctional micelles, that would allow to provide compartments, different in their size, shape and physical or chemical properties.[7] Such structures could be potentially useful as drug carrier systems for two or even more therapeutics, that can be selectively solubilized in specific micellar compartments, or might serve as a scaffold for cascade reactions in catalysis, that would resemble the activity of multienzyme complexes in nature. The major goal of this research was therefore to create two micelles in a distinct environment, that show significant differences in their physical properties. Our approach was based on the self-organization behavior of telechelic polymers, that were composed of a fluorocarbon endgroup ($C_8F_{17}CH_2CH_2$) covalently bound to poly(2-methyl-2-oxazoline) and a hydrocarbon (C) segment of various length (C_6 to C_{18}) as the second endgroup. The fluorinated segment was introduced into the polymer via the initiation method whereas monoalkylated piperazine derivatives were used as termination reagents.[7,8]

$C_8F_{17}CH_2CH_2-OSO_2CF_3$

$+$

(structure: N-methyl-oxazoline ring, CH$_3$)

$\longrightarrow$

$C_8F_{17}CH_2CH_2$ (polymer structure) $N-C_nH_{2n+1}$, with subscript 25, 9, $=O$, CH_3

| fluorophilic | hydrohphilic | lipophilic |
| F | P | L |

$+$

HN (structure) $-N-C_nH_{2n+1}$

Polymers with m = 25, 50 and n = 6, 8, 10, 12, 14, 16, 18 were synthesized and characterized. In addition model polymers of the LP-type and FP-type, composed of either a hydrocarbon end group (LP, $C_{16}H_{33}$- PMeOx$_n$, n = 10, 25, 30, 50, MeOx = 2-methyl-2-oxazoline) or a fluorocarbon one (FP, $C_8F_{17}CH_2CH_2$-PMeOx$_n$, n = 10, 20, 35, 50), respectively, were prepared. Both, LP- and FP-type polymers aggregate above a certain concentration and form micelles. The cmc values increase for the LP and FP polymers with increasing length of the hydrophilic block length. Dynamic and static light scattering measurements of $C_8F_{17}CH_2CH_2$-PMeOx$_{35}$ gave aggregation numbers of 22 ± 2 and a hydrodynamic radius of 5.6 nm above the cmc up to 10 g/L. Only at rather high concentrations (c > 50 g L^{-1}) a strong scattering angle dependency was observed, indicating that the aggregates change their shape most likely by the formation of extended, rod-like micellar structures (Fig. 4).

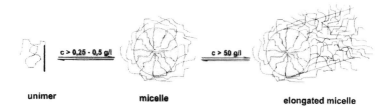

unimer   micelle   elongated micelle

c > 0,25 - 0,5 g/l   c > 50 g/l

Figure 4. From unimers via micelles to higher aggregates for LP- and FP-type polymers.

Afterwards polymer FPL_{16} was investigated by dynamic light scattering measurements at 0.25 g/L, 5.0 g/L and 20.0 g/L in aqueous solution. The results showed clearly that within this concentration range aggregates were formed with a hydrodynamic radius of 5.7 nm, that slightly increased to 6.7 nm at 20 g/L. This result suggested that at very low concentrations above the first critical aggregation concentration, micelles are formed in solution, which will most likely form network structures at very high amphiphile concentrations. In the following studies ^{19}F NMR measurements and the pyrene Ham effect were used to quantify the aggregation of fluorocarbon endgroups within the micellar networks, and to figure out, if fluorocarbon and hydrocarbon endgroups in such micellar networks form mixed micelle or pure fluorocarbon and hydrocarbon-like aggregates. In the first set of experiments we studied the Ham effect of our model compounds LP and FP in order to see the effect of pyrene fluorescence in the presence of micelles with either pure fluorocarbon and hydocarbon-like micellar core. The values I_1/I_3 were 1.28 in the presence of a hydrocarbon-like micelle, which is in good accordance with literature values, whereas in the presence of micelles with a fluorocarbon-like core, the change was much less pronounced compared to pure water with I_1/I_3 found to be 1.56 (pyrene in water $I_1/I_3 \approx 1.80$). Moreover, by using an excess of pyrene, we found pyrene excimer formation in the presence of micelles with a hydrocarbon interior but no pyrene excimer formation in the presence of micelles with a fluorocarbon-like core, suggesting significant differences in the solubilization behavior of pyrene in the presence of these two types hydrophobic core environment. In the next set of experiments, we studied the ABC type polymers with fluorocarbon and hydrocarbon chain ends by fluorescence spectroscopy. As can be seen from Fig. 5, formation of the first aggregate is predominantly determined by the length of the most hydrophobic endgroup. In the case of short alkyl chain with hexyl, octyl, and decyl chain length, the fluorocarbon endgroup $C_8F_{17}CH_2CH_2$ is more hydrophobic and aggregates first. I_1/I_3 values but also excimer formation (I_1/I_E) resembles the results found with micelles composed of a pure fluorocarbon core (FP). In the case of longer alkyl endgroups (C_{14}, C_{16} and C_{18}), the hydrocarbon endgroups start to aggregate before the fluorocarbon do so, and the micellar aggregates show behavior similar to that of micelles based on LP amphiphiles. In addition, it became clear that at higher concentrations of the latter mentioned FPL amphiphiles, pyrene is only able to detect the hydrocarbon like environment and it is not possible to analyze the formation of fluorocarbon-like aggregates at the same time.

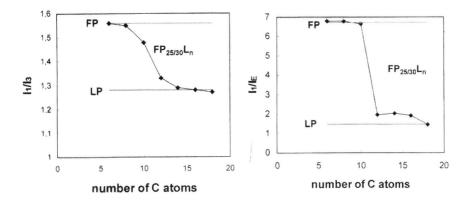

Figure 5. Ham effect of pyrene, lipophilic core (LP): $I_1/I_3 = 1.28$, $I_1/I_E = 1.47$; fluorophilic core (FP): $I_1/I_3 = 1.56$, $I_1/I_E = 6.7$.

In order to overcome this limitation, we applied a second methodology to study the aggregation behavior of the fluorocarbon chain of $FP_{25}L_{16}$ as a function of amphiphile concentration by ^{19}F NMR spectroscopy. Micelles composed of FP were again used as a model compound to study the transverse relaxation of the terminal CF_3-group, that can be described with a biexponential equation.[8]

$$\underbrace{A_{short} \cdot e^{-\frac{\tau}{T_{short}}}}_{A^*} + \underbrace{B_{long} \cdot e^{-\frac{\tau}{T_{long}}}}_{B^*} = 1$$

The short relaxation time corresponds herein to the fraction of FP dissolved in solution and not aggregates in micelles, whereas the long relaxation time is typical for almost solid-state like behavior of the CF_3-group in the micellar core. Based on these findings it was possible to quantify the amount of aggregated fluorocarbon chains versus free one in aqueous solution as a function of amphiphilie concentration. The results are shown in Table 3 for $FP_{25}L_{16}$.

Table 3. Transverse Relaxation times T_{short} and T_{long} and the fractions A_{short} and B_{long} from ^{19}F NMR for $FP_{25}L_{16}{}^{a)}$ in aqueous solution

c_{pol}	A^*_{short}	B^*_{long}	T_{short}	T_{long}
wt-%			µs	µs
0.05	0.12	0.88	292	2120
2.5	0.16	0.84	164	2829
5.0	0.20	0.80	217	1986
10	0.35	0.65	198	2089
20	0.69	0.31	210	2909

$^{a)}$ Polymer **9** with 8 CF_2 groups 25 P units and 16 CH_2 groups.

The results show clearly that the fluorinated endgroups aggregate linearly with increasing amphiphile concentration. The complete aggregation behavior based on our results found by dynamic light scattering, fluorescence measurements and ^{19}F NMR spectroscopy is summarized in the following picture.

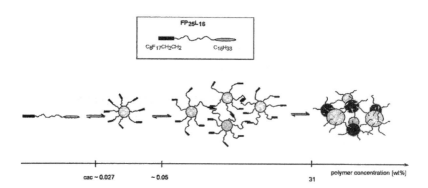

Figure 6. Aggregation diagram of $FP_{25}L_{16}$ versus concentration in aqueous solution.[8]

Amphiphilic AB-block Copolymers for Micellar Catalysis

Homogeneous and heterogeneous catalysis compete in industrial scale. Although homogeneous catalysis has many advantages over its heterogeneous counterpart, such as high reactivity, mild reaction conditions, high selectivity and no diffusion problem, the disadvantages include recycling of the catalyst and product separation. A possibility to combine the benefits of both homogeneous and heterogeneous are liquid-liquid two-phase systems, but also microheterogenization via colloidal assemblies.[9-15]

Limitations for the two-phase catalysis are the large amount of organic solvent which is needed and poor solubility of many starting compounds in water. The idea of micellar catalysis is that the hydrophobic core of a micelle functions as organic phase. Starting compounds are highly soluble in the unpolar phase and the catalyst is immobilized (covalently connected) in this phase, too (Fig. 7).[12]

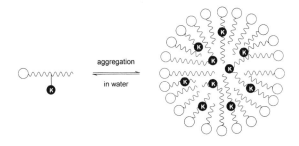

Figure 7. Tentative picture of the aggregation of functionalized ampihpihlies in aqueous solution.

Ideal for such an application are amphiphilic block copolymers from 2-methyl-2-oxazoline (water soluble) and 2-nonyl-2-oxazoline (water insoluble). The hydrophobic part should contain few units with ligands which allow immobilization of interesting transition metals. These ligands can either be introduced via initiation, termination, copolymerization or polymer modification. In principle, all these strategies are possible. For the introduction of triphenylphosphane ligands the polymer modification is favorable. In our case, blockcopolymer **10** was synthesized by sequential addition of suitable monomers. The aryliodide function of **10** was subsequently converted into the triphenylphosphine unit by a palladium catalyzed P-C coupling reaction almost quantitatively, resulting in **11**.

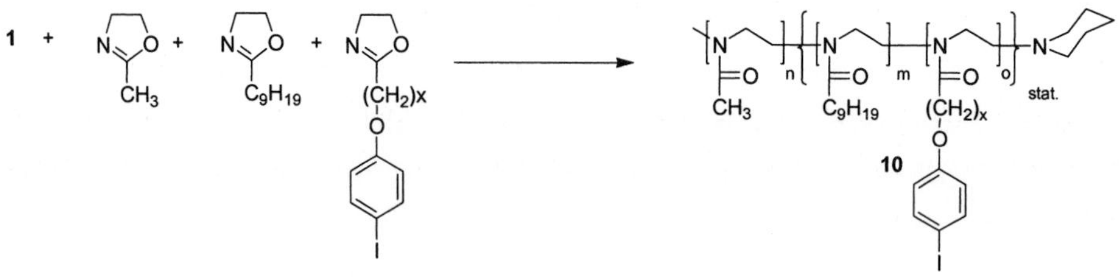

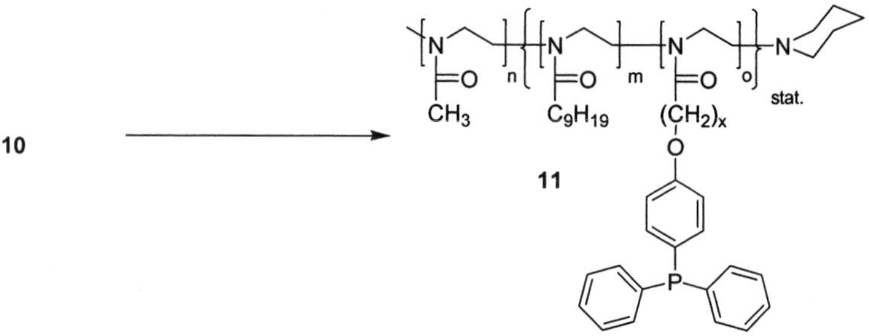

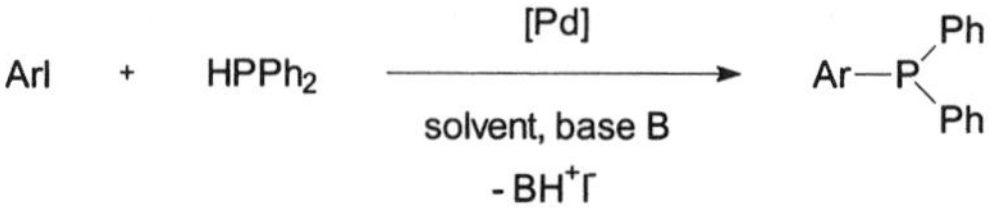

base = NEt₃, KOAc
solvent = DMAc, CH₃CN

As already mentioned for industrial applications of two-phase catalysis it is essential that the solubility of the starting compound is sufficient in water. Therefore, usually higher olefins can not be converted into the corresponding aldehydes by a common two-phase hydroformylation process due to their poor solubility in water. Alternatively, polymer **11** was used for the hydroformylation of 1-octene ($R = C_6H_{13}$).

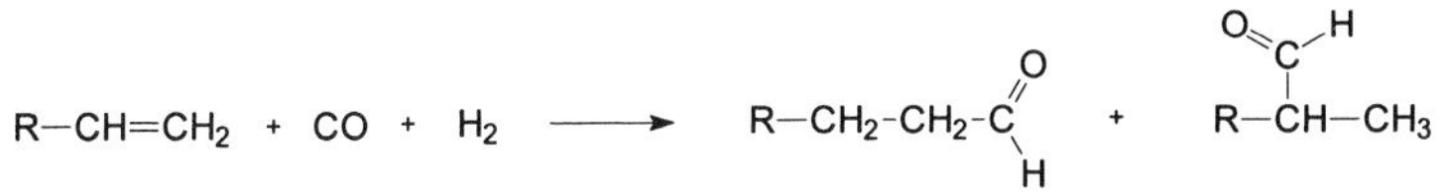

$$R-CH=CH_2 \ + \ CO \ + \ H_2 \ \longrightarrow \ R-CH_2-CH_2-\overset{\displaystyle O}{\underset{\displaystyle H}{C}} \ + \ R-\overset{\displaystyle \overset{O}{\|}C-H}{CH}-CH_3$$

Above the critical micelle concentration, polymer **11** starts to form micelles in water, which have hydrophobic cores with the triphenylphosphane in the core. Consequently, the transition metal ions are immobilized therein. A hydrophilic shell guarantees the water solubility of **11**. One can consider these polymers as nanoreactors with high local concentration of catalyst and starting compound. For comparison experiments were carried out as classical two-phase reactions with Rh-TPPTS (Rh / $P(m\text{-}C_6H_4\text{-}SO_3^-)_3$), with Rh-TPPTS in the presence of sodium dodecylsulfonate (SDS), and finally in the presence of polymer **11**, containing immobilized rhodium. Table 4 summarizes some results.[13,14]

Table 4. Hydroformylation of 1-octene [a]

ligand	c_{PPh_3}	P : Rh	octene : Rh	t	conversion	n : iso	TON	TOF
				h	%			h^{-1}
TPPTS	10^{-3}	25	2000	10	< 5	73 : 27	20	2
TPPTS / SDS	10^{-3}	25	2100	10	< 6	72 : 28	64	2
11[b]	10^{-3}	18	8500	3	95.5	76 : 24	7400	2480

[a] reaction conditions: V = 50 mL, T = 125 °C, p_0 = 51 bar, H_2 : CO = 1 : 1.

[b] n = 30, p = 4, q = 2.

The polymer **11** containing the valuable transition metal can be recycled and used again.

Star-like Polymers as "Unimolecular Micelles"

Another polymer support material currently under investigation in our group are star-like polymers with a hyperbranched core and amphiphilic, water-soluble graft arms. Such materials can be easily synthesized in a size from 20 to 50 nm, which allow for easy separation via ultrafiltration techniques. In addition, the amphiphilic structure of such polymers is very similar to micellar aggregates with the advantage that one molecule shows already many features of micellar aggregates with respect so their amphiphilicity and are thus called "uni-molecular micelles". The synthetic procedure is shown in the following scheme.[15]

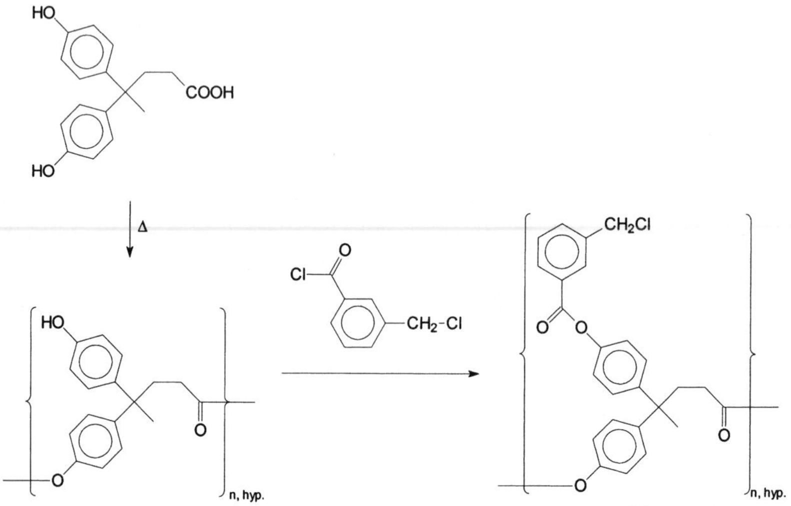

The hyperbranched polymer **12** is soluble in many common solvents – its modification yields **13**, a multifunctional macroinitiator which was applied for the living cationic polymerization of 2-alkyl-2-oxazolines. The following reaction sequence yields the desired unimolecular micelle:[16]

1. copolymerization of 2-(p-iodophenoxyalkyl)-2-oxazoline and 2-nonyl-2-oxazoline
2. 2-methyl-2-oxazoline
3. termination with piperidine
4. polymer analogous conversion of the iodophenyl-groups into triphenylphosphane-groups

These star-like polymers are currently under investigation as support materials in the Heck reaction of iodobenzene with styrene and the hydroformylation of 1-octene. From our experiment it is still not clear which of the two methods (classical micelle or unimolecular micelle) gives better results. In long term the unimolecular micelle concept looks promising because of the great variety in the choice of cores.

Summary and Outlook

In summary we have shown the various applications of 2-oxazoline polymerization in the field of macromonomer synthesis, lipopolymers for stealth liposomes, micellar networks and micellar catalysis. Especially the latter field of research provides enormous opportunities to develop new amphiphilic polymer supported catalysts due to the increasing number of transition-metal catalyzed reactions in synthetic organic chemistry as well as polymer synthesis. And within this research we believe, that poly(2-oxazoline)s can take up a key position to evaluate new synthetic approaches in polymer supported catalysis due to their structural versatility and easy accessibility.[17,18]

[1]   B. L. Culbertson, *Prog. Polym. Sci.* **2002**, *27*, 579 and references therein.
[2]   S. Kobayashi, H. Uyama, *J. Polym. Sci. Polym. Chem.* **2002**, *40* 192 and references therein.
[3]   O. Nuyken, G. Maier, A. Groß, *Macrmol. Chem Phys.* **1996**, *197*, 83.
[4]   A. Groß, G. Maier, O. Nuyken, *Macrmol. Chem Phys.* **1996**, *197*, 2811.
[5]   P. Kuhn, *PhD thesis*, TU Munich, **1997**.
[6]   P. Kuhn, R. Weberskirch, O. Nuyken, C. Cevc, *Desgn. Mon. Polym.* **1998**, *1*, 327.
[7]   R. Weberskirch *PhD thesis*, TU Munich, **1998**.
[8]   R. Weberskirch, J. Preuschen, H. W. Spiess, O. Nuyken, *Macromol. Chem. Phys.* **2000**, *201*, 995.
[9]   G. M. Parhall, S. D. Iffel, *Homogeneous Catalysis*, 2. Ed., J. Wiley, New York, **1992**.
[10] B. Cornils, *Top. Cum. Chem.* **1999**, *206*, 133.
[11] J. Manassen in F. Barsolo, R. L. Burwell, *Catalysis in Research*, Plenum, London, **1973**, p. 183ff.
[12] J. H. Fendler, E. J. Fendler, *Catalysis in Micellar and Macromolecular Systems*, Academic Press, London, **1975**.
[13] P. Persigehl, *PhD thesis*, TU Munich, **2000**.
[14] O. Nuyken, P. Persigehl, R. Weberskirch, *Macromol. Symp.* **2002**, *177*, 163.
[15] R. Weberskirch, R. Hettich, O. Nuyken, D. Schmaljohann, B. Voit, *Macromol. Chem Phys.* **1999**, *200*, 863.
[16] N. West, *PhD thesis*, TU Munich, **2000**.
[17] M. T. Zarka, O. Nuyken, R. Weberskirch, *Chem. Eur.J.* **2003**, *9*, 3228.
[18] T. Kotre, O. Nuyken, R. Weberskirch, *Macromol. Rapid Commun.* **2002**, *23*, 871.

Macromol. Symp. **2004**, *215*, 231-254

Comparison of the Mechanism and Kinetics of Living Carbocationic Isobutylene and Styrene Polymerizations Based on Real-Time FTIR Monitoring[†]

Judit E. Puskas,* Sohel Shaikh

Macromolecular Engineering Research Centre, Department of Chemical and Biochemical Engineering, The University of Western Ontario, London, Canada N6A 5B9
E-mail: jpuskas@uwo.ca

Summary: This paper will compare the mechanism and kinetics of living carbocationic polymerization of isobutylene (IB) and styrene (St), initiated by the 2-chloro-2,4,4-trimethyl-pentane (TMPCl) / TiCl$_4$) system in 60/40 (v/v) methylcyclohexane / methyl chloride mixed solvent at −80 and −75 °C. The rate of initiation was found to be first order in TiCl$_4$ in both systems. While initiation is instantaneous in IB polymerization at $[TiCl_4]_0 \geq [TMPCl]_0$, it is slow in St polymerization. Kinetic derivation showed that initiating efficiency is dependent on [M] in this latter system, which was also demonstrated experimentally. The apparent initiation rate constant was determined from initiator consumption rate data and was found to be $k_{i,app}$ = 1.39 l^2/mol^{2}sec. The rate of St consumption measured using a real time fibre-optic mid-FTIR monitoring technique compared well with gravimetric data and was found to be closer to first order in TiCl$_4$ at $[TiCl_4]_0 < [TMPCl]_0$. However, the rate followed a close to second order in TiCl$_4$ at $[TiCl_4]_0 \geq [TMPCl]_0$. The mechanistic model proposed earlier for living carbocationic IB polymerization, which yielded good agreement with experimental data, seems to apply to carbocationic St polymerization as well. This model reconciles the discrepancy between rate constants published for carbocationic IB and St polymerizations, and accounts for shifting TiCl$_4$ orders. However, independent investigations are necessary to verify the proposed mechanistic model. Optimized conditions led to living carbocationic St polymerization producing high molecular weight PS with 100% initiating efficiency.

Keywords: cationic polymerization; kinetics; mechanism; polyisobutylene; polystyrene; real time FTIR

†This paper is dedicated to the memory of Dr. Michael Lanzendörfer.

 DOI: 10.1002/masy.200451119

Introduction

Living polymerization has been at the frontiers of research in recent years. Living polymerizations have no side reactions such as irreversible termination and chain transfer, and provide great control over the polymerization process[1]. While living anionic polymerizations have been researched extensively and are commercialized, living carbocationic and radical polymerizations are relatively new[2]. Living carbocationic polymerization, the primary interest of our research group, is believed to be a more complex system than the classical living anionic polymerization[2]. The design and effective synthesis of various polymeric structures such as blocks, grafts, stars and arborescent (hyperbranched) polymers require thorough understanding of the mechanism of the polymerization process. Furthermore, optimization of the process conditions needs kinetic understanding. Our group has been investigating the mechanism and kinetics of living carbocationic polymerizations, and has proposed a comprehensive mechanism for living isobutylene (IB) polymerization initiated by 2,4,4-trimethyl pentyl chloride (TMPCl)/TiCl$_4$ [3]. TMPCl was used as a model for the growing chain end thus the rates of initiation and propagation were considered to be equal. In this model, two reaction pathways (Path A and Path B) proceed in competition as shown in Scheme 1. Although the existence of the polymerization-inactive complexes I*LA and P$_n$*LA has not been demonstrated experimentally, this model consolidated experimental evidence that the reaction order in TiCl$_4$ can shift between first and second order[3-12]. According to this mechanism, both pathways may proceed simultaneously, depending on the actual [TiCl$_4$] concentration. The existence of Path A has been demonstrated experimentally: using [TMPCl]$_0$ > [TiCl$_4$]$_0$ conditions were created under which Path A was shown to dominate the process[3,6,9,12]. The mechanism presented in Scheme 1 has been debated, which prompted us to further investigate this model, and compare it with other models. In addition, we intended to investigate the kinetics and mechanism of carbocationic styrene (St) polymerizations, in comparison with our model living IB polymerization.

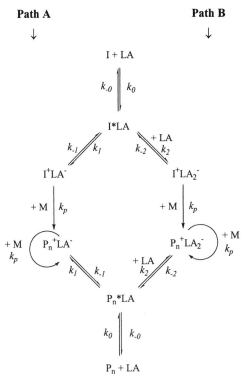

Scheme 1: Proposed comprehensive mechanism for living IB polymerization[3].

This paper will discuss some aspects of the mechanism and kinetics of living IB polymerization. It will also investigate the kinetics and mechanism of carbocationic St polymerizations initiated by the TMPCl/TiCl$_4$ system, and compare them with the kinetics and mechanism of carbocationic IB polymerization.

Experimental

Materials

2-chloro-2,4,4-trimethyl-pentane (TMPCl) was synthesized by HCl addition to 2,4,4-trimethyl-1-pentene[12], and its purity was checked by ^{1}H NMR. TiCl$_4$, dimethyl acetamide (DMA), DtBP (2,6-di-*tert.*-butylpyridine) and methanol were used as received (Aldrich). Methyl chloride (MeCl, BOC gases) was dried by passing it through a column filled with BaO and CaCl$_2$. Hexane (Hx) and methylcyclohexane (MeCHx) were refluxed with CaH$_2$ for at least 24 hours and distilled freshly before use. Styrene (St, Aldrich) was purified by vacuum distillation before use.

Procedures

Polymerizations. The polymerizations were carried out in an MBraun LabMaster 130 glove box equipped with an integral cold hexane bath under dry nitrogen at -78^{0}C, as reported earlier for IB polymerizations[3, 12].

IR Monitoring. The polymerization reactions were monitored in real time using a low temperature fiber-optic probe equipped with a transmission (TR) head[5]. The IR probe (Remspec Inc.), with its source and receiver cables, was interfaced with a Bio-Rad FTS 175C FTIR unit, to which a mid-IR fiber-optic liquid nitrogen cooled 0.5 x 0.5 mm MCT detector was attached. The cable bundle was fed into the glove box through a port, and the probe was immersed into the reactor. FTIR data were analyzed as reported earlier[6].

Polymer Characterization. Polymer molecular weights (MWs) and molecular weight distributions (MWDs) were determined by Size Exclusion Chromatography (SEC) using a Waters system as described earlier[6]. The ASTRA (Wyatt Technology) software was used to obtain absolute MW data with dn/dc=0.183 ml/g for polystyrene (PS) in THF.

Results and Discussion

Living carbocationic IB polymerization

Mechanistic models

Living IB polymerization is governed by a dynamic equilibrium between covalent and ionic species, with the equlibrium strongly shifted towards the former. Several mechanistic models were proposed to describe this system. Our first model is shown in Scheme 2. This model was developed for IB polymerization initiated by TMPCl/TiCl$_4$, which we consider a model system as the structure of the initiator resembles that of the growing PIB chain. In this model, $I_n^+//LA^-$ and $P_n^+//LA^-$ were shown to be ion pairs.

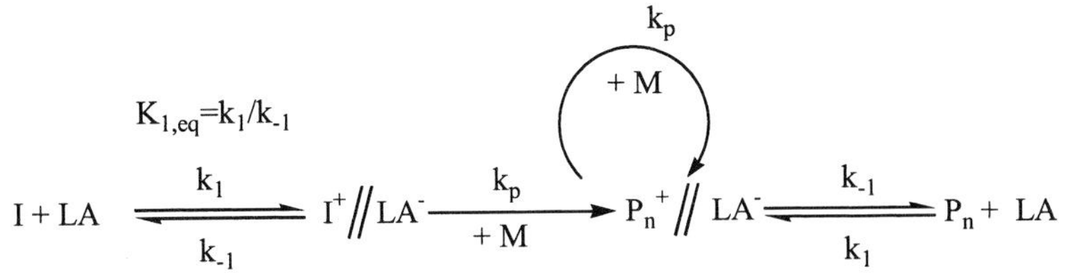

Scheme 2: Model of living IB polymerization initiated by TMPCl/TiCl$_4$.[12,13]

Majoros and Nagy from the Kennedy group developed a closed-loop comprehensive model using the Winstein ionicity spectrum, shown in Scheme 3:

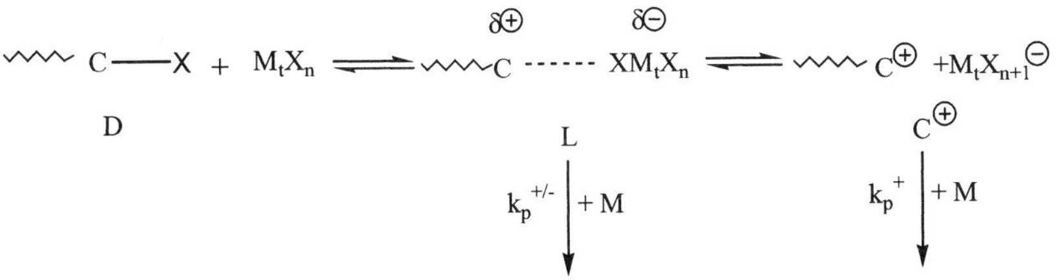

Scheme 3: Comprehensive closed-loop mechanism by Majoros et al.[4]

where D represents a "dormant cationogen" (initiator or polymer), L stands for a sub-spectrum of polarized (stretched or activated, more-covalent-than-ionic) dipole intermediates leading to living polymerization, and C summarizes a further sub-spectrum of ionized entities starting with contact ion pairs through solvent separated ion pairs to fully solvated "free" ion pairs. These authors proposed that L and C can both initiate polymerization, but only L will lead to living conditions.

The reaction mechanism shown in Scheme 4 is based on the model of living IB polymerization proposed by Storey's group[8].

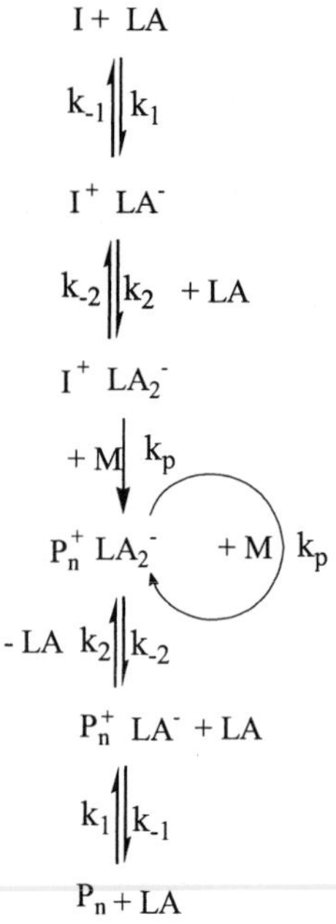

$$I + LA$$

$$k_{-1} \big\|\big\| k_1$$

$$I^+ \ LA^-$$

$$k_{-2} \big\|\big\| k_2 \quad + LA$$

$$I^+ \ LA_2^-$$

$$+ M \big| \ k_p$$

$$P_n^+ \ LA_2^- \qquad + M \big) \ k_p$$

$$- LA \ k_2 \big\|\big\| k_{-2}$$

$$P_n^+ \ LA^- \ + LA$$

$$k_1 \big\|\big\| k_{-1}$$

$$P_n + LA$$

Scheme 4: Mechanism of living IB polymerization based on a model proposed by Storey and Choate[8].

These authors suggested that propagation takes place predominantly through chains possessing dimeric gegenions which form by the reaction of additional $TiCl_4$ with monomeric gegenions, as opposed to direct ionization by neutral, dimeric Ti_2Cl_8. They also presented kinetic derivation that direct ionization by neutral, dimeric Ti_2Cl_8 would lead to first order dependence in $[TiCl_4]$, which would have been in direct contradiction with their experimental results.

The same group later proposed another model; the mechanistic pathway based on this is presented in Scheme 5. This work also reported propagation predominantly by active centers with dimeric gegenions.

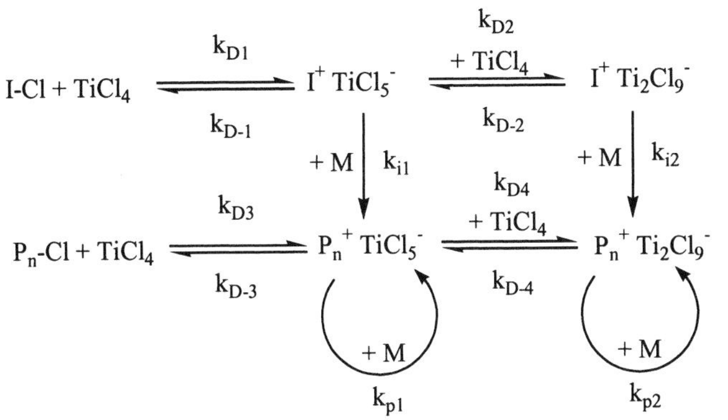

237

$$I\text{-}Cl + TiCl_4 \underset{k_{D\text{-}1}}{\overset{k_{D1}}{\rightleftharpoons}} I^+\,TiCl_5^- \underset{k_{D\text{-}2}}{\overset{k_{D2} \;+\,TiCl_4}{\rightleftharpoons}} I^+\,Ti_2Cl_9^-$$

$$+M \downarrow k_{i1} \qquad\qquad +M \downarrow k_{i2}$$

$$P_n\text{-}Cl + TiCl_4 \underset{k_{D\text{-}3}}{\overset{k_{D3}}{\rightleftharpoons}} P_n^+\,TiCl_5^- \underset{k_{D\text{-}4}}{\overset{k_{D4}\;+\,TiCl_4}{\rightleftharpoons}} P_n^+\,Ti_2Cl_9^-$$

$$\bigcirc\, +M \quad k_{p1} \qquad\qquad \bigcirc\, +M \quad k_{p2}$$

Scheme 5: Mechanism of living IB polymerization based on the model proposed by Storey and Donnalley[11].

Based on the work of Faust et al.,[10, 14-15] the following mechanism can be presented, as shown in Scheme 6:

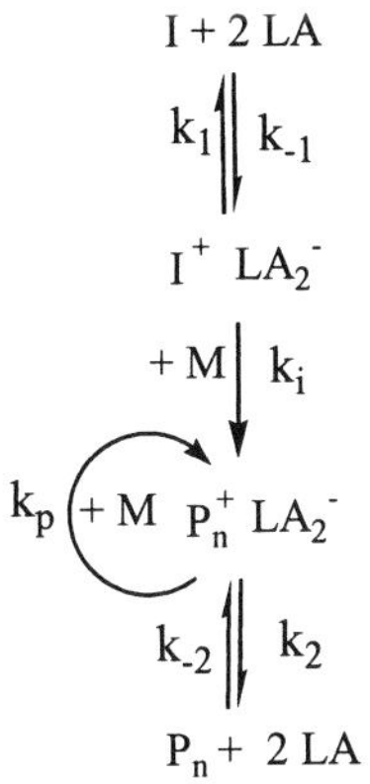

$$I + 2\,LA$$

$$k_1 \updownarrow k_{-1}$$

$$I^+\,LA_2^-$$

$$+M \downarrow k_i$$

$$k_p \left(+M \quad P_n^+\,LA_2^- \right)$$

$$k_{-2} \updownarrow k_2$$

$$P_n + 2\,LA$$

Scheme 6: Model of living IB polymerization based on the work of Faust et al.[10]

The mechanistic models proposed in Schemes 1-6 all agree in the existence of dynamic equilibria between dormant and active species, and that only a fraction of the initiator and polymer chains is ionized at any given time. However, there is disagreement in the mechanism of initiation and propagation. The next section will compare these models based on kinetic investigations.

Kinetics and rate constants

As mentioned above, living IB polymerization initiated by TMPCl/TiCl$_4$ can be considered as a model system, because the structure of the TMPCl initiator resembles the structure of the growing chain. Thus in the first model shown in Scheme 2 it was assumed that $k_i = k_p$, which greatly simplified the kinetic derivations[12]. The conversion dependence of the molecular weight distribution (MWD) in this system was described using the following equation.

$$\frac{DP_w}{DP_n} = 1 + \frac{(2\bar{l}-1)}{DP_n} = 1 + \frac{\bar{l}_0}{DP_n}$$ (1)

where $\bar{l} = (k_p/k_{-1})[M]$ and $\bar{l}_0 = \bar{l}$ at $[M]_0$. From experimental MWD-conversion data $k_p/k_{-1} = 16.5$ l/mol was calculated[13]. A similar treatment was later published by Müller et al.[16]. Roth and Mayr subsequently measured $k_p = 6\pm2\times10^8$ l/molsec using the "diffusion clock" method[17]. Schlaad et al. recently published $k_p = 7\times10^8$ l/molsec, also obtained from competition experiments[14]. These high values agree with industrial experience. In contrast, k_p values obtained from kinetic measurements are several orders of magnitude lower (see Table 1), with the exception of k_p measured in irradiation-induced IB polymerization in bulk[23]. This discrepancy was discussed by Plesch[22, 24] and remains a problem.

Table 1: k_p values in carbocationic IB polymerizations obtained by various methods

k_p (l.mol^{-1}s^{-1})	Initiating system	Solvent	T / ^{0}C	Ref.
6×10^3	AlBr$_3$ / TiCl$_4$	heptane	-14	18
7.9×10^5	Light / VCl$_4$	in bulk	-20	19
1.2×10^4	Et$_2$AlCl/Cl$_2$	CH$_3$Cl	-48	20
9.1×10^3	Ionizing radiation	CH$_2$Cl$_2$	-78	21, 22
1.5×10^8	Ionizing radiation	in bulk	-78	23
6×10^8	R-Cl / TiCl$_4$	CH$_2$Cl$_2$	-78	17*
7×10^8	IB 33-mer / TiCl$_4$	Hexanes / CH$_3$Cl	.-80	14*

* Diffusion clock / competition experiments

Using $k_p = 6\times10^8$ lmol^{-1}sec^{-1} Puskas and Peng[7] calculated $k_{-1} = 3.9\times10^7$ sec^{-1}. Based on the model shown in Scheme 2, $k_1 = 0.22$ lmol^{-1}sec^{-1} was obtained from experimental initiator consumption data[3]. These rate constants were then used to simulate experimental data with the Predici software developed for polymerization kinetic modelling[25], and good agreement was found between simulated and experimental data[7]. However, the model in Scheme 2 does not reflect the experimental fact that living IB polymerization usually exhibits close to second order in [TiCl$_4$]. The model shown in Scheme 1 was proposed to accommodate the fractional and shifting TiCl$_4$ orders found experimentally in living IB polymerizations. Using this model, the following composite rate constants were obtained from experimental data: $K_0 k_1 = 0.22$ lmol^{-1}sec^{-1} from initiator consumption data and $K_0 K_1 k_p = 3.4$ l^2 mol^{-2}sec^{-1} from monomer consumption data in Path A, and $K_0 K_2 k_p = 52$ l^3 mol^{-3}sec^{-1} from monomer consumption data in Path B. Initiator consumption data cannot be measured experimentally in Path B as initiation practically is instantaneous. With $k_p = 6 \times 10^8$ l mol^{-1}sec^{-1} we calculated $K_0 K_1 = 5.7 \times 10^{-9}$ l mol^{-1} and $K_0 K_2 = 8.7 \times 10^{-8}$ l^{2}mol^{-2}. Individual rate constants were obtained by parameter estimation using the model in Scheme 1 and experimental data[3, 26] (the details of the parameter estimation will be published separately). Table 2 lists the values of rate constants obtained by parameter estimation. Remarkably, $k_p = 8.94 \times 10^8$ lmol^{-1}sec^{-1} was obtained, which is very close to values measured by

Table 2: Values of rate constants obtained by parameter estimation

Rate constant	Value
k_0 (l mol^{-1}sec^{-1})	9.3×10^7
k_{-0} (sec^{-1})	3.95×10^7
k_1 (l mol^{-1}sec^{-1})	5.69×10^{-2}
k_{-1} (sec^{-1})	3.90×10^7
k_2 (l mol^{-1}sec^{-1})	8.12×10^{-1}
k_{-2} (sec^{-1})	3.89×10^7
k_p (l mol^{-1}sec^{-1})	8.94×10^8

diffusion clock/competition experiments and in irradiation-initiated IB polymerization (Table 1). The capping rate constants k_{-0}, k_{-1} and k_{-2} were nearly identical (3.95, 3.90 and 3.89×10^7 sec^{-1}) and were also very close to experimentally measured capping rate constants. Schlaad et al. reported $k_{-1} = 5 \times 10^7$ sec^{-1} and $k_{-2} = 5 \times 10^7$ sec^{-1}, and $k_1 = 6.5$ L^{2}mol^{-2}sec^{-1} $k_2 = 16.4$ L^{2}mol^{-2}sec^{-1} (see Scheme 6)[10, 14]. The values of $k_p/k_{-1} = 16.4$ and $k_i/k_{-1} = 6.5$ L/mol were obtained using

TMPCl and a PIB 36-mer as initiators, thus $k_p \approx 3\ k_i$.

Since all rate constants were not available with the models shown in Schemes 3,4 and 5, Predici simulations were carried out using the model in Scheme 1 with the parameters listed in Table 2. Figures 1 and 2 demonstrate very good agreement between simulation and experimental data both for Paths A and B.

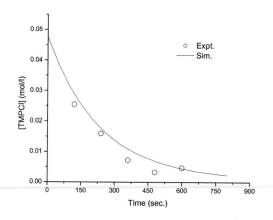

Figure 1. Comparison of predicted and experimental initiator consumption data in living IB polymerization. Path A.
[TMPCl]$_0$=0.050 mol/l, [TiCl$_4$]$_0$=0.025 mol/l, [IB]$_0$=2 mol/l, [DtBP]$_0$=0.007 mol/l, Hx/MeCl=60/40 v/v, T= -80°C.
Simulation: k_0=9.3×10^7 l mol^{-1}sec^{-1}, k_{-0}=3.95×10^7 sec^{-1}, k_1= 5.69×10^{-2} l mol^{-1}.sec^{-1}, k_{-1}=3.9×10^7 sec^{-1}, k_2=8.12 l.mol^{-1}.sec^{-1}, k_{-2}=3.89×10^7 sec^{-1}, k_p=6×10^8 l.mol^{-1}.sec^{-1}.

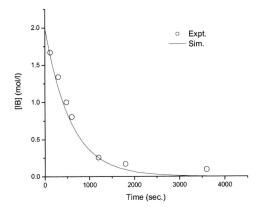

Figure 2. Comparison of predicted and experimental monomer consumption data in living IB polymerization. Path B.

$[TMPCl]_0=0.004$ mol/l, $[TiCl_4]_0=0.064$ mol/l, $[IB]_0=2$ mol/l, $[DtBP]_0=0.007$ mol/l, Hx/MeCl=60/40 v/v, T: - 80°C

Simulation: $k_0=9.3\times10^7$ l mol^{-1}sec^{-1}, $k_{-0}=3.95\times10^7$ sec^{-1}, $k_1=5.69\times10^{-2}$ l mol^{-1}.sec^{-1}, $k_{-1}=3.9\times10^7$ sec^{-1}, $k_2=8.12$ l.mol^{-1}.sec^{-1}, $k_{-2}=3.89\times10^7$ sec^{-1}, $k_p=6\times10^8$ l.mol^{-1}.sec^{-1}.

Simulation was also carried out with the model shown in Scheme 6. Figure 3 shows that good fit with experimental data can be obtained with $k_p=3\times10^7$ l.mol^{-1}.sec^{-1}, but $k_p=7\times10^8$ l.mol^{-1}.sec^{-1} leads to too fast propagation.

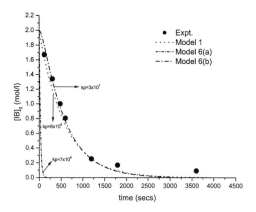

Figure 3. Comparison of predicted and experimental monomer consumption data in living IB polymerization. Path B.
[TMPCl]$_0$=0.004 mol/l, [TiCl$_4$]$_0$=0.064 mol/l, [IB]$_0$=2 mol/l, [DtBP]$_0$=0.007 mol/l, Hx/MeCl=60/40 v/v, T: - 80°C
Simulation: Model 1: k_0=9.3×10^7 lmol^{-1}sec^{-1}, k_{-0}=3.95×10^7 sec^{-1}, k_1= 5.69×10^{-2} l mol^{-1}.sec^{-1}, k_{-1}=3.9×10^7 sec^{-1}, k_2=8.12 l.mol^{-1}.sec^{-1}, k_{-2}=3.89×10^7 sec^{-1}, k_p=6×10^8 l.mol^{-1}.sec^{-1}; Model 6: k_1= 6.5 l mol^{-1}.sec^{-1}, k_{-1}=5×10^7 sec^{-1}, k_2=16.4 l.mol^{-1}.sec^{-1}, k_{-2}=3.4×10^7 sec^{-1}, k_p=7×10^8 for Model 6(a) and 3×10^7 l.mol^{-1}.sec^{-1} for model 6(b).

The good fits demonstrate that various mechanistic models can fit the same experimental data, thus achieving good match between experimental and simulated data alone cannot be used to prove any mechanistic models. Independent experiments are crucial to establish the mechanistic pathways. In the case of living IB polymerization independent experiments demonstrated TiCl$_4$ orders shifting between 1 and 2, depending on experimental conditions. It should be mentioned here that the model in Scheme 5 would dictate propagation predominantly by active centres possessing monomeric counteranions as $k_p \gg k_D$. This is in direct contradiction with the authors' experimental data.

The Puskas model also reconciles the discrepancy between k_p data published for IB polymerizations[14, 17, 18-23]. The lower values (10$^{4-5}$ L/molsec) were obtained from kinetic measurements that would yield composite rate constants including the pre-equilibrium, while the close to diffusion limited ($k_p \cong 10^8$ l/ mol^{-1}sec^{-1}) values were obtained by diffusion clock methods that yield the true rate constants[17, 14]. For instance, the rate constant obtained by Marek[18] (Table 1) used the assumption that $[P^+] = [AlBr_3]_0$, thus the rate equation would read as follows:

$$\frac{-d[M]}{dt} = k_p[P^+][M] = k_p[LA]_0[M] \tag{2}$$

Thus the initial slope of the ln ([M]$_0$/[M])/[LA]$_0$ plot yielded k$_p$. In comparison, the rate equation using Model 1 can be written as follows:

$$\frac{-d[M]}{dt} = k_p[P^+][M] = k_p[I]_0 LA_0^2[M] \qquad \text{(Path B)} \tag{3}$$

and the initial slope of the ln ([M]$_0$/[M])/[LA]$_0$ plot would now be k$_p$[I]$_0$[[LA]$_0$. Considering the usual initiator concentration of 10^{-3} mol/l and TiCl$_4$ concentration of 10^{-2} mol/l in living IB polymerizations could explain the several order of magnitude difference in k$_p$ measured by the diffusion clock method or kinetic measurements. In addition, complexation between initiator and Lewis acid, or monomer and Lewis acid suggested earlier[18-22, 24] could also influence kinetically obtained rate constants. Remarkably, irradiation-initiated IB polymerization in bulk with no coinitiator[23] yielded k$_p$ = 1.4x10^8 l.mol^{-1}.sec^{-1}.

Kinetics of carbocationic St polymerization initiated by TMPCl/TiCl$_4$

Several research groups have investigated the kinetics and mechanism of carbocationic St polymerizations. Polymerizations initiated with TMPCl are most important, since this reaction mimics blocking St with living PIB in polyisobutylene-polystyrene (PIB-PS) block copolymer synthesis. In contrast to living IB polymerization, in the case of St polymerization initiated by TMPCl/TiCl$_4$ the structure of the initiator and the growing chain is different, thus k$_i$ ≠ k$_p$. Indeed, slow initiation was invariably reported in this system. Kaszas *et al.* achieved living St polymerization in the presence of very strong electron pair donors (EDs) such as hexamethylene phosphoramide, or in the combined presence of N,N-dimethyl acetamide (DMA) as ED and 2,6-di-*tert.*-butylpyridine (D*t*BP) as a proton trap. Carbocation stabilization by EDs was claimed to be the reason for the living conditions[27]. Majoros and Nagy from the Kennedy group analyzed experimental data of St polymerization initiated with TMPCl/TiCl$_4$ using the model shown in Scheme 2[4]. The rate of initiation was found to increase with increasing [TiCl$_4$]$_0$, and 100%

initiation efficiency was reached. However, chain transfer to monomer was present and prevented the synthesis of controlled high MW PS. The relatively broad MWD was attributed to slow initiation. Fodor et al. from the Faust group reported that living conditions could be achieved in the sole presence of DtBP, although initiation was slow [28]. Chain-chain coupling by intermolecular alkylation was avoided by quenching the reactions prior to complete monomer conversion. MWDs were rather broad. The reaction order in terms of $TiCl_4$ was found to be 2. Storey et al. reported that St polymerization with the TMPCl/$TiCl_4$ system proceeds with slow initiation[29].

The fact that $k_i \neq k_p$ further complicates the already complex mechanism shown in Scheme 1, and simplification is needed. In the simple model shown in Scheme 2, only three parameters (k_p, k_1 and k_{-1}) are needed. However, k_1 would represent a composite "rate constant" which includes $[TiCl_4]_0$ and K_0 as well as depicted in Scheme 1. The simplified model was used to simulate IB polymerizations, with the understanding that k_1 was dependent on $[TiCl_4]_0$ and good agreement was found with experimental data[7].

For St polymerization, the simplified scheme (Scheme 7) is shown below:

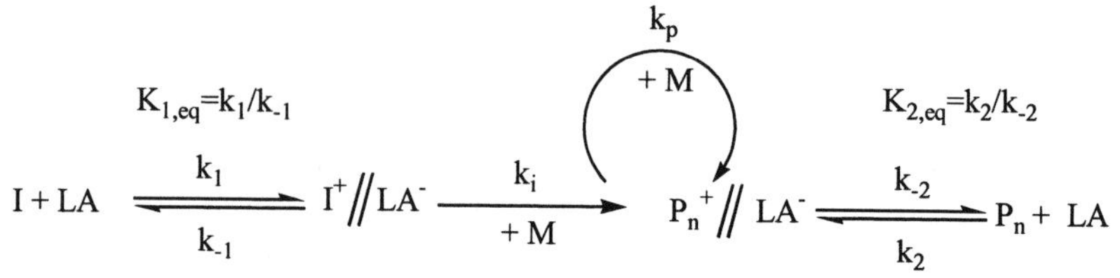

$$K_{1,eq}=k_1/k_{-1} \qquad \overset{k_p}{\underset{+M}{\bigcirc}} \qquad K_{2,eq}=k_2/k_{-2}$$

$$I + LA \underset{k_{-1}}{\overset{k_1}{\rightleftharpoons}} I^+ /\!/ LA^- \xrightarrow[+M]{k_i} \quad P_n^+ /\!/ LA^- \underset{k_2}{\overset{k_{-2}}{\rightleftharpoons}} P_n + LA$$

Scheme 7: Simple model of living St polymerization.

$K_{1eq} = k_1/k_{-1}$ and $K_{2eq} = k_2/k_{-2}$ are apparent equilibrium constants and k_i and k_p are the initiation and propagation rate constants, respectively. In the St polymerization experiments, we used DtBP as proton trap and DMA as a strong donor to prevent nucleophilic substitution of the St rings. The concentration of these additives was kept constant. It is known that these additives lower the polymerization rate[8,11,27]. However, their presence at a constant concentration would not interfere with our investigation of the effect of $[TiCl_4]$ and [IB] on the polymerization rate. The effect of temperature also needs to be addressed here. It is very difficult to ensure isothermal condition in carbocationic polymerizations, even at high dilution. The effect of temperature on the rate of polymerization would manifest itself in non-linear semilogarithmic monomer consumption rates.

The St polymerizations in this study showed linear rate plots both in Path A and B, shown in Figures 4 and 5:

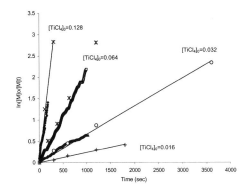

Figure 4. ln $[M]_0/[M]$ vs time plots in TMPCl/TiCl$_4$ initiated St polymerization at -75^{0}C. Path B. $[TMPCl]_0$=0.004 mol/l, $[DMA]$=0.002 mol/l, $[DtBP]$=0.007 mol/l, $[St]_0$=2 mol/l, MeCHx/MeCl=60/40 (v/v) FTIR: based on –CH$_2$=C-H wagging vibration at 1690 cm^{-1}.

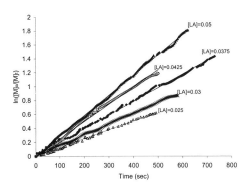

Figure 5. ln $[M]_0/[M]$ vs time plots in TMPCl/TiCl$_4$ initiated St polymerization at -75^{0}C. Path A. $[TMPCl]_0$=0.05 mol/l, $[St]_0$=2 mol/l, $[DMA]$=0.002 mol/l, $[DtBP]$=0.007 mol/l, MeCHx/MeCl=60/40 (v/v) FTIR: based on –CH$_2$=C-H wagging vibration at 1690 cm^{-1}.

Figures 4 and 5 demonstrate that non-isothermal conditions did not effect the polymerization rate. This can be explained as follows: when the dormant-active eqilibrium is strongly shifted and the steady-state approximation $d([I^+]+[P_n^+]/dt \approx 0$ can be evoked, and $[I] + [P_n] = [I]$, the rate will be directly proportional to $[I]_0$ and $[LA]_0$, in spite of changing $[I]$ and $[P_n]$. This was shown by gravimetry in living IB polymerization[12], and in living vinyl ether polymerization.[30] Another approach to achieve Path A and reduce the exotherm would be to simply reduce the $[TiCl_4]$ concentration. We attempted this with $[TMPCl]_0 = 0.004$ mol/l. However, we could not find appropriate conditions and at $[TiCl_4]_0 = 0.02$ mol/l polymerization stopped completely as we reached the DMA concentration.

In contrast to IB polymerization, initiator consumption data can be obtained in both Path A and B since initiation is slow in Path B as well. The experimental ln $[I]_0/[I]$ plots were not linear. Thus the kinetics of initiation were evaluated using the initial rate method, which is recommended for systems with unknown mechanism[31].

Kinetics of Initiation

Using the simplified model in Scheme 7, the initiator consumption can be written as follows:

$$\frac{-d[I]}{dt} = k_1[I][LA] - k_{-1}[I^+] \tag{4}$$

Since K_1 is very small and $k_1 \ll k_i$,[7, 12] the steady-state approximation can be used:

$$\frac{d[I^+]}{dt} = k_1[I][LA] - k_{-i}[I^+][M] - k_{-1}[I^+] \cong 0$$

$$\frac{d[I^+]}{dt} = k_1[I][LA] - k_{-i}[I^+][M] - k_{-1}[I^+] \cong 0 \tag{5}$$

Therefore from (5), $[I^+] = \dfrac{k_1[I][LA]}{k_i[M] + k_{-1}}$ $\tag{6}$

Substituting (6) into (4) we get

$$\frac{-d[I]}{dt} = \frac{k_1[I][LA]\dfrac{k_i}{k_{-1}}[M]}{1 + \dfrac{k_i}{k_{-1}}[M]} \tag{7}$$

For living IB polymerizations initiated by TMPCl, $k_i \approx k_p$, thus

$$\frac{k_i}{k_{-1}}[M] \gg 1$$

and

$$\ln\frac{[I]_0}{[I]} = k_1[LA]_0 t \tag{8}$$

The rate of initiator consumption was found to be directly proportional to k_1 and $[TiCl_4]_0$[3,12] and $k_1 = 0.22$ lmol^{-1}sec^{-1} was obtained from experimental initiator consumption data[3]. For St polymerization using TMPCl as initiator $k_i \ll k_{-1}$ can be assumed. This is a reasonable assumption since $k_{-1} \approx 3.9 \times 10^7$ l/molsec[11] and slow initiation was reported by several authors for the TMPCl/TiCl$_4$/St system[4, 27-29]. Thus

$$\frac{k_i}{k_{-1}}[M] \ll 1$$

and

$$\frac{-d[I]}{dt} = \frac{k_1 k_i}{k_{-1}}[I][LA][M] \tag{9}$$

Thus, based on kinetic derivation, in the case of St polymerization the rate of initiator consumption will depend not only on $[TiCl_4]_0$ but $[M]$ as well. This was proven experimentally, as shown in Figure 6. Simply increasing $[M]_0$ from 0.5 to 2.0 M increased the initiator efficiency from 20% to 100% when the conversion reached about 80% conversion.

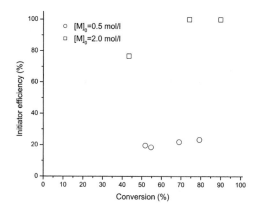

Figure 6. Effect of [St]$_0$ on I$_{eff}$ in polymerizations initiated by TMPCl/TiCl$_4$. [TMPCl]$_0$=0.004 mol/l, [TiCl$_4$]$_0$=0.04 mol/l, [DMA]=0.002 mol/l, [DtBP]=0.007 mol/l, MeCHx/MeCl=60/40, T= -75^{0}C.

Equation (6) can be integrated analytically for low conversion where [M] $\approx$ [M]$_0$

$$\ln\frac{[I]_0}{[I]} = k_{i,app}[LA]_0[M]_0 t \tag{10}$$

where $k_{i,app} = \dfrac{k_1 k_i}{k_{-1}}$

Thus k$_{i,app}$ data were obtained from the initial slope of experimental initiator consumption plots. In carbocationic St polymerizations initiated by TMPCl/TiCl$_4$/DMA slow initiation was found not only at [TMPCl]$_0$ > [TiCl$_4$]$_0$, but also at [TMPCl]$_0$ < [TiCl$_4$]$_0$ with relatively low [TiCl$_4$]$_0$ as shown in Figure 7.

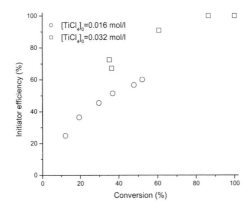

Figure 7. Slow initiation in carbocationic St polymerizations initiated by TMPCl/TiCl₄. Path B. [TMPCl]₀=0.004 mol/l, [St]₀=2 mol/l [DMA]=0.002 mol/l, [DtBP]=0.007 mol/l MeCHx/MeCl=60/40, T= -75^{0}C.

The initial slope of the $\ln[I]_0/[I]$ plots, $[d[I]/dt]_0$, was obtained from the nonlinear initiator consumption plots and the values are listed in Table 3.

Table 3: $k_{i,app}$ values in carbocationic St polymerizations initiated by TMPCl/TiCl₄.
[St]₀=2 mol/l [DMA]=0.002 mol/l, [DtBP]=0.007 mol/l MeCHx/MeCl=60/40, T= -75^{0}C

$[I]_0$ mol/l	$[TiCl_4]_0$ mol/l	$(d[I]/dt)_0$	$k_{i,app}$ $l^2mol^{-2}s^{-1}$
0.05	0.050	0.125	1.25
0.05	0.030	0.092	1.53
0.05	0.025	0.077	1.55
0.05	0.013	0.035	1.39
0.004	0.016	0.040	1.23

It can be seen that $k_{i,app}$ values are independent of $[TiCl_4]_0$, indicating that initiation is first order in TiCl₄. Interestingly, Storey *et al.* found similar values (k_i= 1.3 $l^2mol^{-2}s^{-1}$) using the rapid monomer consumption (RMC) technique[32]. The TiCl₄ order was also obtained from the dependence of the initial rate on $[TiCl]_0$:

$$\left(\frac{d[I]}{dt}\right)_0 = k_{i,app}[TiCl_4]_0^n \tag{11}$$

$$\ln\left(\frac{d[I]}{dt}\right)_0 = \ln k_{i,app} + n\ln[TiCl_4]_0 \tag{12}$$

where $k_{i,app}$ is an apparent rate constant. The $\ln(d[I]/dt)_0$ vs $\ln[TiCl_4]_0$ plot was linear, with a slope of n = 0.98, verifying first order. This finding is similar to the living IB system[3]. The kinetic treatment presented here may be applicable to other polymerizations governed by dormant-active equilibria[33].

Kinetics of Propagation

The $TiCl_4$ order of propagation was investigated by varying the $TiCl_4$ concentration, while keeping all other variables constant, as discussed above:

$$\ln\left(\frac{d[M]}{dt}\right)_0 = \ln k_{p,app} + n\ln[TiCl_4]_0 \tag{13}$$

where $k_{p,app}$ is an apparent rate constant lumping all the other variables. In case of a complex reaction several variables may influence the rate and hence the reaction order. These may be factors such as complex formation between products (for instance, the complex forming between $TiCl_4$ and DMA), the difference between K_i and K_{eq}, or even nonisothermal conditions. The initial rate method is designed to circumvent these effects and will yield correct order data. The accuracy of the data will depend on the accuracy of initial rate measurements, which are greatly improved with *in-situ* FTIR monitoring methods[5]. Both *in-situ* FTIR and gravimetric analysis were used in our experiments. Figures 4 and 5 show that the plots are linear, so the initial slope method does not have to be used. Figure 8 shows ln $(k_{p,app})$ values obtained from the slopes of the lines in Figure 4, versus ln $[TiCl_4]_0$. The $TiCl_4$ order is n =1.83, similar to that reported for living IB polymerizations[3, 7].

Figure 8 shows FTIR linear rate plots for St polymerizations conducted with excess TMPCl over TiCl₄ (Path A). The reaction was relatively fast and therefore gravimetry was not conducted. Conditions were not isothermal, but this did not seem to effect polymerization rates. Figure 9 shows the order plot.

From Figure 9, the TiCl₄ reaction order is 1.2. This finding indicates that, similarly to IB polymerization initiated by TMPCl/TiCl₄, the reaction order is closer to 1 at $[TMPCl]_0/[TiCl_4]_0 \geq 1$.

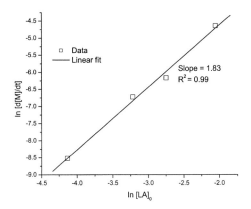

Figure 8. ln $(k_{p,app})$ vs ln$[TiCl_4]_0$ plot for $[TiCl_4]_0 >$ $[TMPCl]_0$ in TMPCl/TiCl₄ initiated St polymerization at -75°C. $[TMPCl]_0$=0.004 mol/l, $[St]_0$=2 mol/l, [DMA]=0.002 mol/l, [DtBP]=0.007 mol/l, MeCHx/MeCl=60/40 (v/v).

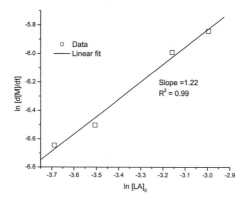

Figure 9. $\ln(k_{p,app})$ vs $\ln[LA]_0$ plot for $[LA]_0 \leq [I]_0$ in TMPCl/TiCl$_4$ initiated St polymerization at -75^0C. [TMPCl]$_0$=0.05 mol/l, [St]$_0$=2 mol/l, [DMA]=0.002 mol/l, [DtBP]=0.007 mol/l, MeCHx/MeCl=60/40 (v/v).

In summary, the carbocationic polymerization of St initiated by the TMPCl/TiCl$_4$/DMA system was found to be a complex reaction. Analysis of kinetic data showed that the rate of initiation is directly proportional to [TiCl$_4$]$_0$ and [M], and $k_{i,app}$ values were obtained. The polymerization reaction order with respect to [TiCl$_4$]$_0$ was found to be 1.83 at [TMPCl]$_0$/[TiCl$_4$]$_0$ < 1 and 1.2 at [TMPCl]$_0$/[TiCl$_4$]$_0$ > 1. Judicious selection of the reaction conditions led to living conditions; living polymerization was achieved at [TMPCl]$_0$ = 0.004 mol/L and [TiCl$_4$]$_0$ = 0.032 – 0.064 mol/L, yielding high MW PS (M$_n$ = 45,000 g/mol). High [TiCl$_4$] (0.128 M) led to alkylation, but only after the monomer was depleted as shown in Figure 10.

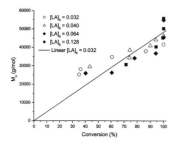

Figure 10. [TMPCl]$_0$=0.004 mol/l, [St]$_0$=2 mol/l, [DMA]=0.002mol/l, [DtBP]=0.007mol/l MeCHx/MeCl=60/40, T= -75^{0}C.

Conclusion

In conclusion, living carbocationic IB and St polymerizations exhibited similar kinetic behaviour. The rate of initiation was first order in $TiCl_4$ in both systems, but was shown to be also dependent on $[St]_0$ – in this system initiation was slow. The rate of propagation was found to be closer to first order in $TiCl_4$ at $[TiCl_4]_0 < [TMPCl]_0$ and followed a close to second order in $TiCl_4$ at $[TiCl_4]_0 \geq [TMPCl]_0$. By optimization of polymerization conditions, high molecular weight PS was produced at a high initiator efficiency.

Acknowledgement

The authors would like to thank W. Luo and A. Naskar for their contribution to this work. Financial support from NSERC (Canada) and Bayer Polymers (Bayer Inc., Canada) is also acknowledged.

[1] J. P. Kennedy, *J. Polym. Sci., Part A: Polym. Chem.* **1999**, *37*, 2285
[2] J. E. Puskas, G. Kaszas, *Prog. Polym. Sci.* **2000**; *25*, 403
[3] J. E. Puskas, M. G. Lanzendörfer, *Macromolecules* **1998**; *31*, 8684
[4] I. Majoros, A. Nagy, J. P. Kennedy. *Adv. Polym. Sci.* **1994**, *112*, 1
[5] J. E. Puskas *et al.*, in "*In Situ Monitoring of Monomer and Polymer Synthesis*", J. E. Puskas, T. E. Long, R. F. Storey, Eds., Kluver Academic/Plenum Pub., **2003**, p.37
[6] C. Paulo, J. E. Puskas, S. Angepat, *Macromolecules* **2000**, *33*, 4634
[7] J. E. Puskas, H. Peng, *Polym. React. Eng.* **1999**, *7(4)*, 553
[8] R. F. Storey, K. R. Choate Jr., *Macromolecules* **1997**, *30*, 4799
[9] Y. Wu, Y. Tan, G.Wu, *Macromolecules* **2002**, *35*, 3801
[10] H. Schlaad, Y. Kwon, R. Faust, M. Mayr, *Macromolecules* **2000**, *33*, 743
[11] R. F. Storey, A. B. Donnalley, *Macromolecules* **2000**, *33*, 53
[12] G. Kaszas, J. E. Puskas, *Polym. React. Eng.* **1994**; *2(3)*, 251
[13] J. E. Puskas, G. Kaszas, M. Litt *Macromolecules* **1991**, *24*, 5278
[14] H. Schlaad, Y. Kwon, L. Sipos, R. Faust, B Charleaux *Macromolecules* **2000**, 33, 8225
[15] H. Schlaad, K. Erentova, R. Faust, B Charleaux, M Moreau, J. P. Vairon, H. Mayr *Macromolecules* **1998**, 31, 8058
[16] A.H.E.Müller, G.Litvinenko, D. Yan, *Macromolecules* **1996**, *29*, 2339
[17] M. Roth, H. Mayr, *Macromolecules* **1996**, *29*, 6104
[18] M. Marek, M. Chmelir, *J. Polym Sci.* **1967**, *22*, 177
[19] L. Toman, M. Marek, *Makromol. Chem.* **1976**, *22*, 3325
[20] P. L. Magagnini, S. Cesca; P. Giusti, A. Priola, M. Di Maina, *Makromol. Chem.*, **1977**, *178*, 2235
[21] K. Ueno, H Yamaoka, K. Hayashi, S. Okamura, *Int. J. Appl. Radiat. Isotopes*, **1966**, *17*, 595
[22] P. H. Plesch, *Prog. Reaction Kinetics* **1993**, *18*, 1
[23] R. B. Taylor, F. Williams, *J. Am. Chem. Soc.*, **1969**, *C22*, 177
[24] P. H. Plesch, *Macromolecules* **2001**, *34*, 1143
[25] Wulkow, M. PREDICI®, *Simulation Package for Polyreactions*, Ver. 5.18.5, CiT GmbH, **2001**
[26] S. Shaikh, H. Peng, J. E. Puskas, *Polym. Prepr.* **2002**, *43(1)*, 260
[27] G. Kaszas, J. E. Puskas, J. P. Kennedy, W. G. Hager. *J. Polym. Sci., Part A: Polym. Chem.* **1991**, *29*, 421

[28] Z. S. Fodor, M. Györ, H. Wang, R. Faust, *J. Macromol. Sci., Pure Appl. Chem.* **1993**, *A30(5)*, 349
[29] R. F. Storey, S. J. Jeskey, *Polym Prepr.* **2000**, *41(2)*, 1895
[30] M. Kamigaito, K. Yamaoka, M. Sawamoto, T. Higashimura, Macromolecules **1992**, 25, 6400
[31] O. Levenspiel, *"Chemical Reaction Engineering"*, 3rd ed., J. Wiley & Sons, New York, 1999
[32] R. F. Storey, Q. A. Thomas, *Macromolecules* **2003**, 36, 5065
[33] J. E. Puskas, W.Luo, *Macromolecules* **2003**, 36, 6942

Macromol. Symp. **2004**, *215*, 255-265

Novel Cationic Ring-Opening Polymerization of Cyclodextrin: A Uniform Macrocyclic Monomer with Unique Character

Masato Suzuki,* Osamu Numata, Tomofumi Shimazaki

Department of Organic and Polymeric Materials, and International Research Center of Macromolecular Science, Tokyo Institute of Technology, Tokyo 152-8552, Japan
E-mail: msuzuki@polymer.titech.ac.jp

Summary: Cyclodextrin (CD) derivatives were found to act as uniform macrocyclic monomers for cationic ring-opening polymerization, yielding poly[(1→4)-D-glucopyranoside]s. Interestingly, as demonstrated in the polymerization of *O*-permethylated α-, β-, and γ-CDs, more strained monomer (α- > β- > γ-) proved less reactive. This finding comes from the unique molecular shape of CD; the monomer having larger cavity (γ- > β- > α-) is favorable for the polymerization. In place of ordinary initiators such as oxonium salts and MeOTf, a combination of HI with an activator of $ZnCl_2$ was found to work as the initiator effectively controlling the polymerization; the glucan obtained from *O*-permethylated γ-CD showed a unique molecular weight distribution, which has large, regular intervals corresponding to the molecular weight of the uniform macrocyclic monomer (= 1634).

Keywords: cationic polymerization; cyclodextrin; polysaccharides; ring-opening polymerization; uniform macrocyclic monomer

Introduction

A molecule of cyclodextrin (CD) has a unique cylinder-like shape; the wall consists of cyclic oligo[α-(1→4)-D-glucopyranoside], which is usually hexa-, hepta-, or octa-mers termed α, β, and γ, respectively, and the inside cavity is relatively hydrophobic in water to form inclusion complexes with various molecules.[1] Thus there have been a large number of studies using CD derivatives as host molecules, disclosing unique features of CDs, and recently, supramolecular chemistry of CDs is an attractive subject.[2] Such works are promoted by the facile availability of three CDs having different cavity sizes as well as by biochemical interest

DOI: 10.1002/masy.200451120

256

in cyclic oligosaccharides. Industrial application of CDs is also extensive especially for food, medicines, and cosmetics.[1]

Herein, we disclose another new aspect of CD; in this study, CD derivatives are examined as uniform macrocyclic monomers for ring-opening polymerization. The glycoside bond forming CD is a kind of acetal linkage, so that CD is topologically identical to a macrocyclic acetal. Since cyclic acetal is known to be polymerizable with the aid of cationic initiators, CD could be expected to act as a macrocyclic monomer, producing linear glucan.

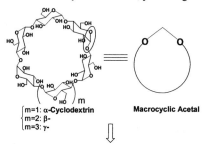

[m=1: α-Cyclodextrin
 m=2: β-
 m=3: γ-]   **Macrocyclic Acetal**

Polymerizable?

Polysaccharide has been prepared by ring-opening polymerization of sugar anhydride or orthoester as well as by enzymatic or chemical polycondensation.[3] In this project, three unique features arising from the character of CD are discussed:

1) special polymerization behavior due to the unique character of CD, which forms inclusion complexes with various molecules;

2) molecular weight distribution of the product polymer with large, regular intervals, because a CD monomer is a uniform oligomer possessing a large, exact molecular weight;

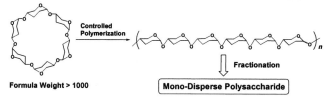

3) production of a sequentially regulated glucan, when one of the glucose components in a CD monomer is modified to achieve regioselective cleavage of the ring;

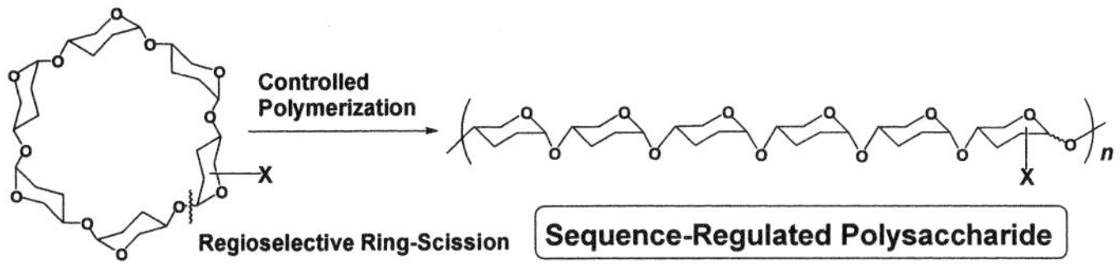

A literature survey disclosed that enzymatic and chemical ring-opening reactions of CD derivatives are known. Free CDs are converted by cyclodextrin glucanotransferase into amylose via the reverse process of the formation.[4] Restricted acetolysis or thiolysis of OH-protected CD derivatives gives the corresponding maltooligosaccharide derivatives, one of which was modified and subjected to polycondensation.[5] However, the direct chemical ring-opening polymerization of CD derivatives has never been investigated.

Ring-Opening Polymerization of O-Permethylated CD with Ordinary Initiator

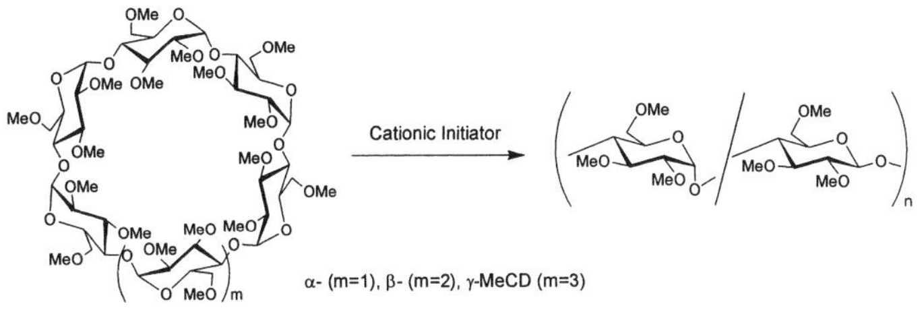

Because the free OH groups of CD disturb the cationic polymerization, all of them should be transformed to inert groups. For the initial attempt, we employed O-permethylated CDs (MeCDs) as the monomers, since the methyl ether is the smallest protective group and its use minimizes the steric hindrance. Three MeCDs were prepared from α-, β-, and γ-CDs, respectively, and subjected to polymerization with ordinary initiators.[6] As shown in Table 1, MeCDs were found to be capable of acting as macrocyclic monomers for cationic ring-opening polymerization. Oxonium salts such as $Et_3O^+X^-$ (X = PF_6, $SbCl_6$, BF_4) and MeOTf are effective initiators in CH_2Cl_2, and a protic acid of HOTf initiates the polymerization even in a less polar solvent of toluene. Increasing the amount of the initiator from 10 mol%

Table 1. Cationic ring-opening polymerization of O-permethylated cyclodextrins (MeCDs) with ordinary initiators ([MeCD] = 0.1 M)

Run	MeCD	Initiator (mol%)	Solv.	Temp /°C	Time /h	Yield[a] /%	M_w[b]	M_n[b]	α-Linkage[c] /%
1	α–	Et$_3$OPF$_6$ (5)	CH$_2$Cl$_2$	r.t.	182	94	13900	5500	76
2	α–	Et$_3$OPF$_6$ (10)	CH$_2$Cl$_2$	r.t.	68	87	15000	8500	78
3	α–	Et$_3$OPF$_6$ (20)	CH$_2$Cl$_2$	r.t.	38	88	8900	5500	72
4	α–	Et$_3$OPF$_6$ (10)	CH$_2$Cl$_2$	0	63	0	-	-	-
5	β–	Et$_3$OPF$_6$ (10)	CH$_2$Cl$_2$	r.t.	18	83	20200	10100	80
6	β–	Et$_3$OPF$_6$ (20)	CH$_2$Cl$_2$	r.t.	21	91	9500	5600[d]	72
7	β–	Et$_3$OPF$_6$ (10)	CH$_2$Cl$_2$	0	135	83	30200	15700	76
8	γ–	Et$_3$OPF$_6$ (10)	CH$_2$Cl$_2$	r.t.	21	95	19400	9800	76
9	α–	Et$_3$OSbCl$_6$ (10)	CH$_2$Cl$_2$	r.t.	39	71	8900	5600	73
10	β–	Et$_3$OSbCl$_6$ (10)	CH$_2$Cl$_2$	r.t.	19	89	10500	6400	80
11	β–	Et$_3$OBF$_4$ (10)	CH$_2$Cl$_2$	r.t.	40	83	21200	11700	74
12	γ–	Et$_3$OBF$_4$ (10)	CH$_2$Cl$_2$	r.t.	18	75	16100	8900	-
13	α–	MeOTf (5)	CH$_2$Cl$_2$	r.t.	64	46	9200	4900	88
14	α–	MeOTf (10)	CH$_2$Cl$_2$	r.t.	62	66	7500	5100	81
15	α–	MeOTf (20)	CH$_2$Cl$_2$	r.t.	67	83	6500	4300	84
16	β–	MeOTf (10)	CH$_2$Cl$_2$	r.t.	118	66	10000	6200	85
17	γ–	MeOTf (10)	CH$_2$Cl$_2$	r.t.	188	75	8800	5400	83
18	α–	Et$_2$OBF$_3$ (10)	CH$_2$Cl$_2$	40	33	(40)	-	-	-
19	α–	MeI/AgBF$_4$ (10)	CH$_2$Cl$_2$	r.t.	137	63 (60)	10800	7000	83
20	α–	HOTf (10)	PhMe	r.t.	24	28 (45)	4900	3500	87
21	β–	HOTf (10)	PhMe	r.t.	15	67 (89)	8700	5400	87
22	γ–	HOTf (10)	PhMe	r.t.	1.4	53 (74)	12100	7400	95

[a] The hexane-insoluble polymer; the monomers were almost quantitatively converted except for runs 4 and 18-22, where the monomer conversions are shown in parentheses. [b] Estimated by GPC (polystyrene standards, CHCl$_3$). [c] The proportions of the α-glycoside linkage, which were evaluated by the ^{1}H NMR spectra. [d] M_n=4500 (VPO, benzene, 40°C).

to 20 mol%, as expected, produces lower molecular weight polymer, while decreasing it to 5 mol% does not cause the molecular weight to increase (runs 1-3, 5-6, and 13-15). This finding suggests that there is a chain transfer reaction involved, which becomes prominent when the propagation reaction is slow.

The ^{1}H NMR spectra showed that the resultant poly[(1→4)-D-glucopyranoside]s consist of not only α- but also β-glycoside linkages, whose relative contents are given in Table 1.[6] In the propagation reaction, one of six, seven, or eight α-glycoside linkages of MeCD monomers is cleaved and then re-formed as the α- or β-glycoside linkage in the polymer main chain. Assuming that the glycoside linkage formed newly by the propagation exclusively has the β-form, the proportion of the α-glycoside linkage in the product polymer could be theoretically evaluated to be 83, 86, and 89% for α-, β-, and γ-MeCD, respectively. The observed values shown in Table 1 are 72-95% and independent of the monomer. Thus an undesired reaction, which involves the inversion of the α-glycoside linkage to the β-form, could take place during the polymerization (*vide infra*).

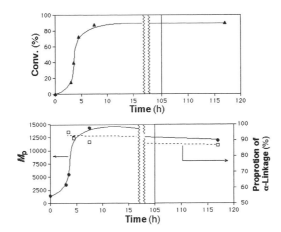

Figure 1. Monomer conversion, molecular weight (at a peak top of a GPC profile in CHCl₃, calibrated with PSt standards; M_p) of the polymer fraction, and the proportion of the α-glycoside linkage contained in the polymer, as a function of time in the polymerization of β-MeCD initiated with Et₃OPF₆; reaction conditions: [β-MeCD] = 0.1 M, [Et₃OPF₆] = 0.01 M, in CH₂Cl₂, at r.t.

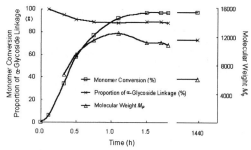

Figure 2. Monomer conversion, molecular weight (at a peak top of a GPC profile in CHCl$_3$, calibrated with PSt standards; M_p) of the polymer fraction, and the proportion of the α-glycoside linkage contained in the polymer, as a function of time in the polymerization of β-MeCD initiated with HOTf; reaction conditions: [β-MeCD] = 0.1 M, [HOTf] = 0.01 M, in CH$_2$Cl$_2$, at r.t.

GPC and ^{1}H NMR analyses of the reaction mixtures revealed the time-dependence of the monomer conversion, the molecular weight of the polymer fraction, and the relative content of the α-glycoside linkage in the product polymer (Figures 1 and 2). There is observed somewhat of an induction period in Figure 1, but not in Figure 2. It is speculated that Et$_3$OPF$_6$ could not directly initiate the polymerization due to its steric bulkiness (*vide infra*) and HPF$_6$ generated during the induction period could act as the actual initiator. The molecular weight of the polymer fraction increases with increasing monomer conversion and afterward decreases during the prolonged reaction time. The proportion of the α-glycoside linkage in the polymer also gradually decreases with time even after the monomer has been consumed. These findings suggest the chain transfer reactions shown in the following schemes, which decrease the content of the α-glycoside linkage and the molecular weight, respectively.

<u>Initiation and Propagation Reactions</u>

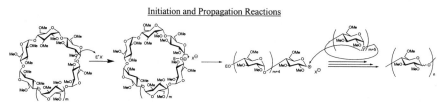

Chain Transfer Reactions to Polymer

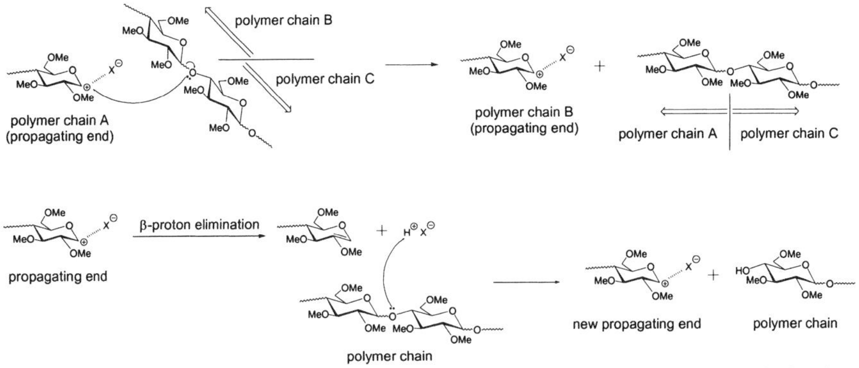

These chain transfer mechanisms are supported by another experiment; O-permethylated amylose (Mw = 54000, Mn = 26000) that is prepared from natural amylose underwent reduction in molecular weight (Mw = 7400, Mn = 4400) and contamination with the β-glycoside linkages, when it was treated with Et_3OBF_4 in CH_2Cl_2 at r.t. for 71 hr.

It is interesting to investigate the relative reactivity of three MeCDs. As shown in Table 1, β-MeCD is polymerizable even at 0°C but α-MeCD is inert under the same conditions (runs 4 and 7). The relative reactivity was examined by subjecting the mixture of two MeCDs to the competitive polymerization in a reaction tube. As seen in Figure 3, the polymerizability increases in the order of α- < β- < γ-, which is identical with increase of the ring size of MeCD and with decrease of the ring strain.[6] This finding is interesting because, in ring-opening polymerizations of analogous monomers having difference in ring size, a more strained monomer usually shows higher polymerizability. To elucidate why the more strained monomer (α- > β- > γ-) is less reactive in the polymerization of MeCDs, the characteristic shape of the CD molecule should be regarded.

As shown in the above scheme, the CD ring is opened by electrophilic attack onto the glycosyl oxygen with an initiator or a propagating end of the oxycarbocation. The CPK space-filling molecular models of MeCDs reveal that the glycosyl oxygen atoms are located at the inside of the cavity of CD. Accordingly, *electorophiles have to go into the cavity of MeCD to open the CD ring for the polymerization*, which means that the larger size of the CD ring is more favorable.

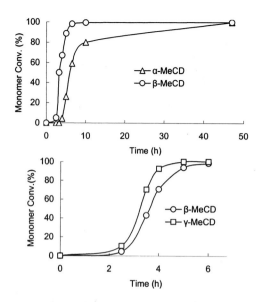

Figure 3. Monomer conversion, which was evaluated from ¹H NMR spectra of the reaction mixtures, as a function of time in the competitive polymerizations between α- and β- and between β- and γ-MeCDs initiated with Et₃OPF₆; reaction conditions: [α-MeCD] = [β-MeCD] = [γ-MeCD] = 0.05 M, [Et₃OPF₆] = 0.01 M, in CH₂Cl₂, at r.t.

Controlled Polymerization of MeCD

In order to achieve the second feature expected for this polymerization (*vide supra*), which gives unique glucan having large, regular intervals in the molecular weight distribution, it should be required to exclude the chain transfer reactions mentioned above. Thus we planned to use the knowledge about living cationic polymerization of vinyl ether, since MeCD and vinyl ether are polymerized *via* analogous propagating ends of an oxycarbocation. Herein the combinations of HI with an activator have been applied to control the cationic ring-opening polymerization of MeCD.

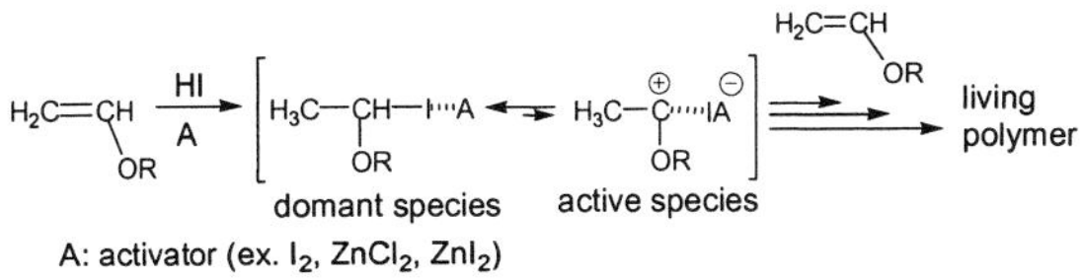

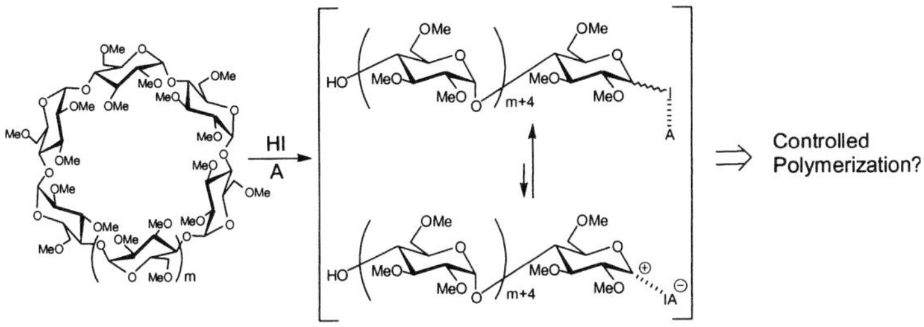

Various reaction conditions were examined: activator: I_2, $ZnCl_2$, or ZnI_2; solvent: CH_2Cl_2 or PhMe; temp.: r.t. or 0°C. The resultant polymers were analyzed with GPC and MALDI-TOF-MS.[7] Consequently, HI-$ZnCl_2$ (10 mol%) in CH_2Cl_2 at 0°C proved most effective to control the polymerization, which however, should still be stopped at moderate monomer conversion to avoid chain transfer to polymer. When the conversion of γ-MeCD was 45%, the high resolution GPC elution profile showed the formation of the "macro" oligomers (DP = 1-5) that were separately detected due to a relatively large difference in molecular weight. Using the MALDI-TOF-MS analysis, the obtained glucan was found to have the unique molecular weight distribution; there are large, regular intervals, each of which were identical to the molecular weight of γ-MeCD.[7]

Ring-Opening Polymerization of Other CD Derivatives

The poly[(1→4)-D-glucopyranoside] obtained from MeCD could not be transformed further, since the O-methyl groups cannot be restored to the hydroxyl groups without affecting the polymer chain. Thus the polymerizability of O-peracetylated and O-perbenzylated CDs has been investigated, since the acetyl and benzyl groups are easily deprotected back to the hydroxyl groups. Unfortunately, however, these CD derivatives proved inert under the same polymerization conditions as mentioned above for MeCDs and reactions forced by heating resulted in polymers consisting of unidentified structures. This is ascribable to the steric hindrance and the reactivity of the acetoxy and benzyloxy groups; the bulky substituents cover the cavity mouth of CD and obstruct the ring-opening reaction *via* the electrophilic attack onto the glycoside oxygen located inside, and the acetoxy and benzyloxy groups are more reactive toward an electrophile to induce side reactions.

Thus partly modified MeCDs were examined for the polymerization. The primary hydroxyl groups at C-6 positions of the glucose units in CD are more reactive than other secondary hydroxyl groups and can be selectively transformed to various functional groups. Additionally, the C-6 positions are located at the smaller mouth of the CD cavity, so that the polymerization would be less affected by the steric bulkiness at the C-6 positions. Therefore, several CD derivatives, composed with 6-*O*-modified 2,3-di-*O*-methylglucose units, were prepared and subjected to polymerization. Consequently, the CD derivatives shown below proved polymerizable, but less reactive than MeCD.[8] The product glucan could undergo further transformation at the C-6 positions.

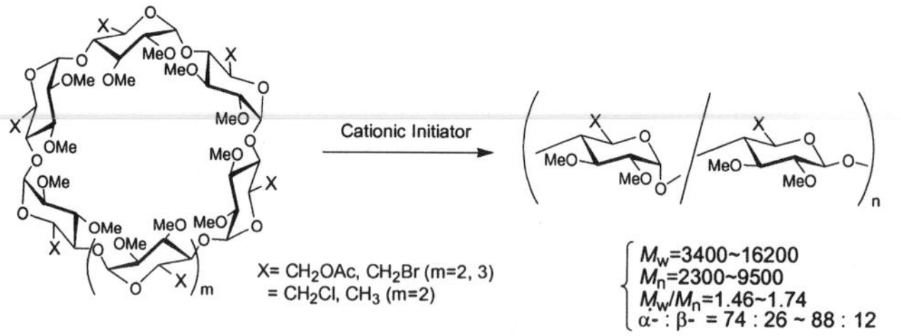

$X = CH_2OAc, CH_2Br \ (m=2, 3)$
$= CH_2Cl, CH_3 \ (m=2)$

$M_w = 3400 \sim 16200$
$M_n = 2300 \sim 9500$
$M_w/M_n = 1.46 \sim 1.74$
$\alpha\text{-} : \beta\text{-} = 74 : 26 \sim 88 : 12$

As mentioned in the Introduction, there is a third unique feature expected for the polymerization of CD derivatives: production of a sequentially regulated glucan. In order to realize this expectation, a CD derivative, one of whose glucose components is selectively modified at the C-2 position, was designed, since a substituent at the C-2 position exerts the strongest influence on the reactivity of the adjacent glycoside bond. The electronic, steric, and neighboring effects of the C-2 substituent could induce regioselective bond cleavage at the vicinal C-1 position, resultantly yielding a sequentially regulated glucan. We have already succeeded in preparing several mono-2-*O*-modified *O*-permethylated cyclodextrins.[9] The substituent effects upon the polymerization are under investigation.

Conclusions

OH-Modified cyclodextrin derivatives have proved polymerizable; they undergo cationic ring-opening polymerization to yield linear glucan. Representative study using *O*-permethylated cyclodextrins has disclosed unique features of this polymerization: unusually, the less strained monomer is more reactive; there are large, regular intervals in the molecular weight distribution of some product polymers. The former finding comes from the unique molecular shape of cyclodextrin, which has a cavity; the larger cavity of the less strained monomer is favorable for electrophile to enter the cavity for the ring-opening reaction. The latter finding has been achieved by the controlled polymerization of *O*-permethylated γ-cyclodextrin, which is a uniform, giant monomer.

[1] J. Szejtli, *"Cyclodextrin Technology"*, Kluwer Academic Publishers, Boston, 1988.
[2] C. J. Easton, S. F. Lincoln, *"Modified Cyclodextrins: Scaffolds and Templates for Supramolecular Chemistry"*, Imperial College Press, River Edge, 1999.
[3] C. Schuerch, in: *"Encyclopedia of Polymer Science and Engineering"*, 2nd ed., H. F. Mark, N. M. Bikales, C. G. Overberger, G. Menges, Eds., John Willey & Sons, New York, 1988, Vol. 13, p 147. H. Sumitomo, M. Okada, in: *"Ring-Opening Polymerization"*; K. J. Ivin, T. Saegusa, Eds, Elsevier Applied Science, London, 1984, Vol. 1, p 299. M. Hori, H. Kamitakahara, F. Nakatsubo, *Macromolecules* 1997, *30*, 2891. S. Kobayashi, H. Uyama, S. Kimura, *Chem. Rev.* 2001, *101*, 3793.
[4] J. A. Rendleman, C. A. Knutson, *Biotechnol. Appl. Biochem.* 1998, *28*, 219.
[5] N. Sakairi, L.-X. Wang, H. Kuzuhara, *J. Chem. Soc. Chem. Commun.* 1991, 289, and *J. Chem. Soc. Perkin Trans. 1* 1995, 437. N. Sakairi, K. Matsui, H. Kuzuhara, *Carbohydr. Res.* 1995, *266*, 263. N. Sakairi, H. Kuzuhara, *Chem. Lett.* 1993, 1093, and *Carbohydr. Res.* 1996, *280*, 139.
[6] M. Suzuki, O. Numata, T. Shimazaki, *Macromol. Rapid Commun.* 2001, *22*, 1354.
[7] M. Suzuki, T. Shimazaki, *Org. Biomol. Chem.* 2003, *1*, 604.
[8] The detail will be presented elsewhere.
[9] M. Suzuki, Y. Nozoe, *Carbohydr. Res.* 2002, *337*, 2393.

Macromol. Symp. **2004**, *215*, 267-280

Wavelength Flexibility in Photoinitiated Cationic Polymerization

Yusuf Yagci

Istanbul Technical University, Department of Chemistry, Maslak, Istanbul 34469, Turkey

E-mail: yusuf@itu.edu.tr

Summary: The spectral sensitivity of onium salt photoinitiators in cationic polymerization can be tuned from the short wavelength region of the UV spectrum to wavelengths up to the visible region by using direct and indirect activation, respectively. Indirect activation is based on the electron transfer reactions between onium salts and free radical photoinitiators, appropriate sensitizers and compounds capable of forming charge transfer complexes. Bisacylphosphine oxides, dimanganese decacarbonyl in conjunction with alkyl halides and titanocene type photoinitiators such as Irgacure 784 were shown to be useful free radical promoters providing the possibility of performing cationic polymerization in the long wavelength and visible region. The synthetic routes to prepare block copolymers by using electron transfer photosensitization and free radical promoted cationic polymerization are also described.

Keywords: electron transfer; long wavelength; onium salts; photoinitiated cationic polymerization; spectral sensitivity

Introduction

Photoinitiated cationic polymerization is an important industrial process widely used in different applications such as coatings, inks, adhesives and photolithography[1, 2]. Photoinitiated cationic polymerization can be initiated by onium salts such as diaryliodonium[3,4] triarylsulphonium[5] and alkoxypyridinium[6-8]. Photolysis of these salts leads to the formation of reactive cations or Brønsted acids which react efficiently with monomers such as vinyl ethers and cyclic ethers.

Photoinitiated cationic polymerization by direct irradiation of simple onium salts can be performed usually below 300 nm. This brings some limitations on the potential uses of cationic polymerization particularly when visible light emitting light sources are used. Chemical attachment of chromophoric groups onto aromatic groups of the salts is one way to broaden the spectral sensitivity of the onium salts. However, it requires multistep synthetic procedures and the

DOI: 10.1002/masy.200451121

desired shift in the absorption characteristics of the salts could still not be achieved. Several indirect ways to overcome this problem have been described[9]. All of these pathways involve electron transfer reactions either with photoexcited sensitizer [10-14] or free radicals[15-18] and with the electron donor compounds in the excited charge transfer complexes[19].

This paper provides an overview of indirect ways for providing working conditions for photoinitiated cationic polymerization at a broad wavelength range. Special emphasis will be placed on the use of organic and organic-inorganic hydride-free radical sources such as bisacylphosphine oxides, and manganese decacarbonyl and titanocene type photoiniators, respectively.

Photosensitized Cationic Polymerization

The absorption characteristics of the onium salt based photoinitiating systems can be adapted for particular applications requiring activation at wavelengths between 330-650 nm by the use of various photosensitizers. Numerous photosensitizers were found to be effective in inducing onium salt decomposition by the electron transfer reaction as depicted in reaction (1) (Table 1). Notably, aromatic carbonyl photosensitizers which undergo electron transfer via a radical route are not included[15, 20].

$$(PS^*) + On^+ \longrightarrow PS^{+\bullet} + On\bullet \tag{1}$$

The photosensitizer radical cations initiate the cationic polymerization directly. Alternatively, they undergo hydrogen abstraction yielding substituted photosensitizer and Brønsted acid capable of initiating cationic polymerization.

$$PS^{\overset{\bullet}{+}} \begin{array}{c} \xrightarrow{\text{Monomer}} \text{Polymer} \\ \\ \xrightarrow{\text{R-H}} PS-R + H^+ \xrightarrow{\text{Monomer}} \text{Polymer} \end{array} \tag{2,3}$$

Table 1. Electron transfer photosensitizers used in photoinitiated cationic polymerization

Photosensitizer	Reference
Anthracene	10-12
Perylene	21-23
Phenothiazine	14, 21-24
Pyrene	21
Carbazole and derivatives	25-26

Photoinitiated Cationic Polymerization by Charge Transfer Complexes

It has been shown[19] that charge transfer complexes formed by mixing certain pyridinium ions with aromatic electron donors act as photoinitiators for cationic polymerization. It was suggested that radical cations of donors, formed by excitation of CT complex according to the following mechanism, are responsible for the polymerization. Once the radical cations are formed, the initiaton than proceeds in a similar way to that described above for the photosensitizer radical cations.

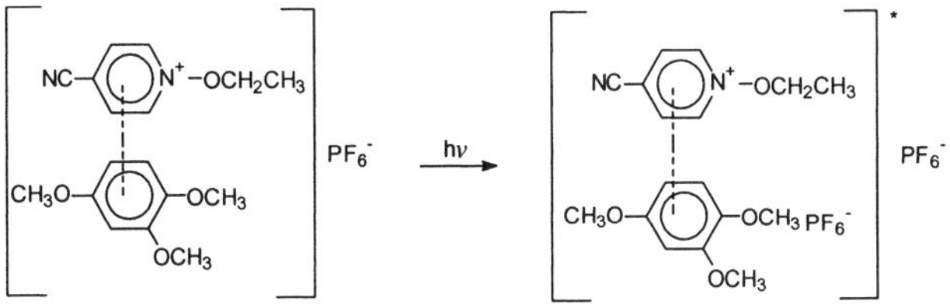

$$(4)$$

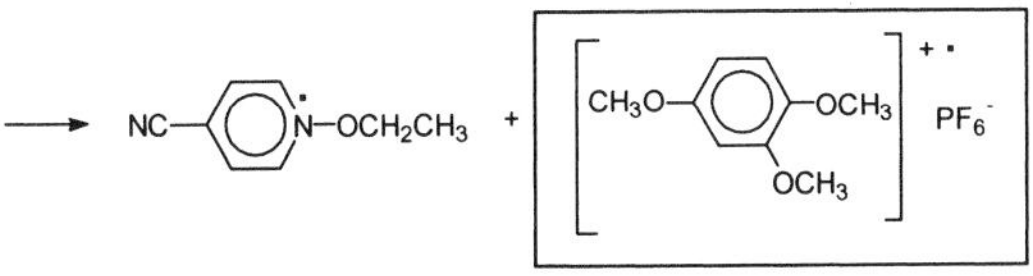

$$(5)$$

Depending on the structure of the electron donor compound and the acceptor pyridinium salt the absorption bands of the charge transfer complexes lie in the range between 350-650 nm. Therefore, the polymerizations can practically be performed at these wavelengths. Interestingly,

CT initiation can be realized only with alkoxy pyrdinium salts. The iodonium and sulphonium salts either do not form such complexes or their complexes absorb at lower wavelengths.

Free Radical Promoted Cationic Polymerization

Among the indirectly acting initiating systems, free radical promoted cationic polymerization is the most flexible route, since free radical photoinitiators with a wide range of absorption characteristics are available. Many photochemically formed radicals[27-32] can be oxidized by onium salts. The cations thus generated are used as initiating species for cationic polymerization according to the following reactions.

$$PI \xrightarrow{h\nu} R\cdot \tag{6}$$

$$R\cdot \xrightarrow{On^+} R^+ + On\cdot \tag{7}$$

$$R^+ \xrightarrow{M} Polymer \tag{8}$$

The use of visible light free radical photoinitiators

Although the use of many photoinitiators in free radical promoted cationic polymerization has been described, the sensitivity of the initiating system could be only rarely extended to the region of visible light. In a recent study[33], radicals formed by the irradiation system containing a xanthen dye and an aromatic amine, were oxidized by a diaryliodonium salt. By using the dye, the wavelengths of incident light chosen were between 500 and 650 nm. The initiation mechanism is assumed to involve the oxidation of α-amino radicals formed after hydrogen abstraction to the respective cations, which initiate the polymerization.

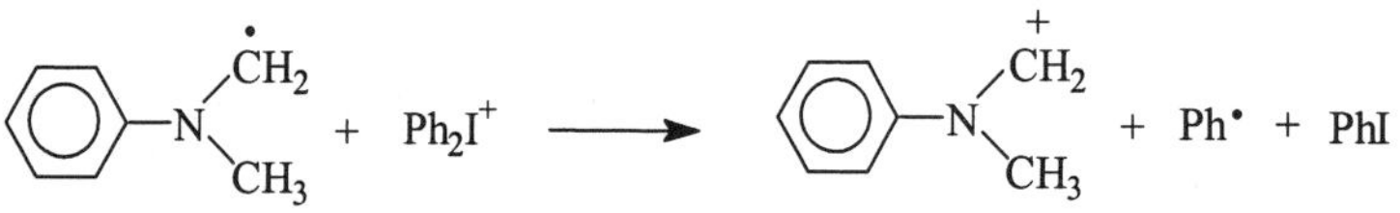

$$(9)$$

More recently, a novel visible light initiating system for the cationic polymerization of cyclic ethers such as cyclohexene oxide (CHO) and alkyl vinyl ethers such as butyl vinyl ether (BVE) was described[34]. This system consists of an organic halide, namely halogen containing solvents, dimanganese decacarbonyl, $Mn_2(CO)_{10}$, and an onium salts such as diphenyl iodonium salt, $Ph_2I^+PF_6^-$. Radical generation was achieved upon irradiation of $Mn_2(CO)_{10}$ in the presence of organic halide at $\lambda = 436$ nm where the onium salt is transparent. The proposed initiation mechanism, in which solvent methylene chloride and diphenyl iodonium salt were used as organic halide and oxidant, respectively, is shown below.

$$Mn_2(CO)_{10} \xrightarrow{h\nu} 2\,Mn(CO)_5 \tag{10}$$

$$CH_2Cl_2 + Mn(CO)_5 \longrightarrow \dot{C}H_2Cl + Mn(CO)_5Cl \tag{11}$$

$$\dot{C}H_2Cl + Ph_2I^+\,PF_6^- \longrightarrow \overset{+}{C}H_2Cl\ PF_6^- + PhI + Ph\cdot \tag{12}$$

Both systems consisted of three components for the generation of reactive cations, and the initiation mechanisms are rather complex. However, titanocene derivatives represent one of the few examples of the one component systems directly activated upon photolysis to initiate free radical polymerization[36]. Titanocene type photoinitiator, namely Irgacure 784, was also tested for its suitability as a visible light free radical source in free radical promoted cationic polymerization. For this purpose, cyclohexene oxide (CHO) was polymerized with combination of Irgacure 784 and an onium salt. CHO was polymerized effectively with iodonium and N-alkoxy pyridinium salts. Simple triaryl sulphonium salts do not undergo radical induced decomposition due to the unfavorable redox potentials. Notably, the two components of the initiating system are indispensable for the polymerization to occur; no polymer is formed in the absence of one of the compounds under our reaction conditions. When a proton scavenger, 2,6-di-tert-butyl-4-methylpyridine, was present in the reaction mixture, no polymer was formed during irradiation indicating that protons act as initiators. Photodecomposition behavior of titanocene derivatives is related to their structure. Radical trapping experiments revealed that non-fluorinated diaryl titanocenes photodecompose via homolytic cleavage of the metal-aryl-ligand bond to generate aryl and titanocene radicals. It is known that fluorinated titanocene derivatives such as Irgacure 784 yield no primary organic radicals, but rather, titanium centered diradicals according to the following reaction.

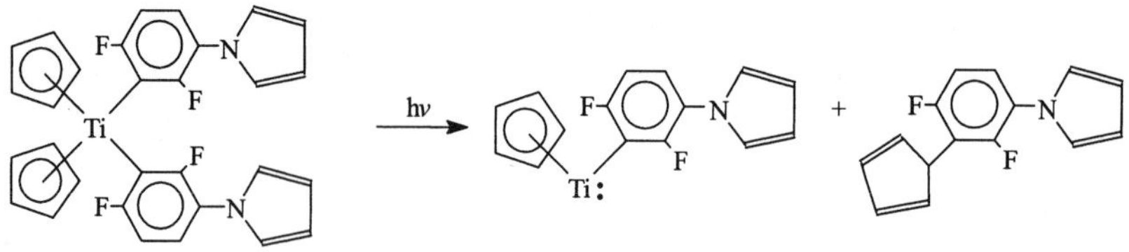

$$ \text{(13)} $$

In the presence of suitable oxidants such as onium salts, these biradicals may undergo electron transfer reaction to yield radical cations.

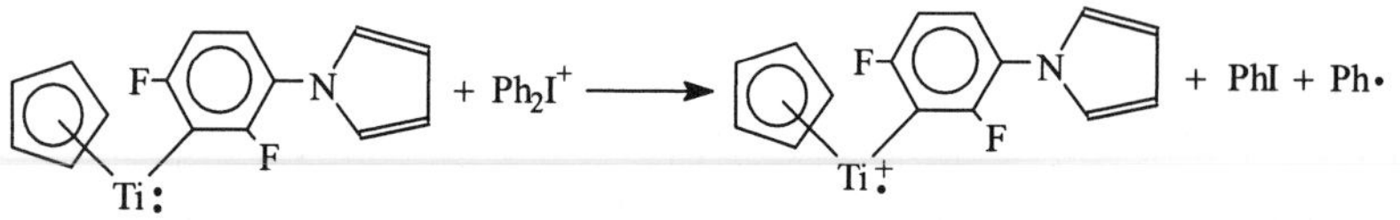

$$ \text{(14)} $$

The possibility of direct initiation by radical cations thus formed is precluded by spectroscopic investigations and absence of polymerization in the presence of a proton scavenger. Although how the protons are generated is not known at present, these results suggest that they played an important role in the initiation

The use of bisacylphosphine oxides

The value of acylphosphineoxides absorbing UV light at around 380 nm has been reported for free radical promoted cationic polymerization[27]. However, in spite of their high performances, these photoinitiators are only active when the photochemically generated benzoyl and phosphonyl radicals undergo addition or abstraction reactions[28] with monomer. The resulting carbon centered radicals are converted to carbocations by reaction with Ph_2I^+ ions.

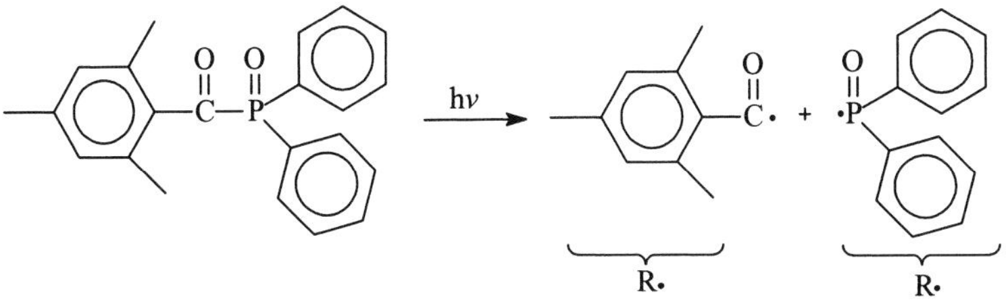

$$\text{(15)}$$

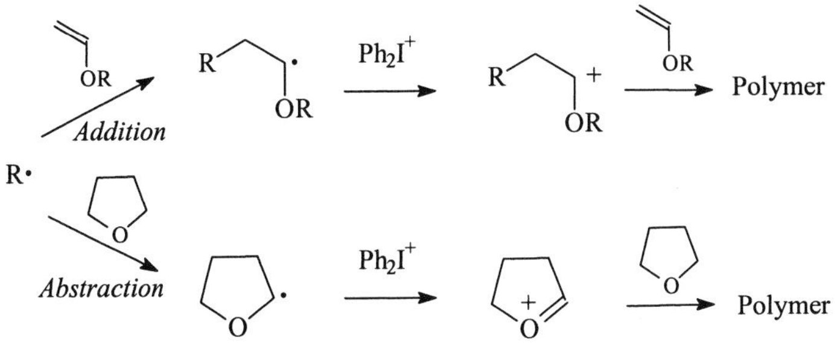

$$\text{(16)}$$

$$\text{(17)}$$

Therefore, their use in technical applications requiring monomers, which are unreactive towards photochemically generated radicals, is impossible. In an effort to overcome these limitations, we have recently reported[37] the use bisacylphosphine oxides that are especially appropriate in free radical promoted cationic polymerization. By introducing another benzoyl substituent adjacent to the phosphonyl group, the absorption is shifted to longer wavelengths. Moreover, the radicals formed from biscaylphosphine oxides were shown to be oxidizable directly without the necessity of the reaction with the monomer. The structure of bisacylphosphineoxides used successfully in free radical promoted cationic polymerization is given below.

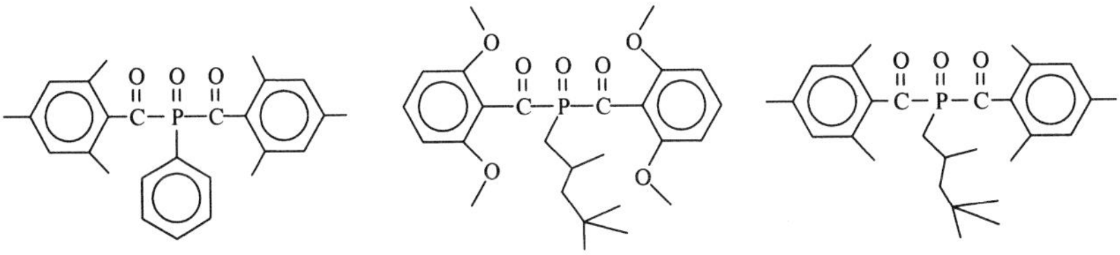

Chart 1. The structures of bisacylphosphineoxides used in free radical promoted cationic polymerization

In general, radicals with electron donating groups undergo oxidation more favorably and are therefore more efficient in free radical promoted cationic polymerization. However, phosphinoyl radicals formed from bisacylphosphine oxides according to reaction (18) possess electron withdrawing benzoyl substituents, and their reactivity should be lower than those formed from monoacylphosphineoxide (see reaction 5). Contradictorily, the bisacylphosphineoxides perform significantly better than with monoaclyphosphine oxide. This better performance was correlated with p-character on the phosphorous atom as reflected by the ^{31}P hyperfine coupling constant.

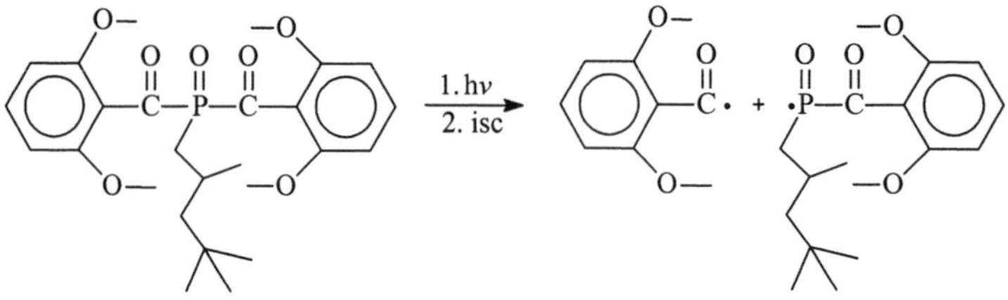

$$(18)$$

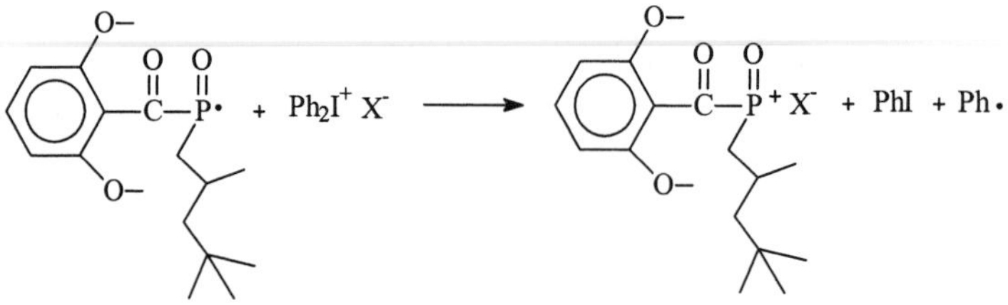

$$(19)$$

It follows that phosphonium ions formed according to the reaction (19) must be capable of initiating the cationic polymerization, as illustrated in reactions (20) and (21), using cyclohexene oxide (CHO) as an example monomer. In this case, poly(cyclohexene oxide) chains with attached benzoylphosphinoyl groups should be formed. Indeed, ^{1}H-NMR measurements revealed, apart from aliphatic protons, also existence of aromatic protons in the polymer.

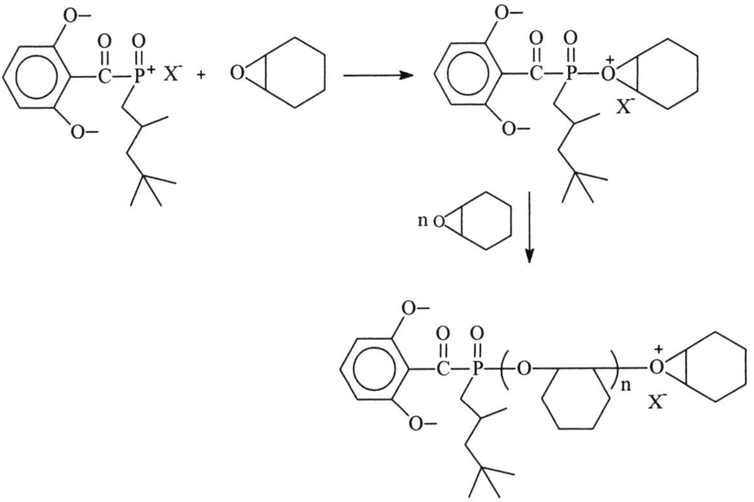

$$\text{(20)}$$

$$\text{(21)}$$

The use of free radicals generated by hydrogen abstraction

Oxidizable radicals can also be formed from the hydrogen abstraction reaction of photoexcited aromatic carbonyl compounds. In this case, the singlet photoexcited sensitizer undergoes successive intersystem crossing, hydrogen abstraction and electron transfer reactions to yield Brønsted acids which are responsible for the initiation.

$$Ar_2C=O \xrightarrow{h\nu} \left[Ar_2C=\bar{O}\right]^1 \xrightarrow{i.s.c.} \left[Ar_2C=\bar{O}\right]^3 \qquad (22)$$

$$\left[Ar_2C=\bar{O}\right]^3 + R\text{-}H \longrightarrow Ar_2\overset{\bullet}{C}\text{-}OH + R \qquad (23)$$

$$Ar_2\overset{\bullet}{C}\text{-}OH + On^+ \longrightarrow Ar_2\overset{+}{C}\text{-}OH + On\bullet \qquad (24)$$

$$Ar_2\overset{+}{C}\text{-}OH \longrightarrow Ar_2C=O + H^+ \qquad (25)$$

$$H^+ + Monomer \longrightarrow Polymer \qquad (26)$$

By using appropriate sensitizers such as benzophenone, thioxanthone, camphorquinone, benzil and antraquinone derivatives, the wavelength flexibility for the initiation could be achieved in the range of 350-485 nm.

Table 2. Hydrogen abstraction type photosensitizers used in free radical promoted cationic polymerization

Photosensitizer	λ_{max} (nm)[a]	Ref.
Benzophenone	340	7, 15, 16
Thioxanthone	380	10, 11
Benzil	480-486.5	7, 20
Camphorquinone	478	20
2-ethylanthraquinone	325	20

[a]In CH_2Cl_2

Block Copolymers by Photoinitiated Cationic Polymerization Involving Electron Transfer Reactions

Although photoinitiating systems described above have the lack of polymerization control, they still contribute to the macromolecular engineering of synthetic polymers. For example, block copolymers can be prepared by using polymeric photoactive molecules, i.e., molecules capable of acting as electron transfer sensitizer or free radical initiator[38]. These systems consist of a photochemical reaction followed by an electron transfer by which active cations are produced at the chain ends, which themselves initiate the polymerization of a second monomer. Low temperature conditions, usually room temperature, prevent side reaction leading to the formation of homopolymers, and high block yields are attained. Table 3 contains a summary of the types of block copolymers prepared using photoinduced cationic polymerization based on electron transfer reactions.

Early examples were undertaken to show that different polymerization techniques could be combined with photoinduced cationic polymerization. However, recent examples illustrate that living methods like nitroxide mediated radical polymerization (NMP), atom transfer radical polymerization (ATRP) and ring-opening polymerization (ROP) could also used in the controlled synthesis of block copolymers.

Table 3. Summary of block copolymers prepared using photonoinitiated cationic polymerization involving electron transfer reactions

Type of photoinitiated cationic polymerization	Block Copolymer	Ref.
Free radical promoted	Polytetrahydrofuran-*b*-polystyrene	18
Free radical promoted	Polystyrene-*b*-poly(butyl vinyl ether)	39
Free radical promoted	Poly(phenylmethylsilane)-*b*-poly(cyclohexene oxide)	32, 40
Free radical promoted	Polystyrene-*b*-poly(cyclohexene oxide)	41, 42
Free radical promoted	Poly(ε-caprolactone)-*b*-poly(cyclohexene oxide)	42
Free radical promoted	Poly(*p*-methoxystyrene)-*b*-poly(cyclohexene oxide)	43
Free radical promoted	Poly(methyl methacrylate)-*b*-poly(butyl vinyl ether)	40
Photosensitized electron transfer	Polytetrahydrofuran-*b*-poly(cyclohexene oxide)	11
Photosensitized electron transfer	Polytetrahydrofuran-*b*-poly(methyl methacrylate)	44
Photosensitized electron transfer	Poly(cyclohexene oxide)-*b*-polystyrene	45
Photosensitized electron transfer	Polyurethane-*b*-polytetrahydrofuran	46

For example, photoexcited anthracene was reacted[45] with an alkoxy pyridinium salt, which produced a radical cation that could be trapped with TEMPO to create a dual initiating species capable of both cationic polymerization and NMP. After the polymerization of CHO, the macroinitiator was purified and used for the polymerization of styrene. The resulting block copolymers had increased molecular weights (Mn=3200 to 40000) with no increase in the molecular weight distribution (Mw/Mn=1.5).

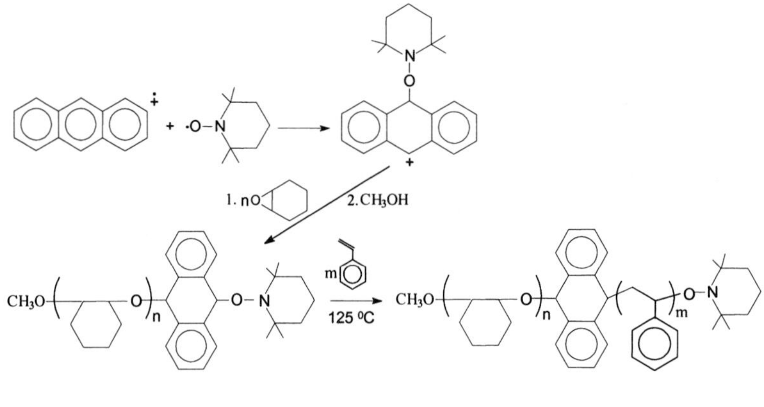

$$\text{(27)}$$

$$\text{(28)}$$

$$\text{(29)}$$

Like low molar mass analogs, electron donating polymeric radicals may conveniently be oxidized to polymeric carbocations to promote cationic polymerization of cyclic ethers. We have recently demonstrated[42] that irradiation of end-chain or midchain benzoin-functionalized polymers, in conjunction with onium salts as oxidants in the presence of cationically polymerizable monomers, makes it possible to synthesize block copolymers as illustrated below for the example of poly(ε-caprolactone)-b-poly(cyclohexene oxide).

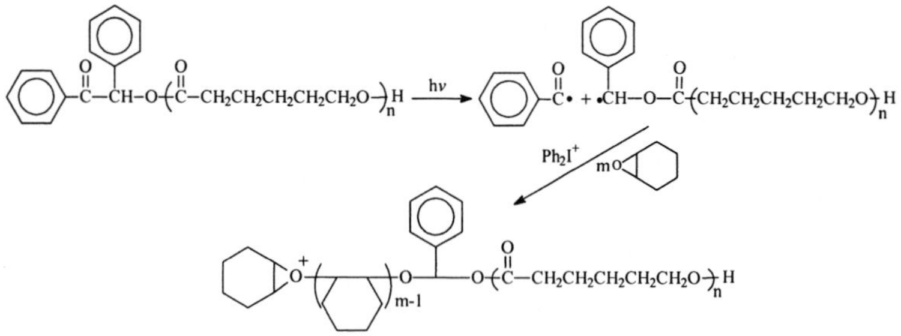

$$\text{(30)}$$

$$\text{(31)}$$

Because the initial photoactive polymers were prepared by controlled polymerization methods such as ATRP or ROP, and the fact that the side reactions are limited in the photoinitiated cationic polymerization step, block copolymers with relatively low polydispersites were formed.

Conclusion

It has been shown that the indirect activation of onium salts based on electron transfer with various additives can be used to initiate cationic polymerization, with the decisive advantage of a photosensitivity range extending up to the visible range. Moreover, the initiating characteristics of such systems favors of the development of novel synthetic procedures for block copolymers with complex structures formed from different monomers.

Acknowledgement

This work was supported by Istanbul Technical University Research Fund and the Turkish Academy of Sciences.

[1] K.Dietliker, in "Chemistry & Technology of UV&EB Formulation for Coatings, Inks& Paints" SITA Technology Ltd., London, (1991) Vol.III.
[2] S.P.Pappas, in "UV Curing: Science and Technology"; Technology Marketing Corp.: Norwalk, CT, 1978
[3] J.V.Crivello, J.Polym.Sci.,Polym.Chem. **1999**, 37, 4241
[4] J.V.Crivello, J.H.W.Lam, J. Polym. Sci., Polym. Chem. Ed., **1980**, 18, 2677
[5] J.V.Crivello, J.H.W.Lam, Macromolecules, **1977**, 10, 1307
[6] Y.Yagci, A.Kornowski, W.Schnabel, J.Polym.Sci., Polym.Chem.Ed., **1992**, 30, 1987
[7] A.Bottcher, K.Hasebe, G.Hizal, Y.Yagci, P.Stellberg, W.Schnabel, Polymer, **1991** 32, 2289
[8] J.V.Crivello, J.H.W.Lam, J. Polym. Sci., Polym. Chem. Ed., **1979,** 17, 1059
[9] Y.Yagci, I.Reetz, Prog.Polym.Sci., **1998**, 23, 1465
[10] Y.Yagci, I.Lukac, W.Schnabel, Polymer, **1993**, 34, 1130
[11] D.Dossow, Q.Q.Zhu, G.Hizal, Y.Yagci, W.Schnabel, Polymer, **1996**, 37, 2821
[12] J.V.Crivello, J.L.Lee, Macromolecules, **1981**, 14, 1141
[13] Y.Chen, T.Yamamura, K.Igarashi, J.Polym.Sci., Polym.Chem., **2000**, 38, 90
[14] E.W.Neison, T.P.Carter, A.B.Scranton, J.Polym.Sci., Polym.Chem.Ed., **1995**, 33, 247
[15] A.Ledwith, Polymer, **1978**, 19, 1217
[16] F.A.M.Abdul-Rasoul, A.Ledwith, Y.Yagci, Polymer, **1978**, 19, 1219
[17] Y.Yagci, W.Schnabel, Makromol.Chem., Macromol.Symp., **1992**, 60, 133
[18] Y.Yagci, A.Ledwith, J.Polym.Sci., Polym.Chem.Ed., **1988**, 26, 1911
[19] G.Hizal, Y.Yagci, W.Schnabel, Polymer ,**1994** 35 , 2428
[20] M.Sangermo, J.V.Crivello, in "Photoinitiated Polymerization", Eds. K.D.Belfield, J.V.Crivello, ACS Symp. Series, 847, Am.Chem.Soc., Washington, 2003, Ch. 21
[21] J.V.Crivello, J.L.Lee, Macromolecules, **1983**, 16, 684
[22] S.Denizligil, R.Resul, Y.Yagci, C.Mc Ardle, J.P.Foussier, Macromol. Chem. Phys.**1996**, 197, 1233
[23] F.Kasapoglu, Y.Yagci, Macromol. Rapid Commun., **2002**, 23, 567
[24] Z.Gomurashvili, J.V.Crivello, J.Polym.Sci., Polym.Chem.Ed., **2001**, 39, 1187
[25] J.V.Crivello, Y.Hua, Z.Gomurashvilli, Macromol.Chem., **2001**, 202, 2133
[26] Y.Hua, J.V.Crivello, J.Polym.Sci., Polym.Chem.Ed., **2000**, 38, 3697
[27] Y. Yagci, and W.Schnabel, , **1987**, Makromol. Chem., Rapid Commun., 8, 209
[28] Y. Yagci, J.Borbley, W.Schnabel, **1989**, Eur. Polym. J., 25, 129
[29] N.Johnen, S.Kobayashi, Y.Yagci, W.Schnabel, **1993**, Polym. Bull. , 30 , 279
[30] A.Okan, I.E.Serhatlı, Y.Yagci, Polym. Bull., **1996**, 37, 723
[31] Y. Yagci, S.Denizligil, J. Polym. Sci., Polym. Chem., **1995**, 33, 1461
[32] Y.Yagci, I.Kminek, W.Schnabel, Eur. Polym. J., **1992**, 28, 387

[33] Y.Bi, D.C.Neckers, **1994**, Macromolecules, 27, 3683

[34] Y.Yagci, Y.Hepuzer, Macromolecules, **1999**, 32, 6367

[35] M.Degirmenci, A.Onen, Y.Yagci, S.P.Pappas, Polym. Bull., **2001**, 46, 443

[36] J.Finter, M.Riediker, O.Rohde, Rotzinger, B. Proc. 1st Meeting Eur. Polym.Fed. Eur.Symp.Polym.Mat.; Lyon; 1987, paper ED03

[37] C. Dursun, M.Degirmenci, Y.Yagci, S.Jockush, N.J.Turro, Polymer, submitted

[38] Y.Yagci, Macromol.Symp., **2000**,161, 19

[39] Y.Yagci , M.Acar , G.Hizal , H.Yildirim, B.Baysal, Angew. Macromol. Chem., **1987**, 154 , 169

[40] D. Yucesan, H. Hostoygar, S. Denizligil and Y. Yagci, Angew. Makromol. Chem., **1994**, 221, 207

[41] Y. Yagci , A. Onen and W. Schnabel, Macromolecules, **1991**, 24 , 4620

[42] Y.Yagci, M.Degirmenci, in "Controlled Radical Polymerization", Ed. K.Matjyazewski, ACS Symp. Series, Am.Chem.Soc., Washington, 2003, in press

[43] A. Baskan Duz, Y. Yagci, Eur. Polym. J., **1999**, 35, 2031

[44] G. Hizal, Y. Yagci, W. Schnabel, Polymer, **1994**, 35 , 4443

[45] T.Girgin, Y.Hepuzer, G.Hizal, Y.Yagci, Polymer, **1999**, 40, 3885

[46] G. Hizal, A. Sarman, Y. Yagci, Polym. Bull. , **1995**, 35 , 567

Macromol. Symp. **2004**, *215*, 281-293 281

Dual Reactivity of Magnesium Compounds as Initiators for Anionic and Cationic Polymerization

Alexander   Arest-Yakubovich,[1]   Boris   Nakhmanovich,[1]   Irina   Zolotareva,[1]
Alexander Yakimansky,[2] *Natalia Pakuro*[1]

[1]Karpov Institute of Physical Chemistry, 10 Vorontsovo pole, Moscow 105064, Russia
E-mail: arest@cc.nifhi.ac.ru
[2]Institute   of   Macromolecular   Compounds,   31   Bolshoi   Prospect   V.O., St. Petersburg 199004, Russia

Summary: Organomagnesium compounds are well known initiators of anionic polymerization of polar monomers. However, we have found recently that in the presence of compounds with labile halogen atoms, e.g., benzyl chloride, they are also active initiators of cationic polymerization of isobutylene and styrene in hydrocarbon media. The tentative scheme of cationic initiation is suggested assuming the formation of benzyl cation connected with $Mg_2Cl_5^-$ counter-ion. The scheme is confirmed by quantum-chemical calculations and ^1H NMR analysis of polyisobutylene. On addition of a polar monomer, N,N-dimethylacrylamide or 2-vinylpyridine, to Bu_2Mg-BzCl-isobutylene polymerizing mixture, the former readily polymerizes. The mixture of homopolymers rather than block copolymers is formed in this case, however, this fact proves the co-existence of anionic and cationic centers in the system.

Keywords: anionic polymerization; cationic polymerization; initiators; isobutylene; magnesium compounds

Introduction

Magnesium is typical of electropositive metals, i.e. metals with the lowest electron affinity. According to the Pauling's scale, its electronegativity is 1.2 and is close to that of lithium (1.0) and potassium (1.0).[1] According to Schlosser,[2] organomagnesium compounds, along with compounds of other alkali and alkaline earth metals are considered as "polar organometallics". Similar to other metals of these groups, in the presence of electron donor solvents magnesium relatively easily transforms into a cation, donating its electrons to strong electron acceptors, for example, to condensed aromatic hydrocarbons.[3, 4] The capability of alkylmagnesiums to initiate anionic polymerization of such monomers as (meth)acrylic esters, 2- and 4-vinylpyridine, N,N-

   DOI: 10.1002/masy.200451122

dimethyl and diethylacrylamides is well known. [5-8]

More surprising is that, contrary to all other alkali and alkaline earth metals, under certain conditions magnesium is able to participate in reactions of the cationic type. In the course of recent studies we have found that in the presence of compounds with labile halogen atom, organomagnesium compounds initiate polymerization of isobutylene, styrene, p-chloromethylstyrene (p-CMS), and their comonomer mixtures. The most convincing evidence of the cationic nature of the process is the polymerization of isobutylene, which proceeds homogeneously in aliphatic or aromatic hydrocarbons, at room temperature, the quantitative yield being reached in a few minutes.

To our knowledge, in the past the possibility for magnesium to participate in cationic processes has been reported only twice, both in 60s. Bryce-Smith et al., who succeeded in the synthesis of etherless Grignard compounds, found that the products obtained with an excess of alkylhalogenides and in the medium of aromatic hydrocarbons catalysed the Friedel-Crafts alkylation of the latter.[9, 10] Somewhat later, another group reported that finely divided $MgCl_2$ initiated polymerization of isobutylene.[11] In contrast to the present study, in both of these cases reactions proceeded heterogeneously, and the mechanism was suggested to involve electron-deficient structures on the surface of deformed crystals. These works performed about 40 years ago, apparently were not extended.

In the present paper, along with new data on anionic polymerization of N,N-dialkylacrylamides, the results of our recent investigations on homogeneous cationic polymerization in the presence of magnesium compounds are given in detail, and the possible mechanism of initiation of this process is discussed. In addition, the results of quantum chemical calculations are given confirming the possibility of formation of active centers, where magnesium is incorporated into a negative counter-ion, and a probable structure of such counter-ion is proposed.

Experimental

All operations were carried out under high vacuum in an all-glass apparatus using break-seal techniques. The general procedures for reagent purification and polymerization were described earlier.[8, 12, 13] Dibutylmagnesium, Bu_2Mg, (Aldrich) was used as received. Benzyl chloride, BzCl, (Aldrich) was repeatedly dried with calcium hydride and distilled under vacuum into ampoules with break-seals.

Molecular weights and molecular weight distributions of the polymers were measured by size exclusion chromatography (SEC) with a Waters-200 instrument equipped with RI and UV (264 nm) detectors. THF was used as an eluent. Polybutadiene calibration was used for polyisobutylene.

The ^{1}H NMR spectra were obtained with 5% CCl_4 solution at 30 °C using a Gemini-300 (Varian) spectrometer operating at 300 MHz. Chemical shifts in ppm were referenced to hexamethyldisiloxane internal standard.

All Density Functional Theory (DFT) geometry optimizations were performed using the TURBOMOLE quantum chemical program[14] with the BP86 gradient corrected functional and split valence basis set with polarization functions at non-hydrogen atoms (SVP) of 6-31G* quality. All computational details have been published elsewhere.[15]

For the optimized geometries, non-specific solvation energies of all the studied structures were calculated semi-empirically using solvent continuum model [16] implemented into the modified MOPAC 6.0 program. PM-3 parametrization was used. The solvent polarity was characterized by its dielectric constant, ε.

Anionic Polymerization

Organomagnesium compounds are incapable of initiating polymerization of non-polar monomers in the absence of strong solvating agents.[3, 5] However, they readily initiate polymerization of (meth)acrylic esters and other monomers containing polar groups.[5, 6] In particular, it was shown recently that organomagnesium compounds are very reactive in the polymerization of N,N-dialkyl acrylamides.[7, 8] Generally speaking, anionic polymerization of N,N-dimethylacrylamide (DMA) was reported for the first time many years ago.[17] However, the polymer synthesized in the first works performed with lithium initiator, was partially crystalline and water-insoluble and had not attracted much attention. Only recently it was found that other initiators yield water-soluble polymers, which generated a new interest in anionic polymerization of such monomers.[18] During last few years, we studied polymerization of DMA and N,N-diethylacrylamide (DEA) under the action of a number of derivatives of alkaline earth metals including magnesium.[8] Some results of this study are interesting from the viewpoint of general theory of stereoregulation in anionic polymerization.

Data on the stereostructure of PDMA and PDEA obtained with organomagnesium initiators are shown in Table 1. For comparison, the results obtained with compounds of lithium and some other metals are also included.

Table 1. Structure of PDMA and PDEA formed with various anionic initiators at -50 °C

Run No.	Monomer	Initiator	Solvent	Triad content		
				mm	mr	rr
1	DMA	BuLi	Toluene	53	47[a]	
2	DMA	Bu$_2$Mg	THF	24	76[a]	
3	DMA	Bu$_2$Mg	Toluene	29	71[a]	
4	DMA	Tr$_2$Ba[b]	Toluene	60	40[a]	
5	DEA	DPHLi[c]	THF	55	41	4
6	DEA	Bu$_2$Mg	THF	10	87	3
7	DEA	EtMgBr	THF	28	66	6
8	DEA	Tr$_2$Ba[b]	THF	20	77	3

[a]Sum of mr and rr configurations
[b]Bis(triphenylmethyl)barium
[c]1,1-Diphenylhexyllithium

The analysis of the data obtained with these little studied monomers shows several deviations from common rules of stereoregulation established earlier, mainly for poly(acrylic esters). Thus, it is admitted that magnesium, similar to lithium, due to its small ionic radius has to favor the formation of polymers with high content of isotactic structure.[6] As can be seen from Table 1, lithium, indeed, forms predominantly isotactic structure with both monomers whereas magnesium compounds give rise to hetero- and syndiotactic polymers. (For PDMA, unfortunately, quantitative determination of mr and rr triad content is hardly possible due to superposition of peaks in the NMR-spectra.) A low content of isotactic structure in PDEA obtained with magnesium was reported also by Kobayashi et al.[7]

Further, it was shown recently that the isotactic structure content in PDMA's obtained with various alkali metals monotonously decreases with increasing ionic radius of counter-ion from ~ 55 % with lithium to 3 % with Cs.[18] On the contrary, as is seen from Table 1, in the case of alkaline earth metals, the stereospecificity in the formation of iso-structure of PDMA increases with increasing ionic radius, so that the polymer obtained in the presence of counter-ion with the largest radius, barium, is similar, with respect to its structure and some other properties, to the lithium-initiated polymer.[8] On the other hand, in the case of DEA, the compounds of magnesium and barium give polymers of approximately the same structure, thus, the counter-ion size only slightly affects the polymer structure.

The reasons for such sharp difference between alkali and alkaline earth metals with respect to stereoregulation in DMA polymerization, as well as for the difference in the behavior of DMA and DEA with respect to alkaline earth initiators are still not clear.

Cationic polymerization

In the course of recent studies we found that, in the presence of compounds with labile halogen atom such as benzyl chloride (BzCl) or *p*-chloromethylstyrene (*p*-CMS), magnesium compounds also readily initiate cationic polymerization. The most convincing evidence is the polymerization of isobutylene, which is known to polymerize only cationically. Typical experimental results on isobutylene polymerization in hydrocarbon solvents are shown in Table 2. In standard experiments (runs 9-12), solvent is condensed from the vacuum line into the ampoule, after that solutions of Bu_2Mg and chlorine-containing component are introduced. The ampoule is frozen and the monomer is condensed there under vacuum. Then the mixture is melted, shaken vigorously and heated up to room temperature. During 10 to 15 min the pale yellow solution formed shows no signs of reaction, after that an intensive heat evolution is observed and the reaction is completed in a few minutes. At molar ratio $RCl:Bu_2Mg \geq 4:1$, the polymerization proceeds with the quantitative yield, at lower ratios (run 9), a similar qualitative pattern is observed, but heat evolution is less and the monomer conversion is not complete.

The polymers obtained at room temperature have low M_n and a broad MWD, which is explained by intensive chain transfer. Nevertheless, irreversible chain termination of active centers is not very important, and the system remains active for sufficiently long time. For example, in one experiment the second monomer, styrene, was introduced 30 min after isobutylene polymerization was finished and it also was quickly polymerized quantitatively. However, the overall molecular weight did not increase as could be expected at intensive chain transfer.

At -50 °C the polymerization is much slower and high monomer conversion is reached only after several hours (runs 13, 14). In this respect the process differs considerably from most other cationic polymerizations of isobutylene where the polymerization rate increases with decreasing temperature.[19-21] However, now it is difficult to judge whether this retardation is related to slower formation of initiating complex or to the decrease in the propagation rate constant.

Table 2. Isobutylene polymerization initiated by Bu_2Mg mixtures with halogen-containing compounds. $[Bu_2Mg] = 1 \cdot 10^{-2}$ mol·L^{-1}; $[IB]_0 = 5$ mol·L^{-1}

Run No.	Solvent	Additive	Cl:Mg molar ratio	T °C	Polymer yield %	$M_n \cdot 10^{-3}$	M_w/M_n
9	Toluene	BzCl	2:1	r.t.[a]	25[b]	-	-
10	Heptane	BzCl	4:1	r.t.[a]	88[b]	1.4	7.7
11	Toluene	BzCl	6:1	r.t.[a]	92[b]	-	-
12	Toluene	BzCl	16:1	r.t.[a]	90[b]	1.1	11
13	Heptane	BzCl	4:1	-50	90[c]	6.2	15
14	Toluene	p-CMS[d]	22:1	-50	84[c]	0.7	8.5

[a]room temperature
[b]Polymerization time – several minutes (see text)
[c] Polymerization time - 4 hours
[d]p-Chloromethylstyrene

Besides BzCl, p-CMS can be also used as an initiating component. This substance is known to be an inimer in cationic self-condensing vinyl polymerization resulting in the formation of hyperbranched polymers (see paper[22] and references therein). The polymer obtained in copolymerization of isobutylene with a small amount of p-CMS (run 14, [p-CMS]:[IB] = 0.09) had much lower apparent molecular weight in comparison to isobutylene homopolymer, obtained under the same conditions (run 13). This fact gives evidence to the formation of highly branched polymer.

Figure 1 shows ^{1}H NMR spectra of polymers formed in heptane at different temperatures. The spectrum of the polymer obtained at room temperature (line 1) contains signals at 4.62; 4.85, and 5.15 ppm, which are attributed to terminal double bonds[19], whereas in the spectrum of the polymer formed at -50 °C (line 2) these peaks are absent. This agreed well with M_n values given in Table 2 indicating intensive chain transfer in the former case and its suppression in the latter. In addition, the ^{1}H NMR-spectra contain signals at 1.98 ppm ascribed to the protons of methylene group of the terminal $-CH_2-C(CH_3)_2-Cl$ unit[19], as well as signals of aromatic protons near 7.2 ppm. IR spectra of the polymers also contain characteristic bands of aromatic rings (697 and 745 cm^{-1}). The presence of the latter in the polymer formed in heptane may be explained only by the participation of benzyl group of halogen-containing component in initiation. This topic is discussed in more detail in the next section.

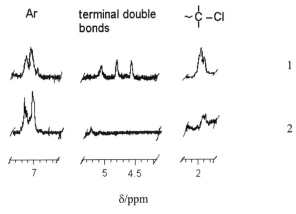

Figure 1. ^{1}H NMR-spectra of polyisobutylene obtained in heptane at room temperature (1) and at -50 °C (2)

Similarly, the Bu$_2$Mg - BzCl mixture initiates polymerization of styrene which cannot be polymerized by organomagnesium compounds in hydrocarbon media. Corresponding data are given in Table 3. At sufficient excess BzCl (run 17), styrene polymerization proceeds similar to isobutylene polymerization: after 10 - 15 min induction period a vigorous reaction starts accompanied with a marked heat evolution. In another experiment (run 18), styrene was added to the Bu$_2$Mg - BzCl mixture in toluene, aged for 20 h at room temperature. In this case, polymerization started immediately after the monomer addition. This confirmed that the induction period is related to the formation of cationic active centers.

The UV-spectum obtained ca. 10 min after completion of styrene polymerization has distinct absorbance maximum at ca. 420 nm. As is known now, this maximum does not correspond to instable polystyrene carbenium ions themselves but to the cyclic products of their secondary transformations.[23,24] Nevertheless, the presence of the maximum confirms cationic character of the processes.

Table 3. Styrene polymerization initiated by Bu_2Mg mixtures with halogen-containing compounds. Toluene, 30 °C. $[Bu_2Mg] = 1 \cdot 10^{-2}$ $mol \cdot L^{-1}$; $[St]_0 = 2.5$ $mol \cdot L^{-1}$

Run No.	Additive	Cl:Mg molar ratio	Polymer yield %	$M_n \cdot 10^{-3}$	M_w/M_n
15	BzCl	1:1	21	1.3	2.0
16	BzCl	2:1	25	1.7	2.0
17	BzCl	14:1	100	0.6	2.5
18[a]	BzCl	14:1	92	0.8	4.0
19	p-CMS[b]	70:1	84	1.5	65

[a]Toluene solution of Bu_2Mg +BzCl was aged for 20 hours at room temperature prior to addition of styrene
[b]p-Chloromethylstyrene; [St]:[pCMS] = 2.8:1

In the presence of Bu_2Mg, styrene also readily copolymerizes with p-CMS. In this case (run 19, [St]:[p-CMS] = 2.8:1), a vigorous polymerization starts without any marked induction period. The polymer obtained has a very broad MWD and is well soluble in THF in spite of high values of $M_w = 94 \cdot 10^3$ and, especially, $M_z = 2.6 \cdot 10^6$. This fact allows to suppose a highly branched structure of the polymer, similar to the case of isobutylene – p-CMS copolymerization.

Tentative scheme of cationic initiation

The whole complex of data presented leaves no doubts that polymerizations of isobutylene and styrene in the presence of Bu_2Mg mixture with labile halogen-containing compounds, in particular, BzCl, proceed via cationic mechanism. The presence of aromatic fragments in polyisobutylene macromolecule as detected by NMR- and IR-spectroscopy, proves the participation of benzyl cation in the initiation. In such a case, magnesium obviously should be incorporated into a negative counter-ion. Therefore, the possible scheme of the generation of cationic active sites should include two main steps: (i) the formation of magnesium halogenides as a result of the exchange between organomagnesium and halogen-containing components, and (ii) the interaction of magnesium halogenides formed with excess BzCl leading to the formation of cationogenic complexes.

The existence of the first step does not need special proofs because reactions of such type are characteristic of organometallic compounds[2].

The possibility of the second step, however, is not obvious and was not considered earlier. In order to estimate the probability of this reaction, quantum chemical calculations were performed.

Relative stability of various forms of chlorine-containing magnesium anions was calculated. A thermodynamically favourable form of the $Mg_2Cl_5^-$ complex anion with symmetry D_{3h} (Figure 2a) was found. It appeared to be 29.5 kcal/mol more stable than the D_{2d}-form of the anion (Figure 2b), which is explained by the fact that Mg atoms are bound via three bridge Cl atoms in the former and only via one Cl atom in the latter case.

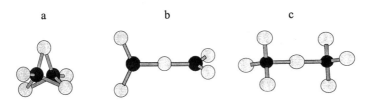

a   b   c

Figure 2. The structures of $Mg_2Cl_5^-$, D_{3h}-symmetry (a), $Mg_2Cl_5^-$, D_{2d}-symmetry (b), $Al_2Cl_7^-$, D_{3d}-symmetry (c) (metal atoms are black, chlorine atoms are gray).

Further, reaction enthalpies of the formation of benzyl cation in the reaction of magnesium chloride with benzyl chloride were calculated both for vacuum (dielectric constant $\varepsilon = 1$) and for toluene medium ($\varepsilon = 2.3$) and compared with enthalpies of similar reactions of such typical cationic catalyst as $AlCl_3$ (Table 4).

Table 4. Calculated reaction enthalpies

Route	Reaction	Reaction enthalpy, kcal/mol	
No		$\varepsilon=1^{a)}$	$\varepsilon=2.3^{b)}$
1	$AlCl_3 + C_6H_5CH_2Cl \rightarrow AlCl_4^- + C_6H_5CH_2^+$	83.5	32.1
2	$MgCl_2 + C_6H_5CH_2Cl \rightarrow MgCl_3^- + C_6H_5CH_2^+$	95.7	46.9
3	$2AlCl_3 + C_6H_5CH_2Cl \rightarrow Al_2Cl_7^- + C_6H_5CH_2^+$	57.3	14.2
4	$2MgCl_2 + C_6H_5CH_2Cl \rightarrow Mg_2Cl_5^- + C_6H_5CH_2^+$	$28.5^{c)}$	$-8.2^{c)}$
5	$2MgCl_2 + C_6H_5CH_2Cl \rightarrow Mg_2Cl_5^- + C_6H_5CH_2^+$	$58.2^{d)}$	$20.5^{d)}$

a) For reactants and products, BP86/SVP total energies are used.
b) For reactants and products, sums of BP86/SVP total energies and PM-3 solvation energies are used.
c) $Mg_2Cl_5^-$ is in the D_{3h}-form with three bridge Cl atoms between Mg atoms.
d) $Mg_2Cl_5^-$ is in the D_{2d}-form with one bridge Cl atom between Mg atoms.

It was found that the ionization of benzyl chloride with the formation of D_{3h}-form of $Mg_2Cl_5^-$ anion (route 4) is energetically possible and even much more favorable than the corresponding reaction of $AlCl_3$ both with the formation of usually assumed $AlCl_4^-$ counter-ion [25] (route 1) and with the formation of more stable $Al_2Cl_7^-$ (route 3) which also contains only one bridge Cl atom (Figure 2c).

Taking into account all said above and the results of calculations, the following scheme of reactions leading to the generation of cationic active centres in the system under consideration may be suggested as a first approximation:

$$Bu_2Mg + BzCl \rightarrow BuMgCl + Bz\text{-}Bu \qquad (1)$$

$$BuMgCl + BzCl \rightarrow MgCl_2 + Bz\text{-}Bu \qquad (2)$$

$$2MgCl_2 + BzCl \rightarrow Mg_2Cl_5^- + Bz^+ \qquad (3)$$

The key point of the Scheme is the binding of $MgCl_2$ formed with the excess BzCl. As a result, magnesium chloride remains in solution and reactions (1)-(3) seem to be of equilibrium character. For example, as is seen from Table 4, in order to obtain a high yield of polyisobutylene one needs 4-fold molar excess of BzCl, whereas according to the stoichiometry of reactions (1)-(3), only 2.5-fold molar excess is necessary for complete binding of Bu_2Mg. It is probable that the equilibrium of the ionization reaction (3) is not completely shifted to the right-hand side. Other considerations concerning the possible reversibility of at least a part of reactions presented by the Scheme will be given in the next section.

An attempt of one-pot synthesis of amphiphilic isobutylene block copolymers

After we have found the ability of magnesium compounds to initiate both anionic and cationic polymerization it seemed attractive to check the possibility of one-pot synthesis of block copolymers of isobutylene with polar monomers. Several test experiments were performed. In all cases, the initiating mixture was BzCl – Bu_2Mg at 4:1 molar ratio, the polymerization temperature -50 °C.

(i) In the first experiment, the DMA – isobutylene mixture (molar ratio ~1:1) was prepared in toluene, then, BzCl was added to the mixture and, after cooling to -50 °C, Bu_2Mg was introduced. Immediately after that the instantaneous polymerization of DMA was observed with precipitation of the polymer. The reaction mixture was allowed to stand for another 4 hours at -50 °C, then the reaction was quenched with methanol and the reaction mixture was washed with heptane, thoroughly separating the extract from the precipitate. According to NMR and IR spectroscopy, the precipitate was pure PDMA at a quantitative yield, and the product isolated from the heptane extract was pure polyisobutylene (yield ca. 20-30 %).

These results may be explained by the fact that due to a high reactivity of DMA[8] it has reacted with Bu_2Mg before the latter had a chance to react with BzCl. Then a slow reaction of unreacted part of Bu_2Mg with BzCl followed resulted in the formation of cationic active centers. Incomplete conversion of isobutylene may be explained by the presence of a large amount of electron donor carbonyl groups which are inhibitors of cationic polymerization.

The next experiments were performed according to another scheme.

(ii) Isobutylene was polymerized in heptane at -50 °C with BzCl – Bu_2Mg mixture. 3 hours after the polymerization started DMA was introduced which was polymerized instantly with the precipitation of the polymer. The mixture was washed with heptane, similar to preceding experiment. Also in this case the precipitate contained pure PDMA (yield 67%), whereas isobutylene homopolymer was isolated from heptane extract (yield ~60%).

(iii) One more experiment was made according to the same scheme. 2-Vinylpyridine (2-VP) was introduced into the system 4 hours after the beginning of isobutylene polymerization. Similar to DMA, 2-VP also polymerized very quickly leading to the formation of partially soluble yellow polymer and yellow-brown precipitate. Ethanol was used for separation of the reaction products in this case, pure polyisobutylene (yield ~80%) being detected by NMR in the precipitate and pure poly-2-VP (yield ~48%) in the ethanol extract.

In two last experiments, the formation of cationic centers initiating isobutylene polymerization obviously took place before the addition of the second monomer. Nevertheless, the system was able to initiate anionic polymerization of polar monomers, i.e. DMA or 2-VP. Probably, due to homogeneity of the system, $MgCl_2$ formed does not precipitate but remains in the solution in the complex-bound state. As a consequence, as was mentioned in the previous section, exchange reactions (1)-(3) are reversible. The equilibrium is not completely shifted to the right hand side so that a part of the compounds with a Mg – C bond, e. g., Bu_2Mg or BuMgCl, remains in the solution and initiates anionic polymerization of polar component, which is terminated shortly thereafter by excess BzCl.

Thus, the attempts to obtain block copolymers via the studied way were unsuccessful. However, the fact of simultaneous proceeding of anionic and cationic polymerization in one and the same system is not trivial and very interesting from scientific point of view.

Conclusion

Therefore, magnesium, apparently, is the only non-transition metal whose compounds are capable of initiating both anionic and cationic polymerization. The possible mechanism of cationic initiation which proceeds in the presence of compounds with labile halogen atoms includes the exchange between organomagnesium and organohalogen components with the subsequent ionization of the latter as a result of their interaction with magnesium halogenides formed. The mechanism is confirmed by the ^{1}H NMR analysis of polyisobutylene obtained in Bu$_2$Mg-BzCl system. According to the data of quantum-chemical calculations the driving force of ionization is the formation of energetically favorable Mg$_2$Cl$_5^-$ counter-ion with D$_{3h}$ symmetry where Mg-atoms are bound via three bridge Cl atoms.

A very interesting feature of the Bu$_2$Mg-BzCl system is co-existence of anionic and cationic centers due to which such different monomers as isobutylene and N,N-dimethylacrylamide can be polymerized simultaneously. However, only homopolymer mixture rather than block copolymer is formed in this case.

Acknowledgment

This work was supported by Russian Foundation of Basic Research, project No. 03-03-32930.

[1]   L. Pauling, "*The Nature of the Chemical Bond*", 3rd ed., Cornell Univ. Press, Ithaka, New York 1960.
[2]   M. Schlosser, in "*Organometallics in Synthesis*", M. Schlosser, Ed., J.Wiley & Sons, Chichester 1994, p.1.
[3]   A. Arest-Yakubovich, *Uspekhi Khim.*, 1981, *50*, 1141.
[4]   B. Bogdanovic, *Acc. Chem. Res.* 1988, *21*, 261.
[5]   H.L. Hsieh, R.P. Quirk, "*Anionic Polymerization: Principles and Practical Applications*", M. Dekker, New York 1996, p. 670.
[6]   T.E. Hogen-Esch, in "*Macromolecular Design of Polymeric Materials*", K. Hatada, T. Kitayama, Eds., M. Dekker, New York 1997, p. 163.
[7]   M. Kobayashi, T. Ishizone, S. Nakahama, *J. Polym. Sci., Part A: Polym. Chem.* 2000, *38*, 4677.
[8]   B.I. Nakhmanovich, Ya.G. Urman, A. Arest-Yakubovich, *Macromol. Chem. Phys.* 2001, *202*, 1327.
[9]   E.T. Blues, D. Bryce-Smith, *Proc. Chem. Soc.* 1961, 245.
[10]   D. Bryce-Smith, *Bull. Soc. chim. Fr.* 1963, 1418.
[11]   K.S.B. Addecott, L. Mayor, C.N. Turton, *Europ. Polym. J.* 1967, *3*, 601.
[12]   B.I. Nakhmanovich, T.N. Prudskova, A.A. Arest-Yakubovich, A.H.E. Müller, *Macromol. Rapid Comm.* 2001, *22*, 1243.
[13]   A. Arest-Yakubovich, B.I. Nakhmanovich, G.I. Litvinenko, *Polymer* 2002, *43*, 7093.
[14]   R. Ahlrichs, M. Bär, M. Häser, H. Horn, C. Kölmel, *Chem. Phys. Lett.* 1989, *162*, 165.

[15]   A.V. Yakimansky, A. H. E. Müller, *J. Am. Chem. Soc.* 2001, *123*, 4932.

[16]   G.E. Chudinov, D. V. Napolov, M. V. Basilevsky, *Chem. Phys.* 1992, *160*, 41.

[17]   K. Butler, P.R. Thomas, G.J. Tyler, *J. Polymer Sci.* 1960, *48*, 357.

[18]   X. Xie, T.E. Hogen-Esch, *Macromolecules* 1996, *29*, 1746.

[19]   Z. Fodor, Y.C. Bae, R. Faust, *Macromolecules* 1998, *31*, 4439.

[20]   C. Paulo, J.E.Puskas, *Macromolecules* 2001, *34*, 734.

[21]   P. Sigwalt, M. Moreau, A. Polton, *Macromol. Symp.* 2002, *183*, 35.

[22]   A.H.E. Müller, D. Yan, M. Wulkow, *Macromolecules* 1997, *30*, 7015.

[23]   V. Bertolli, P.H. Plesh, *J. Chem. Soc. B*, 1968, 1500.

[24]   B. Charleux, A. Rives, J.-P. Vairon, K. Matyjaszevski, *Macromolecules* 1996, *29*, 5777.

[25]   P.H.Plesch, *Macromol. Symp.* 1994, *85*, 1.

Templating Organosilicate Vitrification Using Unimolecular Self-Organizing Polymers Prepared from Tandem Ring Opening and Atom Transfer Radical Polymerizations

T. Magbitang, V. Y. Lee, E. F. Connor, L. K. Sundberg, H.-C. Kim, W. Volksen, C. J. Hawker, R. D. Miller,* J. L. Hedrick*

IBM Almaden Research Center, 650 Harry Road, San Jose, CA 95120, USA

Summary: A unimolecular templating star-shaped polymer with a compatibilizing outer corona, prepared by tandem ROP/ATRP procedures, was dispersed into a thermosetting organosilicate. The organic polymer was thermalized to leave behind its latent image in the matrix with a pore size that reflected the size of the polymer molecule, and provided the expected reduction in dielectric constant.

Keywords: ATRP; nanoporosity; ROP; star-shaped polymer

Introduction

Historically, the semiconductor industry has been able to achieve a two-fold increase in microprocessor performance approximately every three years,[1] primarily achieved by a reduction in device dimension, driven by advances in microlithography allowing on-chip device densities to continue to increase. Even though feature dimensions are projected to decrease to 90 nm or less in the near future, this alone may not be sufficient to continue the historic performance improvements. Other innovative technologies and materials are required to maintain the expected performance enhancements at these dimensions and minimize deviation from the expected performance/density "treadmill." The switch to copper metallurgy,[2] silicon on insulator and silicon-germanium are examples of materials breakthroughs that have significantly bolstered performance.[3] This continual increase in device and wiring densities has therefore placed increasing demands on the insulating material. The integration challenges of any new material place a premium on dielectric generational extendibility (i.e., maintaining the same elemental composition while the dielectric constant progressively decreases). The only route to

DOI: 10.1002/masy.200451123

true dielectric extendibility is the incorporation of porosity, which will require control of pore sizes so that they are at least 10X smaller than the minimal device dimensions.[4]

Over the last few years, we have developed several techniques for the generation of nanoporosity in thermosetting organosilicates using sacrificial polymers or porogens that template the matrix vitrification to produce nanohybrids that are precursors to nanoporous films. These methods include: a) kinetically quenched nucleation and growth processes from miscible porogen/organosilicate combinations that phase separate upon curing, triggered by both chemical and molecular weight changes in the matrix,[5] b) templation using crosslinked single-chain nanoparticle-like porogens that are dispersed in the organosilicate[6] and c) templation using unimolecular self-organizing polymer porogens.[7]   In the first case, the nucleated polymer domains, which consist of many polymer chains, remain nanoscopic in scale with growth inhibited by the crosslinking matrix (kinetically arrested growth), a process that is typically difficult to control.   The last example uses a stimuli-responsive, star-shaped copolymer that creates a nano-sized domain through a matrix-mediated collapse of the interior core of the core-corona polymeric structure. The outer corona of the star compatibilizes the insoluble core in the thermosetting resin and suppresses aggregation or precipitation of the insoluble interior so that a single polymer molecule templates crosslinking and ultimately generates a single hole. The unimolecular micellar star polymers were prepared using ruthenium-catalyzed ring opening metathesis polymerization (ROMP) in a tandem core-in/core out approach that provides pore sizes that are identical to the molecule size in the range of 20-30nm.

Our recent efforts have focused on the reduction in pore size to satisfy the criteria for future device features. This requires a sacrificial template significantly smaller in size, yet retaining the same architecture and function. Towards this goal, we have focused on "bottom-up" or core-out approaches to star-shaped templates.   Specifically, star-shaped polyesters prepared by the controlled ring-opening polymerization (ROP) of ε-caprolactone or lactide[8], initiated from the numerous chain-end hydroxymethyl groups of the analogous dendrimeric[9] and hyperbranched[10] polyesters derived from 2,2'-bis(hydroxymethyl) propionic acid, bis-MPA,[11, 12] provided number average molecular weights per arm that correlated closely to the monomer to initiator ratio. The bis-MPA units also provide exquisite markers for the spectroscopic analysis of the polymers, as the quartenary carbon and the protons on the methyl group are very sensitive to the substitution of the neighboring hydroxyl groups.[13] Along similar lines, dendrimer-like star polymers provide examples of a new macromolecular architecture characterized by a radial geometry where the different layers or generations are comprised of high molecular weight polymer emanating from a central core.[14] This modular approach to polymer design and synthesis provided the opportunity to systematically address some of the issues of branching in macromolecules.[15] Radial or layered block copolymers were generated by compositional variation between generations. In one example, dendrimer-like star block copolymers were prepared by combining ROP with other living/controlled polymerization methods such as atom transfer radical polymerization (ATRP). In some cases, these dendrimer-like star polymers were found to respond to changes in the polarity of the solvent (^{1}H-NMR) and serve as micelle mimics. This tandem ROP/ ATRP "core-out" approach to star-shaped copolymers will be described together with their utility as single molecule templates to direct the crosslinking of organosilicates.

Experimental

Materials

The 1,1,1-tris(p-hydroxyphenyl)ethane (THPE) (Hoechst Celanese) and stannous(II) 2-ethylhexanoate Sn(Oct)$_2$ (Sigma), were used as delivered. 4-(Dimethylamino)pyridinium 4-toluenesulfonate (DPTS) was synthesized according to a literature procedure.[16] The ε-caprolactone was dried over CaH$_2$ (Mallinckrodt), distilled and stored under N$_2$ prior to use. Toluene was dried over Na, distilled and stored under N$_2$. The methyl methacrylate and hydroxyethyl methacrylate (HEMA) were distilled under vacuum and refrigerated under N$_2$ until used. The dendrimers derived from 2, 2-bis(hydroxymethyl)propionic acid (Bis-MPA) were prepared according to a literature procedure.[13] The benzyl 2,2'-bis(hydroxymethyl) propionate was synthesized according to a literature procedure. All other compounds were purchased from Aldrich and used as received.

Measurements

Size-exclusion chromatography (SEC) was carried out on a Waters chromatograph connected to a Waters 410 differential refractometer. Four 5 μm Waters columns (300 X 7.7 mm) connected in series in order of increasing pore size (100, 1000, 10^5, 10^6 Å) were used with THF as eluant. The SEC results were calibrated with polystyrene standards. The thermophysical properties (Tg) were recorded on a Perker-Elmer DSC-7. ^{1}H NMR spectra were recorded in a solution with a Bruker

AM 250 (250 MHz) spectrometer. ^{13}C NMR spectra were recorded at 62.9 MHz on a Bruker AM 250 spectrometer using the solvent carbon signal as an internal standard.

General procedure for the modification of the hydroxy functional end groups of the poly(ε-caprolactone) initiators for ATRP. G-1 (6-OH) (8.00 g, 3.20 mmol) was dissolved in 50 mL of dry THF, and to this solution, triethylamine (1.40 g, 17.75 mmol) was added. The 2-bromo-2-methylpropionyl bromide (1.58 g, 6.85 mmol) was added dropwise over a 15 min period and stirring continued at room temperature for 48 hours. ^{1}H-NMR (CDCl$_3$) δ 1.28–1.40(m, poly, -CH_2CH_2CH_2-), 1.55–1.70 (m, poly, -CH$_2$CH_2CH$_2$-), 1.89(s,6H, CH3), 2.24–2.35 (t, poly, -CH_2CO-), 4.00–4.05 (t, poly, -CH_2O-), 4.11–4.16 (t, 18H, ,-CH_2OH), 4.31 (2, 12H,-CH$_3$(CH_2O)$_2$)), 6.89–7.07 (dd, 12H, Ph–). ^{13}C NMR (CDCl$_3$) 17.74, 24.51, 25.47, 28.29, 30.70, 34.05, 46.71, 51.61, 55.88, 64.05, 65.11, 65.70, 120.69, 129.65, 146.23, 148.62, 171.37, 171.57, 172.77, 173.43.

General procedure for ATRP of methyl methacrylate from functional polycaprolactone. G-1 (6-Br) (0.40 g 0.15 mmol) and dibromo-bis(triphenylphosphine)nickel(II) (7.00 mg, 0.009 mmol) were charged into a flask, which was evacuated for 12h and then purged with nitrogen and evacuated . Dry methyl methacrylate (2.00 g, 20.00 mmol) was added through a rubber septum and allowed to stir at room temperature until the macroinitiator dissolved. Optionally, toluene or THF could be added to facilitate the dissolution of the initiator and/or reduce the viscosity of the polymerization. The reaction flask was placed in a hot oil bath (110°C) and allowed to react for 5–8 hours. The polymers were isolated in hexane, stirred with methanol and isolated by filtration.

Results and Discussion

The dendritic initiators used in this study are the first, second and third generation hydroxy functionalized bis-MPA dendrimers (G-1-3), Scheme 1.[13] The synthesis of the six, twelve and twenty-four arm star polymers was accomplished by the reaction of G-1, G-2 or G-3 respectively, with ε-caprolactone in the presence of a catalytic amount of Sn(Oct)$_2$ in bulk,[11,12] and the characteristics of the polymers are shown in Table 1.   The targeted degrees of polymerization, DP, for each arm of the star polymers ranged from 10 to 50 and the average DP's, calculated by ^{1}H NMR, were comparable to the targeted values (Table 1). Examination of the ^{13}C NMR spectra and comparison with previous studies demonstrated that initiation occurs from each of the hydroxyl groups of the dendritic initiator.[11,12]

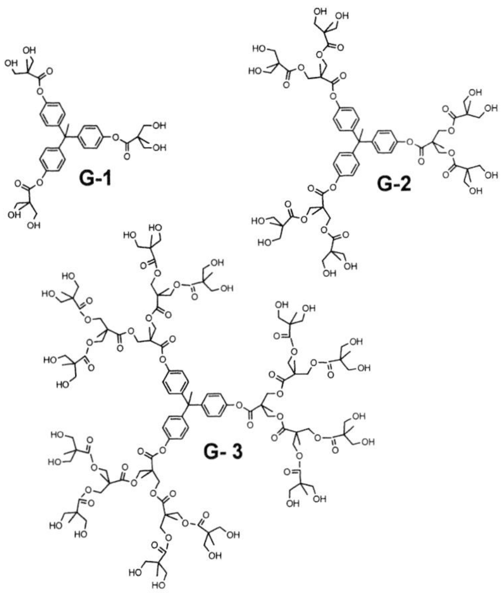

Scheme 1

Table 1. Characteristics of star-shaped polycaprolactones.

Sample Entry	Initiator	Targeted DP, M/I	DP ^{1}H-NMR	Mw/Mn (PDI)	Mw (SEC)
1	G-1	20	19	1.10	14,500
2a	G-2	10	12	1.21	20,000
2b	G-2	20	17	1.12	34,000
2c	G-2	50	48	1.20	62,000
3	G-3	25	24	1.18	70,000

Introduction of initiating centers for ATRP at the chain ends of the star-shaped polycaprolactone was accomplished by esterification of the hydroxyl functional chain ends with 2-bromo-2-methylpropionyl bromide in THF in the presence of triethylamine (Scheme 2).[14g] Isolation of the chain-end functionalized polymers and purification from excess reagents was accomplished by a simple precipitation in methanol. The ^{1}H NMR spectra of the star polymers show a clear shift in the peaks assigned to the methylene group adjacent to the hydroxyl chain end (3.66 ppm) upon the formation of the ester linkage (4.15 ppm). Furthermore, a new peak, from the –CH$_3$ groups of the modified chain end is observed at 1.05 ppm.

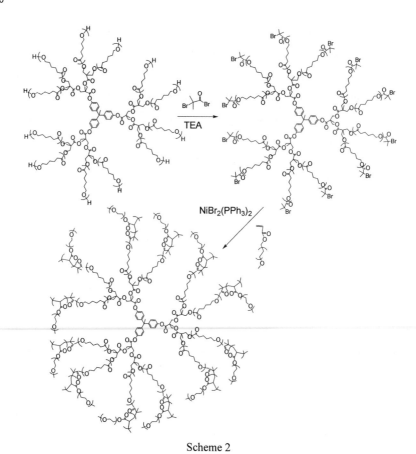

Scheme 2

Table 2. Characteristics of amphiphilic radial block copolymers.

Sample Entry	Core	Shell Type	DP (Shell)	Mw (SEC)	Mw/Mn PDI	D_h (DLS nm)
4a	2a	PEO	12	44,200	1.09	13.2
4b	2b	PEO	16	48,700	1.04	13.4
4c	2c	PEO	8	42,600	1.04	16.0
5	3	PEO	18	99,500	1.08	18.0
6	2b	MMA/HEMA 80/20	-	92,000	1.18	-

ATRP of selected vinyl monomers from the star-shaped macroinitiators was accomplished in solution at 85°C using NiBr$_2$(PPh$_3$)$_2$ as the organometallic promoter.[17] Polymers derived from hydroxyethyl methacrylate (HEMA) and poly(ethylene oxy) methacrylate-functional macromonomer were found to be miscible with organosilicate oligomers and were chosen to provide the corona of the star polymers. The high concentration of HEMA necessary in the copolymer corona required the use of the trimethylsilyl protected HEMA (Si-HEMA) together with dilution with methyl methacrylate in the polymerization to avoid "gelation" and facilitate processing of the mixtures. Upon completion of polymerization, the polymer was dissolved in a 50/50 mixture of methanol/THF, and the trimethylsiloxy group was readily removed under acidic conditions. Generally, ATRP catalysts are used in near stoichiometric amounts relative to the activated bromide, however, controlled polymerization is possible at NiBr$_2$(PPh$_3$)$_2$ levels as low as 10 to 20 mole percent the stoichiometric amount. Reduced catalyst concentrations and diluted polymerization mixtures minimized radical-radical coupling reactions in the preparation of the star polymers. The polymerizations were performed in toluene (20% solids) at 85°C in the presence of 20 mol.% catalyst for 18 hours. These conditions minimize side reactions and limited the conversions so as to minimize the size of the corona. The polymers were isolated in hexane, redissolved in THF and fractionally precipitated by the addition of hexane. This general ATRP procedure, with targeted DP's in the proximity of 10 to 20 for the outer corona (Table 2), was used to survey each of the macroinitiators. Shown in Figure 1 are the SEC traces for the poly(caprolactone) "macroinitiator" and the block copolymers derived from the poly(ethylene oxy) macromonomer. Clearly, from these data, high molecular weight, low polydispersity products are obtained which yield the expected spectroscopic data. The size of the star polymer in THF solution was determined using dynamic light scattering techniques (DLS) (Table 2). In solution, a single star-shaped polymer assumes a solvent-swollen state, and hence the values obtained by DLS represent the upper limit in size. For example, the smaller star **4a** had a Rg of 13.2 nm while that of the larger star **5** was 18.0 nm. The size distribution for both polymers was relatively narrow, consistent with the low polydispersity index measured by size-exclusion chromatography.

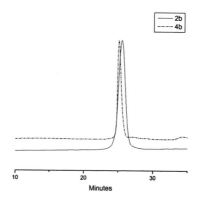

Figure 1. SEC chromatographs of star-shaped polycaprolactone, **2b**, and radial block copolymer, **4b**.

The amphiphilic polymers were dissolved in a solution containing methyl silsesquioxane (MSSQ) prepolymer in propylene glycol monomethyl ether, and the resulting solution was spun on a silicon wafer to produce thin films. Thermal analysis of the micellar star polymers in the hybrid samples by dynamic mechanical analysis (DMA) confirms that at least the internal hydrophobic core of the micellar polymers is phase separated from the MSSQ resin after soft curing (80 and 150 °C), (Figures 2 and 3), consistent with the dispersion and collapse of the core of the amphiphilic pore-generating macromolecule. The compatibilizing PEO arms are ultimately also expelled during further curing (to 200 °C) by frustrated phase separation generating the single polymer templating morphology. Crossectional FESEM and TEM micrographs of porous thin films generated from mixtures of MSSQ with **4a** (40 wt.%) are shown in Figure 4. For a truely templating process, the star polymers should produce foams with hole sizes reflecting the respective porogen dimensions. The dielectric constants of the porous samples decrease predictably from 2.8 to 1.9 with increasing porogen loading (0-40%).

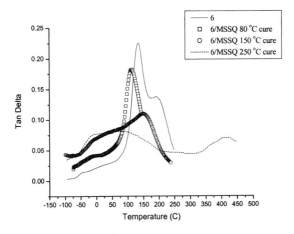

Figure 2. Dynamic mechanical analysis of **6** and mixtures of **6** (40 wt.%) with MSSQ cured to different temperatures.

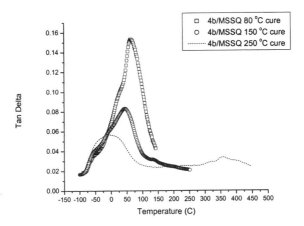

Figure 3. Dynamic mechanical analysis of mixtures of **4b** (40 wt.%) in MSSQ cured to different temperatures.

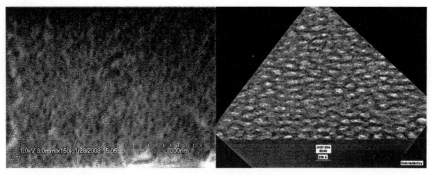

Figure 4. FESEM (left) and TEM (right) micrographs of porous MSSQ derived from **4b** (40 wt.%).

In summary, unimolecular, micellar star polymers, synthesized by tandem ROP/ATRP procedures, provide a controlled templating effect during the thermal cure of organosilicates. The unimolecular nature of the polymeric materials eliminates the complex dynamic assembly characterizing most amphiphilic systems. Porogen burnout results in nanoporous films where the pore sizes and distributions reflect the dimensions of the single polymer molecules.

[1] The National Technology Roadmap for Semiconductors; Semiconductor Industry Association: San Jose, CA 1997.

[2] Wall Street Journal 1997, Sept 22.

[3] New York Times 2001, June 8.

[4] a) a) Hedrick, J. L.; Carter, K. R.; Labadie, J. W.; Miller, R. D.; Volksen, W.; Hawker, C. J.; Yoon, D. Y.; Russell, T. P.; McGrath, J. E.; Briber, R. M. *Adv. Polym. Sci.* **1999**, 141, 1. b)Hedrick, J.L., Labadie, J., Russell, T. P., Hofer, D., Wakharker, V. *Polymer* **1993**, 34, 4717.c) Miller, R. D. *Science* **1999**, 286, 421. d) Hawker, C.J.; Hedrick, J.L.; Miller, R.D.; Volksen, W. *MRS Bull.*, **2000**, 25, 54.

[5] Nguyen, C.; Carter, K. R.; Hawker, C. J.; Hedrick, J. L.; Jaffy, R.; Miller, R. D.; Remenar, J.; Rhee, H.; Rice, P.; Toney, M.; Yoon, D. *Chem. Mater.* **1999**, *11*, 3080. b) Nguyen, C.; Hawker, C. J; Miller, R.; Hedrick, J. L.; Hilborn, J. G. *Macromolecules* **2000**, *33*, 4281. c) Heise, A.; Nguyen, C.; Malek, R.; Hedrick, J. L.; Frank, C. W.; Miller, R. D. *Macromolecules* **2000**, *33*, 2346. d) Mecerreyes, D. Huang, E. Magbitang, T., Volksen, W.; Hawker, C. J.; Lee,V.; Miller, R. D.; Hedrick, J. L. *High Perform. Polym.* **2001**, 13, 11. e) Hedrick, J. L.; Hawker, C. J.; Trollsas, M.; Remenar, J.; Yoon, D. Y.; Miller, R. D. *Mat. Res. Symp. Proc.* **1998**, 519, 65.

[6] Mecerreyes, D.; Lee, V.; Hawker, C. J.; Hedrick, J. L.; Wursch, A.; Volksen, W.; Magbitang, T.; Huang, E.; Miller, R. D. *Adv. Mat.* **2001**, 13,204.

[7] Connor, E. F.; Volksen, W.; Magbitang, T; Hawker, C. J.; Hedrick, J. L.; Miller, R. D. *Angew. Chemie Int. Ed.* **2003**, in press.

[8] Albertsson, A.-C.; Finne, A.; Andronova, N. *J. Poly. Chem. Sci. Part A: Polym. Chem.* **2003**, 41, 2412. b) Storey, R. F.; Mullen, B. D.; Desai, G. S.; Sherman, J. W.; Tang, C. N. *J. Poly. Chem. Sci. Part A: Polym. Chem.* **2002**, 40, 3434. c) Parrish, B.; Quansah, J. K.; Emrick, T. *J. Poly. Chem. Sci. Part A: Polym. Chem.* **2002**, 40,1983. d) Kricheldorf, H. R.; Fechner, B. *J. Poly. Chem. Sci. Part A: Polym. Chem.* **2002**, 40, 1047.

[9] Tomalia, D. A.; Frechet, J. M. J. *J. Poly. Chem. Sci. Part A: Polym. Chem.* **2002**, 40, 2719. b) Hecht, S. *J. Poly. Chem. Sci. Part A: Polym. Chem.* **2003**, 41, 1047. c) Percec, V.; Obata, M.; Rudick, J. G.; De, B. B.; Glodde, M.;

Bera, T. K.; Magonov, S. N.; Balagurusamy, V.; Heiney, P. A. *J. Poly. Chem. Sci. Part A: Polym. Chem. 2002, 40,* 3509.

[10] Bolton, D. H.; Wooley, K. L. *J. Poly. Chem. Sci. Part A: Polym. Chem.* **2002**, 40, 823. b) Orlicki, J.; Thompson, J. L.; Markoski, L. J.; Sill, K. N.; Mooore, J. S. *J. Poly. Chem. Sci. Part A: Polym. Chem.* **2002**, 40, 936. c) Baek, J.-B.; Harris, F. W. *J. Poly. Chem. Sci. Part A: Polym. Chem.* **2003**, 41, 2374.

[11] Trollsås, M.; Hawker, C. J.; Remenar, J. F.; Hedrick, J.L.; Johansson, M.; Ihre, H.; Hult, A. *J. Polym. Sci., Chem. Ed* **1998**, *36,* 2793. b) Atthoff, B.; Trollsas, M.; Claesson, H. Hedrick, J. L. *Macrom. Chem. And Phys. Series* **1998**, 713, 127.

[12] Hecht, S.; Vladimirov, N.; Frechet, J. M. J. *J. Am. Chem. Soc.* **2001**, 123, 18.

[13] a) Ihre, H.; Hult, A. *Polym. Mat. Sci. Eng.* **1997**, *77,* 71. b) Ihre, H.; Hult, A.; Soderlind, E. *J. Am. Chem. Soc.* **1996**, *178*, 6388. c)) Ihre, H.; Hult, A. *Polym. Mat. Sci. Eng.* **1997**, *77*, 71. b) Ihre, H.; Hult, A.; Soderlind, E. *J. Am. Chem. Soc.* **1996**, *178*, 6388. d) Malmstrom, E.; Johansson, M. Hult, A. *Macromolecules*, **1995**, 28, 1698. e) Trollsås, M.; Hedrick, J. L. *Macromolecules* **1998**, *31*,4390.

[14] a) Trollsås, M.; Hedrick, J. L. *J. Am. Chem. Soc.* **1998**, *120*, 4644. b) Trollsås, M.; Claesson, H.; Atthoff, B.; Hedrick, J. L. *Angew. Chem. Int. Ed. Engl.* **1998**, *37*, 3132. c) Trollsås, M.; Hedrick, J. L; Mecerreyes, D.; Dubois, Ph.; Jérôme, R.; Ihre, H.; Hult, A. *Macromolecules* **1998**, *31*, 2756. d) Trollsås, M.; Kelly, M. A.; Claesson, H.; Siemens, R.; Hedrick, J. L. *Macromolecules* **1999**, 32, 4917. e) Gnanou, Y.; Taton, D. *Macromol. Symp.* **2001**, 174, 333. f) Seebach, D.; Herrmann, G. F.; Lengweiler, U.; Bachmann, B. M.; Amrein. W. *Angew. Chem. Int. Ed. Engl.* **1996**, 35, 2795. g) Hedrick, J. L.; Trollsas, M.; Hawker, C. J.; Atthoff, B.; Claesson, H.; Heise, A.; Miller, R. D. *Macromolecules*, **1998**, 31, 8691.

[15] a) Trollsas, M.; Claesson, H.; Atthoff, B.; Hedrick, J. L.; Pople, J. A.; Gast, A. P. *Macromol. Symp.* **2000**, 153, 87. b) Trollsas, M.; Atthoff, B.; Hedrick, J. L. Pople, J. A. Gast, A. P. *Macromolecules* **2000**, 33, 6423.

[16] Moore, J. S.; Stupp, S. I. *Macromolecules* **1990**, 23, 65.

[17] (a) Matyjaszewski, K. Ed., *ACS Symp. Series* **1998**, 685; (b) Patten, T. E.; Xia, J.; Abernathy, T.; Matyjaszewski, K. *Science* **1996**, *272*, 866; (c) Kato, M.; Kamigaito, M.; Sawamoto, M.; Higashimura, T. *Macromolecules* **1995**, *28*, 1721; (d) Granel, C.; Dubois, P.; Jérôme, R.; Teyssié, P. *Macromolecules* **1996**, *29*, 8576; (e) Granel, C.; Moineau, G.; Lecomte, Ph.; Dubois, P.; Jérôme, R.; Teyssié, P. *Macromolecules* **1998**, *31*, 542; (f) Percec, V.; Barboiu, B. *Macromolecules* **1995**, *28*, 7970.

Ring-Opening Polymerization of the Cyclic Ester Amide Derived from Adipic Anhydride and 1-Amino-6-hexanol in Melt and in Solution

Thomas Fey, Helmut Keul, Hartwig Höcker*

Lehrstuhl für Textilchemie und Makromolekulare Chemie der Rheinisch-Westfälischen Technischen Hochschule Aachen, Worringerweg 1, 52056 Aachen, Germany

Summary: The ring-opening polymerization (ROP) of the cyclic ester amide (cEA) **5** (systematic name, 1-oxa-8-aza-cyclotetradecane-9,14-dione) – prepared from adipic anhydride and 1-amino-6-hexanol – in the melt at 165 °C and in solution at 100 °C and 120 °C with $Bu_2Sn(OMe)_2$ or $Ti(OBu)_4$ as initiator yields the alternating poly(ester amide) (PEA) **4** (systematic name, poly(5-(6-oxyhexylcarbamoyl)-pentanoate) with regular microstructure. Kinetic studies for different monomer-to-initiator ratios, different reaction media, initiators and temperatures reveal that the ROP is a first-order reaction with respect to the monomer. Under suitable polymerization conditions termination and transfer reactions are suppressed. The elementary chain growth reaction proceeds by a coordination insertion mechanism in analogy to the polymerization of lactones. By using monohydroxy- and bishydroxy-functional telechelic poly(ethylene oxide) and $Sn(octoate)_2$ as the initiating system poly(ethylene oxide)-*block*-poly(ester amide)s and poly(ester amide)-*block*-poly(ethylene oxide)-*block*-poly(ester amide)s are obtained. The poly(ester amide) **4** is a semicrystalline material with a melting point of 140 °C, the block copolymers are phase separated systems showing two melting points characteristic for the respective homopolymers.

Keywords: cyclic ester amides; kinetics; poly(ester amide)s; ring-opening polymerization

Introduction

Ring-opening polymerization (ROP) of depsipeptides[1] – six-membered cyclic ester amides from α-hydroxy acids and α-amino acids – has been studied intensively in the last decade.[2-12] ROP is usually performed in the melt at temperatures above 100 °C using transition metal initiators and proceeds according to a chain growth mechanism. The ROP of depsipeptides occurs via a coordination insertion mechanism which was studied in detail for the ROP of lactones.[13] The active species is generated upon acyl-oxygen cleavage under exclusive participation of the ester group. By using macro-initiators derived from poly(ethylene oxide) A-B and B-A-B block copolymers are obtained.[14-16]

Reports on ROP of cyclic ester amides with rings larger than six-membered have scarcely been mentioned in the literature. We prepared and polymerized different substituted eleven-membered cyclic ester amides of ε-amino caproic acids and β-hydroxy acids and a thirteen-membered cyclic ester amide of adipic acid and 1-amino-5-pentanol.[17,18]

In this paper results on ROP of a fourteen-membered cyclic ester amide - 1-oxa-8-aza-cyclotetradecane-9,14-dione (cEA **5**) - are reported with special emphasis on kinetic aspects. The possibility to prepare block copolymers by using macroinitiators for the ROP is presented.

Experimental Part

Materials. Adipic acid (Bayer AG) and 1-amino-6-hexanol (Fluka), were used as received. 5-(6-Hydroxy-hexylcarbamoyl)-pentanoic acid (**3**), 1-oxa-7-aza-cyclotridecane-8,13-dione (cEA **6**) and poly(ester amide) **4** were prepared according to the literature.[18,19] Before polymerization the monomer 1-oxa-8-aza-cyclotetradecan-9,14-dione (cEA, **5**) was sublimed at 120 °C and 10^{-2} mbar. The initiators titanium(IV) butoxide (Ti(OBu)$_4$, Acros), dibutyldimethoxytin(IV) (Bu$_2$Sn(OMe)$_2$, Aldrich), and tin(II) 2-ethylhexanoate (Sn(octoate)$_2$, Aldrich), were used without further purification. The monomer, initiators and purified reagents were stored under nitrogen. *N,N*-Dimethylformamide (DMF) and *N*-methyl-2-pyrrolidone (NMP) were refluxed over CaH$_2$ for several hours and distilled before use. The monohydroxy-functional poly(ethylene oxide)s (MPEO 2000, MPEO 5000, and MPEO 10000) and bishydroxy-functional poly(ethylene oxide)s (PEO 2000, PEO 6800 and PEO 11800) of narrow molecular weight distribution ($M_w/M_n \approx 1.1$) from Shearwater Polymers Inc. were used as received.

Polymerizations were carried out in an inert gas atmosphere. Nitrogen (Linde) was passed over molecular sieves (4 Å) and finely distributed potassium on aluminium oxide.

Measurements. ^{1}H NMR and ^{13}C NMR spectra were recorded on a Bruker DPX-300 FT-NMR spectrometer at 300 MHz and 75 MHz, respectively. Chloroform (CDCl$_3$), dimethylsulfoxide (DMSO-d_6) and trifluoroacetic acid (TFA-d) were used as solvents, and tetramethylsilane (TMS) served as an internal standard.

Gel permeation chromatography (GPC) analyses were carried out using a high pressure liquid chromatography pump (Bischoff HPLC pump 2200) and a refractive index detector (Waters 410). The eluting solvent was N,N-dimethylacetamide (DMAc) with 2.44 g·L^{-1} LiCl with a flow rate of 0.8 mL·min^{-1} at 80 °C. Four columns with MZ-DVB gel were applied: length of each column 300 mm, diameter 8 mm, diameter of gel particles 5 μm, and nominal pore widths 100 Å, 100 Å, 10^3 Å and 10^4 Å. Calibration was achieved using polystyrene standards of narrow molecular weight distribution from Polymer Standard Service Mainz.

Differential scanning calorimetric analyses were performed with a Netzsch DSC 204 under nitrogen with a heating rate of 10 K·min^{-1}. Calibration was achieved using indium standard samples. Thermogravimetric analyses were performed on a TG 209 with a TA-System-Controller TASC 412/2 and kinetic software from Netzsch. IR spectra were recorded on a Perkin-Elmer FTIR 1760. C, H, N elemental analyses were performed with a Carlo Erba MOD 1106 instrument.

Preparation of 1-oxa-8-aza-cyclotetradecane-9,14-dione (cEA, 5). *Procedure A:* 5-(6-Hydroxy-hexylcarbamoyl)-pentanoic acid[19] (**3**, 6.36 g, 26.0 mmol) and Ti(OBu)$_4$ (170 mg, 0.5 mmol) were heated in vacuum for 16 d to 170 °C. The cyclic ester amide **5** was collected by sublimation. Yield: 86 % (5.06 g, 22.3 mmol). *Procedure B:* Poly(ester amide) **4** (1.13 g, 5.0 mmol) and Ti(OBu)$_4$ (34 mg, 0.1 mmol) were heated in vacuum for 16 d to 170 °C. The cyclic ester amide **5** was collected by sublimation. Yield: 75 % (856 mg, 3.8 mmol). Mp: 159-161 °C.

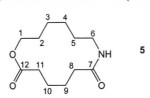

^{1}H NMR (DMSO-d_6): δ = 1.33 (m, 4H, CH$_2$-3/4); 1.41 (m, 2H, CH$_2$-5); 1.51 (m, 2H, CH$_2$-9); 1.59 (m, 4H, CH$_2$-2/10); 2.05 (tr, 2H, CH$_2$-8, ^3J = 6.0 Hz); 2.28 (tr, 2H, CH$_2$-11, ^3J = 6.4 Hz); 3.10 (d/tr, 2H, CH$_2$-6, ^3J = 5.3 Hz, ^3J = 5.7 Hz); 4.00 (tr, 2H, CH$_2$-1, ^3J = 5.3 Hz); 7.71 (br. s, 1H, NH) ppm.

^{13}C NMR (DMSO-d_6): δ = 23.90 (C-10, 1C); 24.80 (C-3/4, 1C); 25.49 (C-9, 1C); 25.53 (C-3/4, 1C); 26.56 (C-2, 1C); 27.89 (C-5, 1C); 34.13 (C-11, 1C); 35.46 (C-8, 1C); 36.94 (C-6, 1C); 63.84 (C-1, 1C); 171.69 (C-7/12, 1C); 172.57 (C-7/12, 1C) ppm.

^{1}H NMR (CDCl$_3$): δ = 1.44 (m, 4H, CH$_2$-3/4); 1.57 (m, 2H, CH$_2$-5); 1.67 (m, 4H, CH$_2$-2/9); 1.77 (m, 2H, CH$_2$-10); 2.26 (tr, 2H, CH$_2$-8, ^3J = 6.0 Hz); 2.37 (tr, 2H, CH$_2$-11, ^3J = 6.4 Hz); 3.36 (d/tr, 2H, CH$_2$-6, ^3J = 5.3 Hz, ^3J = 6.0 Hz); 4.12 (tr, 2H, CH$_2$-1, ^3J = 5.3 Hz); 5.64 (br. s, 1H, NH) ppm.

^{13}C NMR (CDCl$_3$): δ = 24.37 (C-10, 1C); 25.15 (C-3/4, 1C); 25.83 (C-3/4, 1C); 26.14 (C-9, 1C); 27.13 (C-2, 1C); 28.27 (C-5, 1C); 34.86 (C-11, 1C); 36.72 (C-8, 1C); 38.11 (C-6, 1C); 64.40 (C-1, 1C); 172.77 (C-7/12, 1C); 173.44 (C-7/12, 1C) ppm.

^{1}H NMR (TFA-d): δ = 1.60 (m, 4H, CH$_2$-3/4); 1.92 (m, 8H, CH$_2$-2/5/9/10); 2.62 (m, 2H, CH$_2$-8); 2.88 (m, 2H, CH$_2$-11); 3.75 (tr, 2H, CH$_2$-6, ^3J = 5.3 Hz); 4.35 (tr, 2H, CH$_2$-1, ^3J = 5.3 Hz) ppm.

^{13}C NMR (TFA-d): δ = 25.53 (C-10, 1C); 27.29 (C-3/4, 1C); 27.91 (C-3/4, 1C); 27.96 (C-9, 1C); 28.23 (C-2, 1C); 28.74 (C-5, 1C); 35.23 (C-11, 1C); 36.08 (C-8, 1C); 44.39 (C-6, 1C); 69.71 (C-1, 1C); 180.04 (C-7/12, 1C); 181.79 (C-7/12, 1C) ppm.

IR (KBr): 3303 (s); 2926 (s); 2862 (m); 1723 (s, C=O stretching, ester); 1643 (s, C=O stretching, amide I); 1547 (s, amide II); 1458 (m, O-CH$_2$); 1330 (m); 1288 (s); 1262 (m); 1232 (m); 1149 (s, C-O stretching); 1065 (m); 998 (w); 709 (m) cm^{-1}.

Anal. calcd for C$_{12}$H$_{21}$NO$_3$ (227.29): C, 63.41; H, 9.31; N, 6.16. Found: C, 63.35; H, 9.32; N, 6.05.

Ring-opening polymerization of cyclic ester amide 5. All glass vessels were heated in vacuo prior to use, filled with inert gas, and handled in a stream of dry inert gas. *Procedure A:* A mixture of cEA **5** (455 mg, 2.00 mmol) and Bu$_2$Sn(OMe)$_2$ (11.8 mg, 0.04 mmol) was heated to 165 °C. After 15 min the polymerization was terminated by cooling to room temperature (r.t.). The product was dissolved at 60 °C in DMF (8 mL) and precipitated in 150 mL diethyl ether. The polymer was isolated by filtration as a colourless solid. Yield: 93 % (423 mg). GPC: M_n = 18100, M_w = 29700, M_w/M_n = 1.64. *Procedure B*: A solution of cEA **5** (341 mg, 1.50 mmol) in DMF (410 μL) was heated to 100 °C and treated with Bu$_2$Sn(OMe)$_2$ (4.4 mg, 0.015 mmol) for initiation. After 96 h the polymerization was terminated by cooling to r.t. The reaction mixture was diluted with additional DMF (0.6 mL) and precipitated in ether (30 mL). The polymer **4** was isolated by filtration as a colourless solid. Yield: 98 % (335 mg). GPC: M_n = 18400, M_w = 37500, M_w/M_n = 2.04

^{1}H NMR (DMSO-d_6): δ = 1.26 (m, 4H, CH$_2$-3/4); 1.36 (m, 2H, CH$_2$-5); 1.48 (m, 4H, CH$_2$-9/10); 1.54 (m, 2H, CH$_2$-2); 2.04 (m, 2H, CH$_2$-8); 2.27 (m, 2H, CH$_2$-11); 3.00 (d/tr, 2H, CH$_2$-6, ^3J = 6.0 Hz); 3.98 (tr, 2H, CH$_2$-1, ^3J = 6.4 Hz); 7.75 (br. s, 1H, NH) ppm.

^{13}C NMR (DMSO-d_6): δ = 24.03 (C-9, 1C); 24.69 (C-3/4, 1C); 25.03 (C-10, 1C); 25.96 (C-3/4, 1C); 28.02 (C-2/5, 1C); 28.95 (C-2/5, 1C); 33.17 (C-11, 1C); 34.93 (C-8, 1C); 38.19 (C-6, 1C); 63.55 (C-1, 1C); 171.53 (C-7/12, 1C); 172.73 (C-7/12, 1C) ppm.

^{1}H NMR (DMSO-d_6, 100 °C): δ = 1.32 (m, 4H, CH$_2$-3/4); 1.43 (m, 2H, CH$_2$-5); 1.55 (m, 6H, CH$_2$-2/9/10); 2.06 (m, 2H, CH$_2$-8); 2.28 (m, 2H, CH$_2$-11); 3.05 (d/tr, 2H, CH$_2$-6, ^3J = 6.0 Hz); 4.02 (tr, 2H, CH$_2$-1, ^3J = 6.4 Hz); 7.39 (br. s, 1H, NH) ppm.

^{13}C-NMR (DMSO-d_6, 100 °C): δ = 23.52 (C-9, 1C); 24.06 (C-3/4, 1C); 24.39 (C-10, 1C); 25.31 (C-3/4, 1C); 27.49 (C-2/5, 1C); 28.33 (C-2/5, 1C); 32.79 (C-11, 1C); 34.47 (C-8, 1C); 37.84 (C-6, 1C); 62.92 (C-1, 1C); 170.97 (C-7/12, 1C); 171.85 (C-7/12, 1C) ppm.

^{1}H NMR (TFA-d): δ = 1.51 (m, 4H, CH$_2$-3/4); 1.79 (m, 4H, CH$_2$-2/5); 1.87 (m, 4H, CH$_2$-9/10); 2.60 (m, 2H, CH$_2$-8); 2.83 (m, 2H, CH$_2$-11); 3.63 (tr, 2H, CH$_2$-6, ^3J = 7.2 Hz); 4.28 (tr, 2H, CH$_2$-1, ^3J = 6.8 Hz) ppm.

^{13}C NMR (TFA-d): δ = 25.22 (C-9, 1C); 26.98 (C-3/4/10, 2C); 27.96 (C-3/4, 1C); 29.52 (C-2/5, 1C); 29.79 (C-2/5, 1C); 35.04 (C-11, 1C); 35.16 (C-8, 1C); 44.98 (C-6, 1C); 68.74 (C-1, 1C); 180.42 (C-7/12, 1C); 180.98 (C-7/12, 1C) ppm.

IR (KBr): 3314 (s); 3086 (w); 2937 (s); 2863 (s); 1735 (s, C=O stretching, ester); 1640 (s, C=O stretching, amide I); 1546 (s, amide II); 1465 (m, O-CH$_2$); 1420 (m); 1373 (m); 1269 (s); 1176 (s, C-O stretching); 1073 (w); 969 (w); 731 (m); 581 (w) cm^{-1}.

For kinetic measurements after selected reaction times samples were drawn in a stream of dry inert gas and analyzed by means of NMR and GPC.

Ring-opening polymerization of cyclic ester amide 5 using poly(ethylene oxide) MPEO 2000 and Sn(octoate)$_2$ as initiator. Poly(ethylene oxide) MPEO 2000 (80 mg, 0.04 mmol) and Sn(octoate)$_2$ (186 μL of a 0.215 M solution in toluene, 0.04 mmol) were stirred for 45 min at 145 °C. Cyclic ester amide **5** (681 mg, 3.00 mmol) and 500 μL of toluene were added at r.t. to this mixture. To obtain a homogeneous system the mixture was heated to 145 °C, cooled again to r.t. and toluene was removed at reduced pressure. For polymerization the mixture was stirred for 48 h at 165 °C. The polymerization was terminated by cooling to r.t. The product was dissolved in DMAc (5 mL) at 60 °C, precipitated in diethyl ether (150 mL) and isolated by filtration. To remove unconverted macroinitiator the solid product was washed with water; from the resulting suspension the polymer was isolated by centrifugation. Yield: 715 mg (94 %). GPC: M_n = 20600, M_w = 38100, M_w/M_n = 1.85.

R' = Me for A-B block copolymer; R' = R for B-A-B block copolymer

^{1}H NMR (DMSO-d_6, 100 °C): δ = 1.34 (m, 4H, CH$_2$-3/4); 1.44 (m, 2H, CH$_2$-5); 1.56 (m, 6H, CH$_2$-2/9/10); 2.09 (m, 2H, CH$_2$-8); 2.28 (m, 2H, CH$_2$-11); 3.06 (d/tr, 2H, CH$_2$-6, ^3J = 6.0 Hz, ^3J = 6.8 Hz); 3.56 (s, 4H, CH$_2$-13); 4.03 (tr, 2H, CH$_2$-1, ^3J = 6.8 Hz); 7.32 (br. s, 1H, NH) ppm.

^{13}C NMR (DMSO-d_6, 100 °C): δ = 24.03 (C-9, 1C); 24.67 (C-3/4, 1C); 25.03 (C-10, 1C); 25.94 (C-3/4, 1C); 28.02 (C-2/5, 1C); 29.94 (C-2/5, 1C); 33.17 (C-11, 1C); 34.93 (C-8, 1C); 38.19 (C-6, 1C); 63.57 (C-1, 1C); 69.71 (C-13, 1C); 171.53 (C-7/12, 1C); 172.71 (C-7/12, 1C) ppm.

Results and Discussion

Monomer Synthesis. In a previous paper we have studied the catalytic polycondensation of 5-(6-hydroxy-hexylcarbamoyl)-pentanoic acid (3) in the temperature range of 110 °C to 170 °C.[19] Up to 140 °C the polycondensation results in high yields of poly(ester amide) 4, however, at 170 °C the polymer yield reaches a maximum after 90 min, then due to back-biting reactions the yield decreases in favour of the cyclic ester amide 5. We have successfully applied ring-closing depolymerization for the synthesis of cEA 5, a monomer suitable for ring-opening polymerzation. As starting material for the synthesis served adipic anhydride (1) and 1,6-amino hexanol (2) which in a selective reaction in solution result in 5-(6-hydroxy-hexylcarbamoyl)-pentanoic acid (3) (Scheme 1). Preparation of the cyclic ester amide was performed either in a one pot synthesis starting with the α-carboxy-ω-hydroxy amide 3 (procedure A) or starting with the poly(ester amide) 4 (procedure B); Bu$_2$Sn(OMe)$_2$ or Ti(OBu)$_4$ served as catalyst. The yields of cEA 5 were similar for both procedures and reached values from 75 to 86 %. The cyclic ester amide is a colourless solid with a melting point of 159 - 161 °C.

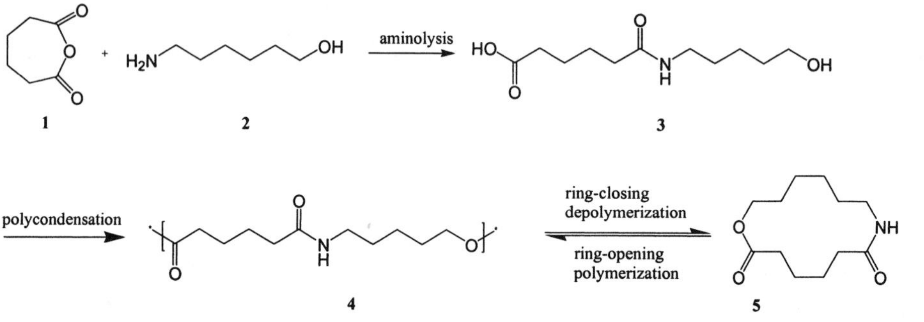

Scheme 1. Synthesis and ring-opening polymerization of cyclic ester amide (cEA) 5.

The ^{1}H and ^{13}C NMR spectra (Figure 1), the IR spectrum and the elemental analysis clearly reveal the cyclic nature of the product. Of special interest in the NMR spectra are the methylene groups adjacent to the functional groups with characteristic resonance lines: CH$_2$O (δ = 4.12 ppm, 64.40 ppm), CH$_2$NH (δ = 3.36 ppm, 38.11 ppm), CH$_2$COO (2.37 ppm, 34.86 ppm) and CH$_2$CONH (δ = 2.26 ppm, 36.72 ppm). Since all these groups show only one resonance signal a uniform microstructure is expected. The NMR spectra of cEA **5** in DMSO-d_6 and TFA-d are given in the experimental part, to be compared with that of the poly(ester amide) **4** which is insoluble in CDCl$_3$.

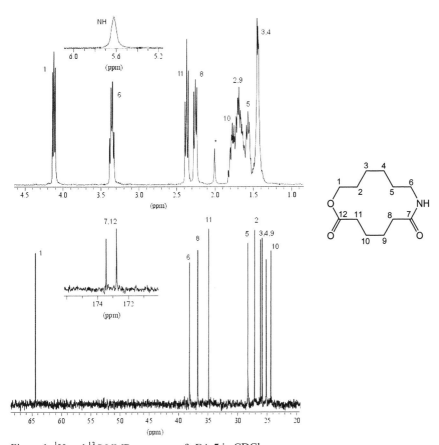

Figure 1. ^{1}H and ^{13}C NMR spectrum of cEA **5** in CDCl$_3$.

Ring-opening polymerization. The ROP of cEA **5** was performed in the melt above the melting temperature of the monomer at 165 °C and in solution of DMF or NMP with $Bu_2Sn(OMe)_2$ or $Ti(OBu)_4$ as the initiator (Table 1).

Table 1. Reaction conditions and results of the ROP of cEA **5** in the melt and in solution

No.	Initiator	$[M]_0/[I]_0$	Solvent	$T/°C$	t/min	M_n[a]	M_w/M_n[a]	$X_p/\%$[b]
1	$Bu_2Sn(OMe)_2$	50	-	165	15	21300	1.79	99
2	$Bu_2Sn(OMe)_2$	100	-	165	60	32800	1.89	96
3	$Bu_2Sn(OMe)_2$	250	-	165	360	22000	1.91	76
4	$Bu_2Sn(OMe)_2$	50	-	145	60	23900	1.81	98
5	$Bu_2Sn(OMe)_2$	50	DMF	100	24	8700	1.62	70
6	$Bu_2Sn(OMe)_2$	50	DMF	120	2	15700	2.10	91
7	$Bu_2Sn(OMe)_2$	50	NMP	100	24	7900	1.60	77
8	$Ti(OBu)_4$	50	DMF	100	0.5	11000	1.69	95

[a] determined by means of GPC in DMAc; [b] determined by ^1H NMR-spectroscopy in TFA-d; [c] $[M]_0 = 3.66$ mol/L.

In the melt with $Bu_2Sn(OMe)_2$ as the initiator the polymer yields obtained at $[M]_0/[I]_0 = 50$ and 100 are > 95 %, at lower initiator concentration ($[M]_0/[I]_0 = 250$) the yields achieve a value of 76% after 6 h. With increasing the $[M]_0/[I]_0$ ratio the molecular weight increases indicating a controlled polymerization, however, the polydispersities of the polymers are relatively high. This is explained by transesterification reactions which eventually lead to a most probable chain length distribution without changing the microstructure of the alternating poly(ester amide). In solution with $Bu_2Sn(OMe)_2$ as the initiator at short reaction times the polymer yields are higher than at longer reaction times. This is interpreted in the sense that in the kinetically controlled regime of the reaction high molecular weight polymer is produced and in the thermodynamic controlled regime due to the monomer - polymer equilibrium cyclic oligomers are produced.

The influence of $[M]_0/[I]_0$ ratio, temperature, and initiator on the course of the reaction was studied by selective kinetic measurements. All poly(ester amide)s (PEA) prepared show a unimodal molecular weight distribution in GPC analyses (Figure 2). The GPC of the PEA prepared in solution clearly shows a single well resolved oligomeric series. The ^1H NMR spectrum shows characteristic resonance lines for an alternating poly(ester amide); the chemical shifts of the characteristic groups adjacent to the functional groups are clearly shifted to higher fields as

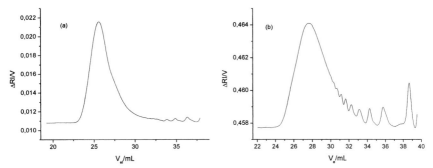

Figure 2. GPC of poly(ester amide)s **4** in DMAc / 2.44 g·L^{-1} LiCl: (a) Prepared in melt at T = 165 °C with Bu$_2$Sn(OMe)$_2$ as the initiator, [M]$_0$/[I]$_0$ = 100 (M_n = 32800, table 1, no. 2). (b) Prepared in DMF solution at T = 100 °C with Ti(OBu)$_4$ as the initiator [M]$_0$/[I]$_0$ = 50 (M_n = 11000, table 1, no. 8).

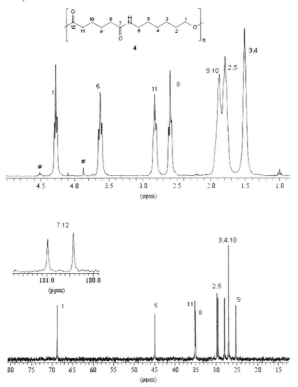

Figure 3. ^{1}H and ^{13}C NMR spectrum of poly(ester amide) **4** in TFA-d (# = end group).

compared to the chemical shifts of the cyclic monomer (Figure 3).

The existence of only one signal in the ^{13}C NMR spectrum for each carbon is indicative for the regular alternating microstructure of the poly(ester amide) **4**. The signals of small intensity in the ^{1}H NMR spectrum at δ = 3.88 ppm and δ = 4.55 ppm were assigned to -COO*CH$_3$* and -*CH$_2$*OCOCF$_3$ end groups.

The poly(ester amide) **4** is a semicrystalline material with a melting point of 140 °C. The endotherm at the melting transition may show two peaks or one peak with a shoulder, indicating the existence of two crystalline modifications. The thermal stability of poly(ester amide) **4** was determined by thermogravimetric analysis: a 5 % mass loss is observed at 278 °C, a 50 % mass loss is observed at 387 °C, and a 90 % mass loss at 432 °C.

Kinetic Aspects. For the kinetic analysis at selected reaction times samples were analyzed; by means of ^{1}H NMR spectroscopy the monomer conversion and by means of GPC the number average molecular weight were determined. The monomer conversion was obtained by comparison of the integrals of the CH$_2$O group of the monomer and the polymer. The influence of different parameters on the kinetics of the ROP of cEA **5** were investigated: The influence of the monomer-to-initiator ratio was studied for $[M]_0/[I]_0$ = 50, 100 and 250 at 165 °C in the melt with Bu$_2$Sn(OMe)$_2$ as the initiator. Polymerization in solution of DMF and NMP was compared with the polymerization in bulk. The influence of the temperature on the polymerization in DMF solution was studied with Bu$_2$Sn(OMe)$_2$ as the initiator. The polymerization rates of the ROP with Bu$_2$Sn(OMe)$_2$ and Ti(OBu)$_4$ as initiators in DMF solution at 100 °C were compared. Finally a comparison of the ROP of the fourteen-membered cEA **5** 1-oxa-8-aza-cyclotetradecane-9,14-dione - with the thirteen-membered cEA **6** - 1-oxa-7-aza-cyclotridecane-8,13-dione - in DMF solution was performed.

Ring-opening polymerization in bulk. Influence of the monomer/initiator ratio. For a controlled polymerization it is expected that with decreasing initiator concentration the degree of polymerization (P_n) increases according to: $P_n = [M]_0 \cdot X_p /[I]_0$, where $[M]_0$ is the initial monomer concentration, $[I]_0$ the initial initiator concentration and X_p the conversion. The first-order plots for the polymerization of cEA **5** in the melt with Bu$_2$Sn(OMe)$_2$ as the initiator are linear for ($[M]_0/[I]_0$ = 50 and 100 up to high conversion (Figure 4a). For $[M]_0/[I]_0$ = 250 the plot is linear up to 120 min, however the conversion increases slowly. From the slope of the straight lines the apparent rate constants were determined to be $k[P^*] = 77.7 \cdot 10^{-4}$ s^{-1} for $[M]_0/[I]_0 = 50$, $24.1 \cdot 10^{-4}$ s^{-1} for $[M]_0/[I]_0 = 100$, and $0.7 \cdot 10^{-4}$ s^{-1} for $[M]_0/[I]_0 = 250$. These results indicate that with decreasing the initiator concentration the efficiency of the initiator decreases. In order to obtain information on transfer reactions the dependence of M_n on conversion was studied (Figure 4b). The values obtained by GPC using PS standards are not absolute values; however, a linear dependence of M_n on conversion is a proof of the absence of transfer reactions. For $[M]_0/[I]_0 = 50$ and 100 we observe a linear dependence up to quantitative conversion. For $[M]_0/[I]_0 = 250$ where the concentration of active species is lowest, deviations form linearity are observed starting with a conversion of 30 %. We have excluded the possibility of a thermal polymerization since at 165 °C the cyclic monomer is stable for at least 3 h. It should be mentioned that the M_n vs. conversion plots do not pass through the origin which is tentatively explained by the method of determination of M_n by GPC using polystyrene standards and the relative high molecular weight of the monomer/initiator adduct.

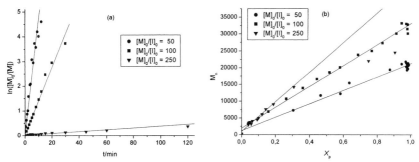

Figure 4. ROP of cEA **5** in bulk for different monomer-to-initiator ratios ($T = 165$ °C, initiator Bu$_2$Sn(OMe)$_2$): (a) First-order plot; (b) M_n vs. conversion.

Ring-opening polymerization in solution. (i) Influence of the solvent. The poly(ester amide) **4** is soluble only in aprotic dipolar solvents. We have performed the polymerization of cEA **5** in DMF and in NMP as a solvent with Bu$_2$Sn(OMe)$_2$ under identical conditions: $T = 100$ °C, $[M]_0 = 3.66$ mol/L and $[M]_0/[I]_0 = 50$. First-order plots show that in DMF as the solvent the conversion increases faster than in NMP as the solvent up to a conversion of 50 % (Figure 5a). The molecular weight increases linear up to this conversion, too (Figure 5b). Both solvents are not really inert solvents. At longer reaction times and decreasing monomer concentration transfer to solvent is to be expected.

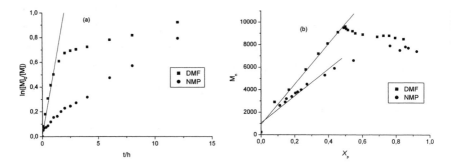

Figure 5. ROP of cEA **5** in solution for different solvents. (initiator: $Bu_2Sn(OMe)_2$, $[M]_0/[I]_0 = 50$, $T = 100$ °C, $[M]_0 = 3.66$ mol·L^{-1}). (a) First-order plot; (b) M_n vs. conversion.

(ii) Influence of the temperature. The influence of temperature on the ROP of the cEA **5** was studied for the polymerization in DMF solution and $Bu_2Sn(OMe)_2$ as initiator at a monomer-to-initiator ratio of $[M]_0/[I]_0 = 50$ for temperatures of 100 °C and 120 °C. For both temperatures the conversion increases rapidly within the first two hours: for $T = 120$ °C after 2h a conversion of 95 % is reached, for $T = 100$ °C the conversion increases fast to ca. 50 % within the first two hours.

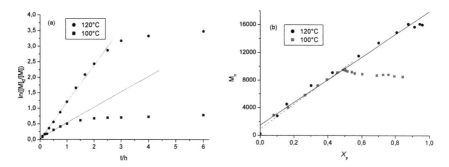

Figure 6. ROP of cEA **5** in DMF-solution at different temperatures. (initiator: $Bu_2Sn(OMe)_2$, $[M]_0/[I]_0 = 50$, $[M]_0 = 3.66$ mol·L^{-1}). (a) First-order plot; (b) M_n vs. conversion.

A first-order plot (Figure 6a) shows a linear dependence at $T = 120\,°C$ up to high conversion and at $T = 100\,°C$ only up to a conversion of 50 %. The conclusion from this result is that with increasing temperature the rate of polymerization increases faster than the rate of adventitious termination caused by the solvent. From the slope of the straight lines the apparent rate constants were determined to be $k[P^*] = 1.5 \cdot 10^{-4}\ s^{-1}$ (at 100 °C) and $3.2 \cdot 10^{-4}\ s^{-1}$ (at 120 °C). The dependence of the number average-molecular weight on conversion is linear up to 95 % conversion at 120 °C and up to 50 % conversion for 100 °C (Figure 6b). This means that at 100°C transfer reactions are observed starting with a conversion of 50 %. At 120 °C there is no indication of transfer reactions. In conclusion, the polymerization of cEA **5** in DMF solution with $Bu_2Sn(OMe)_2$ as the catalyst at 120 °C is a controlled polymerization since the first-order plot is linear and the molecular weight increases linearly up to high conversion. No transfer or termination reactions are observed. At higher temperatures the rate of polymerization is faster than that of transfer and termination reactions.

(iii) Influence of the initiator. The influence of the two initiators $Bu_2Sn(OMe)_2$ and $Ti(OBu)_4$ on the polymerization of cEA **5** was investigated at 100 °C in DMF solution. Within two hours with $Ti(OBu)_4$ as the catalyst complete conversion is obtained while with $Bu_2Sn(OMe)_2$ the conversion is only 50 %. A first-order kinetic plot (Figure 7a) for the linear part reveals an apparent rate constant of $k[P^*] = 22.2 \cdot 10^{-4}\ s^{-1}$ for $Ti(OBu)_4$ and $k[P^*] = 1.5 \cdot 10^{-4}\ s^{-1}$ for $Bu_2Sn(OMe)_2$ as initiator.

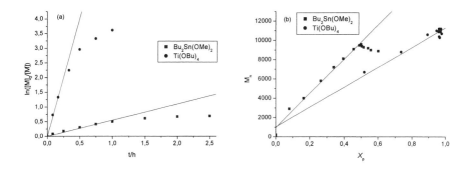

Figure 7. ROP of cEA **5** in DMF-solution with $Bu_2Sn(OMe)_2$ and $Ti(OBu)_4$ as the initiators. ($[M]_0/[I]_0 = 50$, $T = 100\,°C$, $[M]_0 = 3.66\ mol \cdot L^{-1}$). (a) First- order plot; (b) M_n vs. conversion.

The dependence of the molecular weight on conversion is linear up to high conversions for Ti(OBu)$_4$ and up to 50 % conversion for Bu$_2$Sn(OMe)$_2$ as initiator (Figure 7b). At equal conversion the molecular weight obtained with Ti(OBu)$_4$ is lower by a factor of about two. This is expected since the number of chains initiated by Ti(OBu)$_4$ is two times higher.

Polymerization of cyclic ester amides with fourteen- and thirteen-membered rings. In a previous paper[18] we have studied the ring-opening polymerization of cEA **6** in the melt and found for Bu$_2$Sn(OMe)$_2$ as initiator ([M]$_0$/[I]$_0$ = 100) at 160 °C an apparent rate constant of $k[P^*]$ = 20.4 $\cdot 10^{-4}$ s^{-1} which is very close to the value obtained for the cEA **5** at 165 °C, $k[P^*]$ = 24.1$\cdot 10^{-4}$ s^{-1}. In solution we have studied the polymerization of the two monomers at 100 °C in DMF as solvent. Both the first-order plots (Figure 8a) and the M_n vs. conversion plots (Figure 8b) are identical within experimental error.

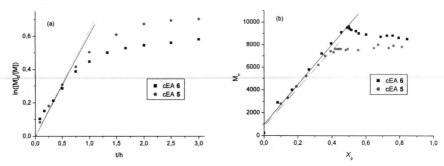

Figure 8. ROP of cEA **5** and **6** in DMF-solution. (initiator: Bu$_2$Sn(OMe)$_2$ [M]$_0$/[I]$_0$ = 50, T = 100 °C, [M]$_0$ = 3.66 mol·L^{-1}). (a) First-order plot; (b) M_n vs. conversion.

Synthesis and characterization of block copolymers. A-B and B-A-B block copolymers with A the poly(ethylene oxide) block and B the poly(ester amide) block have been prepared before by polymerization of cyclic ester amide **6** with an initiating system comprising hydroxytelechelic poly(ethylene oxide) and Sn(octoate)$_2$.[19] For the synthesis of A-B and B-A-B block copolymers comprising an alternating poly(ester amide) block with adipic acid and 1-amino-6-hexanol repeating units and a poly(ethylene oxide) block (cf. below) the cyclic ester amide **5** was polymerized with monohydroxy-functional poly(ethylene oxide)s (MPEO) and bishydroxy-functional poly(ethylene oxide)s (PEO) of various molecular weights in combination with Sn(octoate)$_2$ as catalyst ([Sn(octoate)$_2$]/[OH] = 1). In order to activate all hydroxyl groups and to initiate all chains at the same time poly(ethylene oxide) was first treated with Sn(octoate)$_2$ to generate the active species, then the monomer was added ([cEA]$_0$/[OH]$_0$ = 75) and the mixture was heated to 165 °C for polymerization. For termination and purification the mixture was cooled to room temperature dissolved in DMF and precipitated in diethyl ether (Table 2).

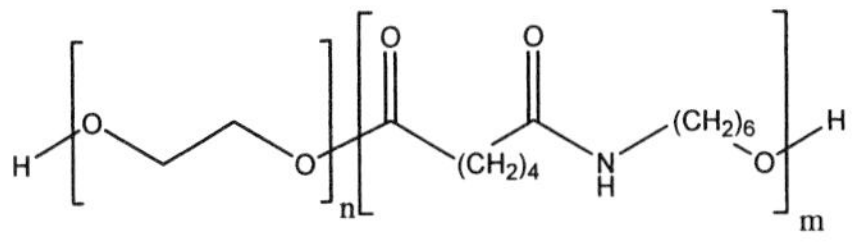

poly(ethylene oxide)-*block*-poly(ester amide)

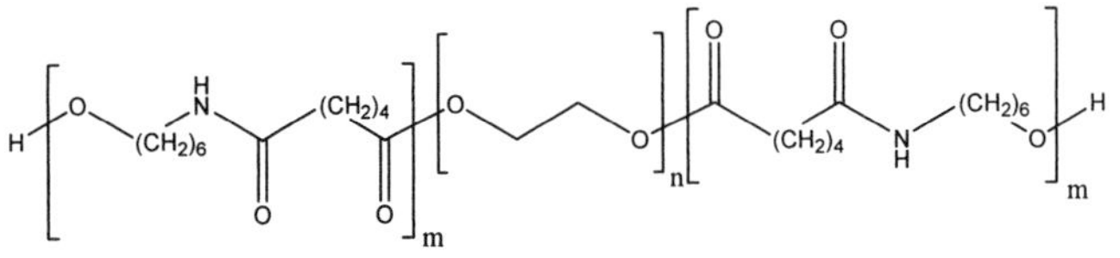

poly(ester amide)-*block*-poly(ethylene oxide)-*block*-poly(ester amide)

Table 2. Reaction conditions and results for the ROP of cEA **5** with poly(ethylene oxide) macroinitiators and Sn(octoate)$_2$ as catalyst in melt at T = 165 °C, t = 48 h, [cEA]/[OH] = 75, [Sn(octoate)$_2$]/[OH] = 1

No.	PEO	X_p of cEA in %[a]	PEO in wt.%[b]	PEO in wt.%[c]	M_n[d] (M_w/M_n)	Yield in %[e]
1	MPEO 2000	99	11	8	20600 (1.85)	94
2	MPEO 5000	95	23	11	18300 (1.76)	76
3	MPEO 10000	93	37	38	24100 (1.72)	65
4	PEO 2000	98	6	5	22400 (1.83)	95
5	PEO 6800	95	17	10	18500 (1.94)	83
6	PEO 11800	93	26	13	20600 (1.96)	74

[a] determined by means of ^{1}H NMR spectroscopy of the crude product; [b] percent by weight of PEO in the feed; [c] percent by weight of PEO in the block copolymer determined by means of ^{1}H NMR spectroscopy after purification of the product; [d] determined by means of GPC in DMAc; [e] determined by weight.

The conversion of the cEA **5** is higher than 90 % in all cases; however, the efficiency of the initiating PEO block decreases with increasing molecular weight of the macroinitiator. This is reflected by the decreasing block copolymer yield and the composition of the block copolymer. The composition of the purified copolymers was determined by ^{1}H NMR spectroscopy (Figure 9). Beside the resonances for the poly(ester amide) blocks assigned before, the singlet at $\delta = 3.56$ ppm is assigned to the ethylene oxy repeating units of the poly(ethylene oxide) block.

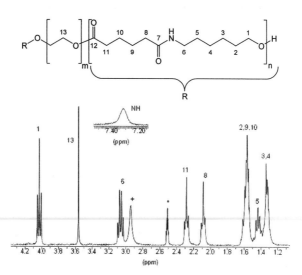

Figure 9. ^{1}H NMR spectrum of the B-A-B triblock copolymer (table 2, no. 4) (DMSO-d_6/100°C).

The GPC analysis of the block copolymers (Figure 10) shows a monomodal distribution of the molecular weight with no residual macroinitiator in the purified product.

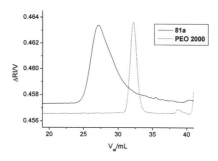

Figure 10. GPC elution curve of the B-A-B triblock copolymer (table 2, no. 4) and of the corresponding macroinitiator PEO 2000 in DMAc / 2.44 g·L^{-1} LiCl.

The thermal characteristics of these block copolymers as determined by DSC measurements reveal a phase separated system with two melting points. For the block copolymer - entry 5 of table 2 - the poly(ester amide) domains show a melting point of $T_m = 136$ °C with a melting enthalpy of $\Delta H_m = 42.5$ J/g which is very close to the melting point of the homopolymer ($T_m = 140$ °C, $\Delta H_m = 53.6$ J/g), and in addition show a melting point of 60 °C for the poly(ethylene oxide) domains.

Conclusions

Ring-opening polymerization of cEA **5** occurs with high rate up to high conversion above the melting point of the monomer. In solution of aprotic dipolar solvents transfer and termination reactions can not completely be excluded. However, DMF should be favoured over NMP, higher temperatures (120 °C) should be preferred over lower temperatures (100 °C) and Ti(OBu)$_4$ should be given preference compared to Bu$_2$Sn(OMe)$_2$ as initiator. Further, the kinetic data show that the polymerization in bulk and under certain conditions in solution is controlled. As a consequence block copolymers were prepared.

Acknowledgement

The financial support of the *Deutsche Forschungsgemeinschaft* (HO 772/34-1 and 2) is gratefully acknowledged.

324

[1] M. M Shemyakin, A. S. Khohlov, *Die Chemie der Antibiotika*, Moskau 1953.
[2] J. Helder, F. E. Kohn, S. Sato, J. W. A. van den Berg, J. Feijen, *Makromol. Chem., Rapid Commun.* **1985**, *6*, 9.
[3] N. Yonezawa, F. Toda, M. Hasegawa, *Makromol. Chem., Rapid Commun.* **1985**, *6*, 607.
[4] P. J. A. in't Veld, P. J. Dijkstra, J. H. van Lochem, J. Feijen, *Makromol. Chem.* **1990**, *191*, 1813.
[5] Eur. Pat. Appl. EP 322154 (1989) Pfizer Inc., invs.: F. N. Fung, R. C. Glowaky.
[6] J. Helder, J. Feijen, S. J. Lee, S. W. Kim, *Makromol. Chem., Rapid Commun.* **1986**, *7*, 193.
[7] C. Samyn, M. van Beylen, *Makromol. Chem., Makromol. Symp.* **1988**, *19*, 225.
[8] T. Ouchi, M. Shiratani, M. Jinno, M. Hirao, Y. Ohya, *Makromol. Chem., Rapid Commun.* **1993**, *14*, 825.
[9] T. Ouchi, T. Nozaki, Y. Okamoto, M. Shiratani, Y. Ohya, *Macromol. Chem. Phys.* **1996**, *197*, 1823.
[10] V. Jörres, H. Keul, H. Höcker, *Macromol. Chem. Phys.* **1998**, *199*, 835.
[11] H. R. Kricheldorf, K. Hauser, *Macromol. Chem. Phys.* **2001**, *202*, 1219.
[12] Y. Feng, D. Klee, H. Keul, H. Höcker, *Macromol. Chem. Phys.* **2000**, *201*, 2670.
[13] (a) H. R. Kricheldorf, *Macromol. Symp.* **2000**, *153*, 55. (b) D. Mecerreyes, R. Jérôme, *Macromol. Chem. Phys.* **1999**, *200*, 2581.
[14] (a) J. M. Schakenraad, P. Nieurenkuis, I. Molenaar, J. Helder, P. J. Dijkstra, J. Feijen, *J. Biomed. Mater. Res.* **1989**, *23*, 1271. (b) J. Helder, P. J. Dijkstra, J. Feijen, *J. Biomed. Mater. Res.* **1990**, *24*, 1005.
[15] V. Jörres, *Dissertation*, RWTH Aachen **1997**.
[16] (a) Y. Feng, D. Klee, H. Höcker, *Macromol. Chem. Phys.* **1999**, *200*, 2276. (b) Y. Feng, D. Klee, H. Keul, H. Höcker *Macromol. Biosci.* **2001**, *1*, 30.
[17] (a) B. Robertz, H. Keul, H. Höcker, *Macromol. Chem. Phys.* **1999**, *200*, 1034. (b) B. Robertz, H. Keul, H. Höcker, *Macromol. Chem. Phys.* **1999**, *200*, 1041. (c) B. Robertz, H. Keul, H. Höcker, *Macromol. Chem. Phys.* **1999**, *200*, 2100.
[18] T. Fey, H. Keul, H. Höcker, *Macromolecules* **2003**, *36*, 3882.
[19] T. Fey, H. Keul, H. Höcker, *Macromol. Chem. Phys.* **2003**, *204*, 591.

Recent Developments in the Ring-Opening Polymerization of ε-Caprolactone and Derivatives Initiated by Tin(IV) Alkoxides

Ph. Lecomte, F. Stassin, R. Jérôme*

Center for Education and Research on Macromolecules (CERM), University of Liège, Sart-Tilman, B6a, 4000 Liège, Belgium
E-mail: rjerome@ulg.ac.be; http://www.ulg.ac.be/cerm

Summary: The macromolecular engineering of aliphatic polyesters by ring-opening polymerization (ROP) initiated by aluminum alkoxides is now well established. Tin (IV) alkoxides are less popular, mainly because of a poorer control of the chain growth. This paper discusses some recent examples from CERM, showing that tin (IV) alkoxides can advantageously replace the aluminum counterparts as ROP initiators. For instance, they can initiate successfully the Ring-Opening Polymerization of α-chloro-ε-caprolactone and dioxepane-2-one. They are also very promising initiators for ROP in supercritical CO_2 and for the synthesis of clay/aliphatic polyester nanocomposites.

Keywords : nanocomposites; polyesters; ring-opening polymerization; supercritical carbon dioxide; tin alkoxides

Introduction

Due to the remarkable properties of biodegradability and biocompatibility, aliphatic polyesters are very promising either as biomaterials or as environmentally friendly materials to address growing ecological concerns.[1,2] During the last decades, ring-opening polymerization (ROP) of lactones and lactides has proven to be a powerful tool to produce high molecular weight aliphatic polyesters. When ROP is initiated by suitable organometallic species, i.e., metal alkoxides with d-orbitals of a favourable energy, irreversible termination and transfer reactions are negligible, and ROP turns living. Many examples are known in which the polymer molecular weight is predetermined by the monomer to initiator molar ratio and the concentration in propagating species is constant. Moreover, the chemical structure of the chain-ends can be controlled, and the polymerization of a second monomer by a living polyester can be initiated with formation of block copolymer. Whenever the initiation step is fast compared to propagation, a low polydispersity index ($M_w/M_n \approx 1.1$) is reported. Nowadays, real macromolecular engineering is carried out as witnessed by a steadily increasing number of papers reporting on the synthesis of diblock, triblock, star-shaped, graft, hyperbranched and dendritic polyesters.[2]

Aluminum alkoxides are very popular ROP initiators because of the very efficient control of the molecular weight parameters.[2] The polymerization proceeds through the well-known "coordination-insertion" mechanism. Polydispersity is low, and transesterification reactions are not significant (kinetic control) until the quasi complete conversion of the monomer. Although tin (IV) alkoxides are easily available, their use has been limited for a long time, because of less efficiently controlled polymerization. In this respect, Penczek et al. reported that selectivity of propagation versus transesterification was lower with tin(IV) than with aluminum alkoxides.[3] Among possible reasons, Kricheldorf et al. proposed slow initiation, possible equilibrium between aggregated/unaggregated species and occurrence of transesterification reactions.[4]

This paper reports on particular examples in which tin(IV) alkoxides are more attractive than their aluminum counterparts, despite a poorer control. The first part is devoted to the polymerization of ε-caprolactone (εCL) derivatives substituted by functional groups. ROP in supercritical carbon dioxide is the topic of the second part. Finally, special attention is paid to the production of PCL nanocomposites, by polymerization in the presence of nanofillers, in both organic media and supercritical CO_2.

Synthesis of poly(ε-caprolactone)s with pendent functional groups

There are two major reasons for attaching functional groups to polyester chains. Firstly, it is highly desirable to tune important properties, such as crystallinity, hydrophilicity, biodegradation rate, bioadhesion, etc. Moreover, functional groups may be used to attach active molecules, e.g., drug, recognition agent, adhesion promoter, probe, etc. Synthesis and polymerization of lactones substituted by a functional group is a straightforward route to new aliphatic polyesters. Figure 1 shows a series of substituted-ε-caprolactones that have been successfully synthesized and polymerized by aluminum alkoxides.[5-22]

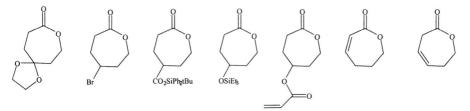

Figure 1. Substituted ε-caprolactones polymerizable by Al(OiPr)$_3$

Synthesis of copolymers of oxepane-1,5-dione and ε-caprolactone

Poly(oxepane-1,5-dione) (POPD) is a highly crystalline polymer, which exhibits a high T_m (147°C) and a T_g of 37°C.[23,24] Both PCL and POPD have an orthorombic unit cell. Remarkably, poly(OPD-co-εCL) random copolymers show a linear increase of T_m with the molar content of OPD units, as a consequence of the isomorphism of PCL and POPD (Figure 2). To the best of our knowledge, this behavior has never been reported before for aliphatic polyesters. It is remarkable that only 30 mol% oxepane-1,5-dione (OPD) copolymerized with εCL is sufficient to increase T_m from 60°C up to 90°C, which opens up new applications, e.g., in the packaging field. The hydrolytic and thermal behavior of these copolymers has been reported elsewhere.[25]

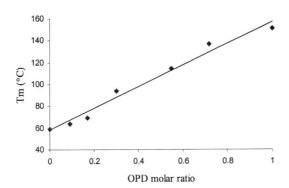

Figure 2. Dependence of the T_m of poly(OPD-co-εCL) with the OPD molar ratio

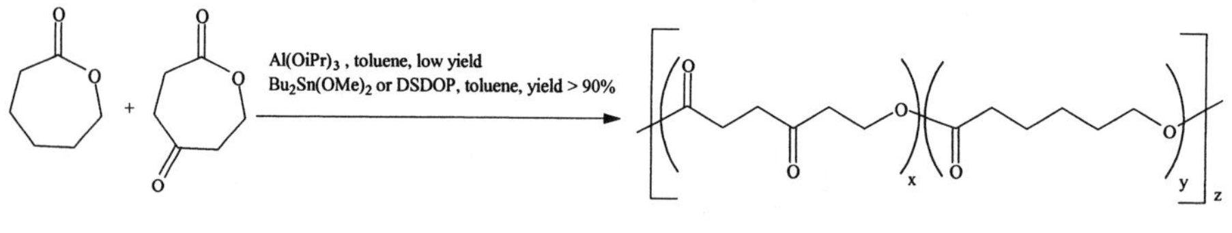

Figure 3. Copolymerization of OPD and εCL into poly(OPD-co-εCL)

Aluminum alkoxides are not proper initiators for copolymerization of OPD and εCL, because they do not tolerate the ketone group (Figure 3).[24] Table 1 shows indeed that the reaction of εCL with aluminum isopropoxide is perturbed by cylcohexanone as soon as the [cyclohexanone]$_0$ / [Al(OiPr)$_3$]$_0$ molar ratio is 20 and higher, as assessed by exceedingly high M_n and polydispersity. As a rule, the yield of the OPD/εCL copolymerization is very low when Al(OiPr)$_3$ is used as an initiator (Figure 3, Table 2). The reason for this disappointing observation is the preferential complexation of Al to the carbonyl group of the ketone rather than to the one of the lactone, which is detrimental to ROP of the monomer.

Table 1. Influence of the [cyclohexanone]$_0$ / [Al(OiPr)$_3$]$_0$ molar ratio on the εCL ROP

Entry	[Cyclohexanone]$_0$ / [Al(OiPr)$_3$]$_0$ [a]	Yield (%)	M_n [b]	M_w / M_n
1	0	86	10.5 K	1.3
2	2	88	10 K	1.3
3	20	80	18 K	1.3
4	200	45	400	2.8

(a) conditions : [εCL] = 0.9M, toluene, room temperature, M_n,th at complete conversion = 12 K.
(b) PMMA calibration.

Table 2. Bulk copolymerization of εCL and OPD ($f_{OPD}=0.3$)

Entry	Initiator	$[M]_0/[In]_0$	f_{OPD}	Time	Yield	F_{OPD}	M_n (SEC)[c]	M_w/M_n
1	Al(OiPr)$_3$[a]	400	0.05	24 h	4 %	-	-	-
2	Al(OiPr)$_3$[a]	400	0.05	1 week	38 %	0.01	7 K	1.3
3	Bu$_2$Sn(OMe)$_2$[b]	300	0.3	3 min	97 %	0.27	21 K	2.1
4	Bu$_2$Sn(OMe)$_2$[b]	600	0.3	5 min	95 %	0.31	39 K	1.9
5	DSDOP[b]	300	0.3	3 min	92 %	0.31	40 K	1.6
6	DSDOP[b]	600	0.3	5 min	94 %	0.31	65 K	1.6

(a) conditions: toluene, 90°C, [M]=2M. (b) bulk, 110°C. (c) PMMA calibration

Substitution of dibutyltin dimethoxide for aluminum isopropoxide is an efficient strategy to solve this problem. Table 3 shows that cyclohexanone no longer has a detrimental effect on the εCL polymerization. High yields of poly(OPD-co-CL) are obtained when ROP is initiated by tin(IV) alkoxides, such as dibutyltin dimethoxide and 2,2-dibutyl-2-stanna-1,3-dioxepane (DSDOP) (Table 2).

Table 3. Influence of the [cyclohexanone]$_0$ / [Bu$_2$Sn(OMe)$_2$]$_0$ molar ratio on the ROP of εCL

Entry	[Cyclohexanone]$_0$ / [Bu$_2$Sn(OMe)$_2$]$_0$[a]	Yield (%)	M_n[b]	M_w/M_n
1	0	93	20 K	1.5
2	30	93	18 K	1.4
3	100	92	19 K	1.6
4	500	94	18 K	1.4

(a) conditions : [εCL] = 0.9M, toluene, room temperature, M_n,th at complete conversion = 22 K.
(b) PMMA calibration

Synthesis of copolymers of α-chloro-ε-caprolactone

DSDOP proved to be a very efficient initiator for ROP of αClεCL.[26] Table 4 shows that molecular weight is predetermined by the monomer to initiator molar ratio. The polydispersity (ca. 1.2) is lower than that observed for the εCL polymerization, whereas the electrowithdrawing chloro group increases the polymerization rate of αClϵCL compared to ϵCL. Although, block and random copolymers of ϵCL and αClϵCL can be synthesized with DSDOP, these reactions are a failure in the presence of aluminum isopropoxide.

Table 4. Ring-Opening Polymerization of αClϵCL inititiated by DSDOP

Entry	[αClϵCL]$_0$ / [DSDOP]$_0$ [a]	Time	Yield	M_n,th	M_n (SEC) [b]	M_w / M_n
1	16.9	90 min	100 %	2500	2500	1.4
2	33.8	90 min	100 %	5000	6000	1.3
3	67.6	10 min	89%	9000	8000	1.2

(a) Conditions: toluene, room temperature. (b) polystyrene calibration

Poly(ϵ-caprolactone) containing α-chlorinated units is very promising because of easy derivatization (Figure 4). Indeed, reaction of αClϵCL with 1,5-diazabicyclo(4.3.0)non-5-ene (DBN) results in an unsaturated lactone, which can thereafter be polymerized into unsaturated polyester. The α-chloro esters units are initiators for atom transfer radical polymerization (ATRP). So, the radical polymerization of MMA has been initiated by a random copolymer of ϵCL and αClϵCL with the traditional CuCl/CuCl$_2$/HMTETA ATRP catalyst. Moreover, reaction of benzoyl-3-butenyl with poly(ϵCL-co-αClϵCL) (molar fraction of αClϵCL ($F_{\alpha Cl\epsilon CL}$) = 0.5) at 60°C in N,N-dimethylformamide (DMF) in the presence of CuBr and tris[2-(dimethylamino)ethyl]amine (Me$_6$TREN), results in the quantitative insertion of olefin by atom transfer radical addition (ATRA). This strategy is very versatile because many functional groups can be attached to the same poly(ϵCL-co-αClϵCL) chains, merely by changing the structure of the olefin involved in the ATRA process. PCL's with pendent alcohol, carboxylic acid, amine and epoxide groups have been prepared, as will be reported in a forthcoming paper. This synthetic pathway to α-substituted PCL's is very straightforward because only three steps from commercially available materials are required, i.e., the Baeyer-Villiger oxidation of α-chlorocyclohexanone, ring-opening polymerization of αClϵCL and finally ATRA of functional olefins (Figure 4).

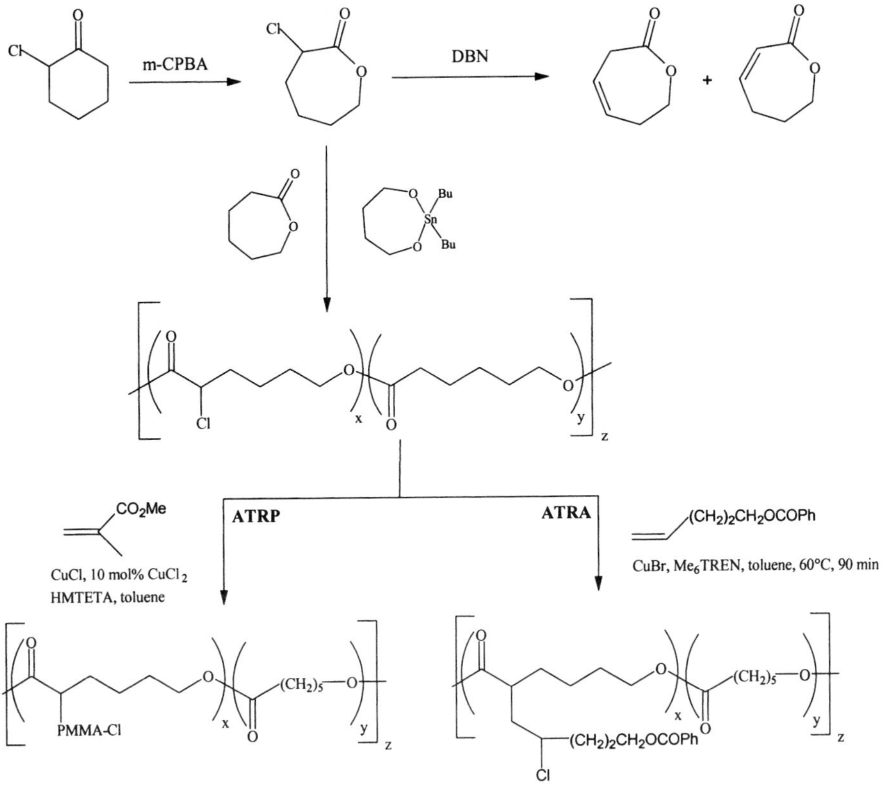

Figure 4. Synthesis and chemical transformation of poly(εCL-*co*-αClεCL) by ATRA and ATRP

Ring-Opening Polymerization of cyclic (di)esters in Supercritical CO_2

Environmental concerns have recently prompted the scientific community to investigate new ways to decrease the use of volatile organic solvents. For this purpose, the use of supercritical fluids appears to be a valuable alternative. Indeed, supercritical fluids combine gas-like and liquid-like properties (solvation power and density). Amongst available supercritical fluids, supercritical carbon dioxide is the best candidate because of low toxicity, low cost, non flammability and easily accessible critical parameters ($T_c = 31.4°C$, $P_c = 73.8$ bar). Moreover, CO_2 is widely available from commercial and industrial supplies and is also easily recycled.

Tin(IV) alkoxides are very efficient initiators for ROP of εCL in supercritical CO_2 (40°C and 210-215 bar).[27] The ^{1}H NMR analysis of the chain ends has confirmed that ROP proceeds through the coordination-insertion mechanism that commonly operates in organic solvents. When polymerization starts, the medium is transparent because the monomer and the very short PCL oligomers are soluble. After a few minutes, the medium turns cloudy as result of the precipitation of the non-soluble PCL chains. Nevertheless, the experimental molecular weight increases regularly with conversion and is predetermined by the monomer-to-initiator molar ratio, at least until 20000 g/mol (Figure 5), on the assumption that the two alkoxides are active. The first order in εCL is observed, and the first order in initiator indicates that tin species are mostly unaggregated in supercritical CO_2. This behavior is different from ROP in toluene, because then the kinetic order in initiator changes with concentration, a value of 1.3 being found at low initiator concentration. Penczek et al. proposed that a kinetic order higher than 1 would be the signature of an equilibrium between more reactive aggregated species and less reactive unaggregated species.[3] This observation is unexpected because, as far as aluminum alkoxides are concerned, unaggregated species are less reactive, which results in a kinetic order lower than 1.

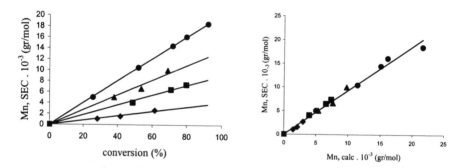

Figure 5. Dependence of M_n (SEC) on the monomer conversion and on theoretical M_n for the εCL ROP initiated by $Bu_2Sn(OMe)_2$ in supercritical CO_2. [εCL]$_0$= 1.39M, [εCL]$_0$ / [Sn]$_0$ = 364 (square), 254 (diamond), 167 (triangle), and 88 (circle)

For the sake of comparison, the apparent rate constants for ROP of εCL have been measured in different media: k_{app} is 56×10^{-3} min^{-1} in toluene, 130×10^{-3} min^{-1} in bulk, 15×10^{-3} min^{-1} in CFC-113 and 3.95×10^{-3} min^{-1} in supercritical CO_2. Thus ROP at 40°C proceeds ca. 14 times faster in toluene and 33 times faster in bulk than in supercritical CO_2. This very slow kinetics has been explained by an equilibrium between propagating species and dormant species. The reversible insertion of CO_2 into the Sn-O bond leads to a carbonated tin compound, as shown in Figure 6.[27] This mechanism has been substantiated by spectroscopy and the activation parameters for the reaction.[28]

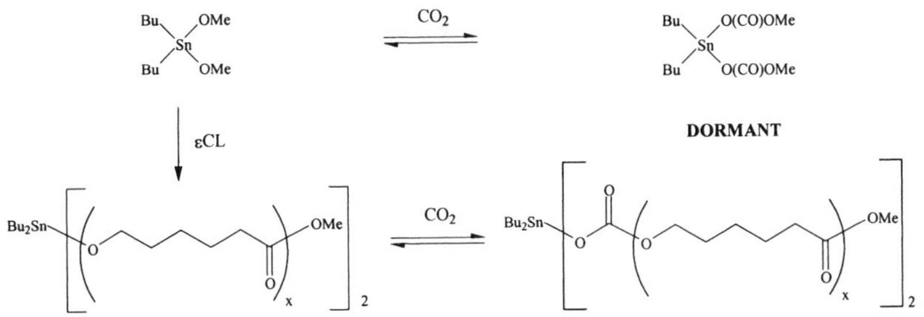

Figure 6. Reversible carbonation of propagating tin (IV) alkoxide by CO_2

When aluminum triisopropoxide is the initiator, ROP of εCL is a failure, as testified by formation of only a few percents of oligomers, as result of the high reactivity of aluminum alkoxide towards CO_2.

It is highly desirable to prepare polyesters free of tin residues in view of biomedical applications. After polymerization, tin remains chemically bonded to the polymer as an ω-alkoxytin moeity and has to be removed from the final polymer. Unreacted monomer has also to be extracted in order to provide polymer with high purity. Purification of PCL by sc CO_2 has been investigated. In a first step, εCL dissolved in the polymerization medium has been extracted by supercritical CO_2. For this purpose, the phase diagram was set up by the cloud point method. εCL exhibits a LCST/UCSP behavior typical of many supercritical CO_2 / solute systems. For instance, 20 v/v % of εCL are dissolved in supercritical CO_2 at 40 °C under a pressure of ca.100 bar, while the solubilization pressure is approximately 135 bar at 50 °C. Figure 7 shows the first-order kinetic profile for the supercritical fluid extraction (SFE) of εCL from a PCL sample containing 15 wt % of monomer. Based on the extraction constant, 95 % of εCL is extracted after ca. 110 minutes, while 99 % of extraction would require ca. 175 minutes. The extraction of tin from PCL is a more difficult task because PCL-bound tin alkoxide has first to be derivatized into species soluble in supercritical CO_2. The strategy consists of reaction of the PCL-alkoxytin end-group with acetic acid and release of dibutyltin diacetate which is extractable by supercritical CO_2. The extraction of tin has been monitored by thermogravimetry and quantified by UV-visible absorption spectrometry. The kinetic profile is quasi-linear (Figure 8) and the slope of the straight line allows the extraction constant to be determined. Extraction of tin is 2.5 times lower than extraction of εCL out of PCL.

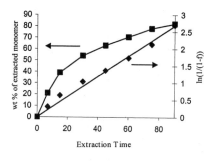 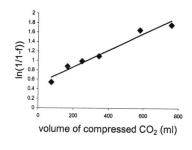

Figure 7. Kinetic profile for the extraction of CL from PCL at 40 °C and 150 bar, at a flow rate of 6 ml/min

Figure 8. Kinetic profile for the extraction of dibutyltin diacetate from PCL at 40 °C and 150 bar, at a flow rate of ca. 4 ml/min

ROP of εCL in supercritical CO_2 has also been carried out in the presence of various poly(tetrahydroperfuorodecylacrylate-b-caprolactone) diblock copolymers. These surfactant contain one CO_2-philic block (PAC8) responsible for steric stabilization and one CO_2-phobic block (PCL) responsible for anchoring to the growing PCL particles. Microspheres have accordingly been prepared with [PCL(20K)-b-PAC8(40K)] surfactant. (conditions: 10 vol% CL, M_n,th = 20 k ; 5 wt% surfactant, 40°C, 300 bar, 400 rpm, 15h) as illustrated in Figure 9.

Figure 9. Preparation by of microspheres dispersion ROP of εCL in supercritical CO_2

Synthesis of nanocomposites

Dispersion of nanoclay sheets into a polymeric matrix is an efficient tool to improve properties, such as thermal stability, mechanical strength, permeability to gases and moisture and enhancement of flame retardancy as result of charring effect. PCL/clay nanocomposites are of special interest not only because of the biocompatibility and biodegradability of PCL but also because of the PCL miscibility with other polymers, including PVC.

A first preparation method for PCL/clay nanocomposites relies on the melt blending of PCL and clay, i.e. natural montmorillonite (MMT) or MMT surface modified by quaternary ammonium salts.[29,30] Moreover, PCL/clay nanocomposites have been prepared by the "in situ intercalative polymerization" process.[31,32] Bulk εCL polymerization has been promoted by dibutyltin dimethoxide in the presence of MMT either native or MMT surface modified by dimethyl 2-ethylhexyl and methyl bis(2-hydroxyethyl) ammonium cations (MMT-C_8H_{17} and MMT-$(CH_2CH_2OH)_2$).[33] The polyester chains are then growing within the clay galleries. The targeted content of filler ranged from 1 wt% up to 10 wt%. In the presence of native MMT, intercalated structures are observed in contrast to exfoliated structures, which are formed when the surface of MMT is modified by cations bearing hydroxyl groups (Figure 10). The hydroxyl groups available at the surface of MMT-$(CH_2CH_2OH)_2$ are reacted with $Bu_2Sn(OMe)_2$ and converted into tin alkoxides. As a rule, M_n decreases upon increasing the MMT-$(CH_2CH_2OH)_2$ content (Table 5, entries 2 to 4), such that the clay to monomer molar ratio allows the molecular weight to be predicted. In parallel, the polydisperisty index decreases from 1.95 down to 1.55 when the MMT-$(CH_2CH_2OH)_2$) content is increased from 1 wt% up to 10 wt% because the propagating chains are confined within the clay interlayer, which is thought to impose limitation to the mobility of the tethered chains, which are less prone to take part to transesterification reactions (Table 5). In contrast, M_n is independent of the MMT-C_8H_{17} content (Table 5, entries 5 and 6). The intercalation process has been extended to PCL nanocomposites with a high clay content (typically in the 25-50 wt% range) in order to prepare masterbatches, which can thereafter be dispersed into molten PCL or PVC, so that the final clay content is lower than 10 wt%.[34]

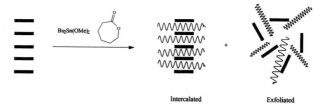

Intercalated   Exfoliated

Figure 10. Intercalation/exfoliation of clays by ROP of εCL

Whenever the "in situ" intercalative polymerization is carried out in the bulk, the recovery of the material is a problem because of an exceedingly high melt viscosity. In order to tackle the problem, the polymerization in the presence of MMT has been conducted in supercritical CO_2.[32] The nanocomposite is then collected as a coarse powder. Table 6 shows that the molecular weight is in line with the theoretical value, except for the entry 2. In this case, chain transfer reaction with hydroxyl groups located at the clay surface would have occurred, so accounting for a lower M_n and a higher polydispersity. Because of the solubility of εCL and the high dispersibility of the clay in supercritical CO_2, high clay loading can be considered at least up to 50wt%.

Table 5. Bulk polymerization of εCL in the presence of MMT

Entry	MMT [a]	Filler content (wt %)	$M_{n,th}$	M_n [b] (SEC)	M_w / M_n
1	-	0	17.1 K	21 K	2.05
2	MMT-$(CH_2CH_2OH)_2$	1	13.7 K	16 K	1.95
3	MMT-$(CH_2CH_2OH)_2$	3	11.2 K	13.5 K	1.80
4	MMT-$(CH_2CH_2OH)_2$	10	5.6 K	4.5 K	1.55
5	MMT-C_8H_{17}	1	17.1 K	14.5 K	2.05
6	MMT-C_8H_{17}	5	17.1 K	17.5 K	2.05

(a) conditions: $[CL]_0/[Bu_2Sn(OMe)_2]_0 = 300$, 24h, room temperature (b) universal calibration

Table 6. Polymerization of εCL in the presence of MMT in supercritical CO_2

Entry	$M_{n, th}$	M_n (SEC) [a]	Time	Conv	M_w / M_n
1 [a]	17.8 K	15.0 K	24 h	85 %	1.65
2 [b]	40.0 K	29.0 K	24 h	100 %	1.80
3 [c]	11.3 K	13.1 K	24 h	70 %	1.60
4 [d]	40.0 K	41.0 K	24 h	100 %	1.60

(a) 4 wt % (MMT-$(CH_2CH_2OH)_2$), 50°C, 190 bar (b) 3 wt % (MMT-$(CH_2CH_2OH)_2$), 50°C, 165 bar (c) 5 wt % MMT-C_8H_{17}, 50°C, 190 bar (d) 3 wt % MMT-C_8H_{17}, 50°C, 165 bar. (e) universal calibration

Conclusions

This paper has shown that tin alkoxides are attractive initiators for the Ring-Opening Polymerization of εCL and derivatives. The efficiency of tin alkoxides has been illustrated for polymerizations in which aluminum alkoxides completely failed. The Ring-Opening Polymerization of α-chloro-ε-caprolactone and dioxepane-2-one is a representative example. Although ROP of cyclic (di)esters in supercritical carbon dioxide cannot be promoted by aluminum alkoxides, tin alkoxides proved to be successful. They have also been used to prepare nanocomposites in the presence of montmorillonite modified by methyl bis(2-hydroxyethyl-ammonium cations.

Acknowledgments

The authors are indebted to the "Services Fédéraux des Affaires Scientifiques, Techniques et Culturelles" for general support to CERM in the frame of the "PAI 5-03: Supramolecular Chemistry and Supramolecular catalysis. Ph. L. is "Chercheur Qualifié" by the FNRS. F. S. is researcher fellow by the FNRS. R. J. thanks the Walloon Region (TECMAVER project).

338

[1] M. Vert, M., J.Feijen, A. C. Albertsson, G. Scott, E. Chiellini, *In Biodegradable Polymers and Plastics*, Royal Society, London, 1992.

[2] D. Mecereyes, R. Jérôme, Ph. Dubois, *Advances in Polymer Science* 1999, *147*,1.

[3] A. Kowalski, J. Libiszowski, J., A. Duda, S. Penczek, *Macromolecules* 2000, *33*, 1964.

[4] H. R. Kricheldorf, A. Stricker, D. Langanke, *Macromol. Chem. Phys.* 2001, *202*, 2525.

[5] C. Detrembleur, M. Mazza, O. Halleux, Ph. Lecomte, D. Mecerreyes, J. L. Hedrick, R. Jérôme, *Macromolecules* 2000, *33*, 14.

[6] D. Mecerreyes, M. Trollsas, J. L. Hedrick, *Macromolecules* 1999, *32*, 8753.

[7] M.Trollsås, M.; V. Y. Lee, D. Mecerreyes, P. Löwenhielm, M. Möller, R. D. Miller, J. L. Hedrick, *Macromolecules* 2000, *33*, 4619.

[8] Ph. Lecomte, V. D'aloia, M. Mazza, O. Halleux, S. Gautier, C. Detrembleur, R. Jérôme, *Polym. Preprint* 2000, *41 (2)*, 1534.

[9] C. G. Pitt, Z. W. Gu, P. Ingram, R. W. Hendren, *J. Polym. Sci, Polym. Chem.* 1987, *25*, 955.

[10] F. Stassin, O. Halleux, Ph. Dubois, C. Detrembleur, Ph. Lecomte, R. Jérôme, *Macromol. Symp.* 2000, *153*, 27.

[11] S. Gautier, V. d'Aloia, V.; O. Halleux, M. Mazza, Ph. Lecomte, R. Jérôme, *J. Biomater. Sci. Polymer Edn.* 2003, *14*, 63.

[12] M. Liu, N. Vladimirov, J. M. J. Fréchet, *Macromolecules* 1999, *32*, 6881.

[13] D. Tian, Ph. Dubois, Ch. Grandfils, R. Jérôme, *Macromolecules* 1997, *30*, 406.

[14] D. Tian, Ph. Dubois, R. Jérôme, *Macromol. Symp.* 1998, *130*, 217.

[15] D. Tian, Ph. Dubois, R. Jérôme, *Macromolecules* 1997, *30*, 2575.

[16] D. Tian, Ph. Dubois, R. Jérôme, *Macromolecules* 1997, *30*, 1947.

[17] D. Mecerreyes, R. D. Miller, J. L. Hedrick, Ch. Detrembleur, R. Jérôme, *J. Polym. Sci., Polym. Chem.* 2000, *38*, 870.

[18] D. Mecerreyes, J. Humes, R. D. Miller, J. L. Hedrick, Ph. Lecomte, Ch. Detrembleur, R. Jérôme, *Macromol. Rapid. Commun.* 2000, *21*, 779.

[19] X. Lou, Ch. Detrembleur, Ph. Lecomte, R. Jérôme, *Macromolecules* 2001, *34*, 5806.

[20] X. Lou, Ch. Detrembleur, Ph. Lecomte, R. Jérôme, *J. Polym. Sci., Polym. Chem.* 2002, *40*, 2286.

[21] X. Lou, Ch. Detrembleur, Ph. Lecomte, R. Jérôme, *e-Polymers* 2002, n°034.

[22] Ch. Detrembleur, M. Mazza, O. Halleux, Ph. Lecomte, D. Mecerreyes, J. L. Hedrick, R. Jérôme, *Macromolecules* 2000, *33*, 7751.

[23] J. P. Latere, Ph. Lecomte, Ph. Dubois, R. Jérôme, *Macromolecules* 2002, *35*, 7857.

[24] J. P. L. Dwan'Isa, Ph. Lecomte, Ph. Dubois, R. Jérôme, *Macromolecules* 2003, 36, 2609.

[25] J. P. L. Dwan'Isa, Ph. Lecomte, Ph. Dubois, R. Jérôme *Macromol. Chem. Phys.* 2003, 204, 1191.

[26] Ph. Lecomte, C. Detrembleur, R. Riva, S. Lenoir , R. Jérôme, under preparation

[27] F. Stassin, O. Halleux, R. Jérôme, *Macromolecules* 2001, *34*, 775.

[28] F. Stassin, R. Jérôme, *J. Chem. Soc. Chem. Commun.* 2003, 232-233.

[29] B. Lepoittevin, M. Devalckenaere, N. Pantoustier, M. Alexandre, D. Kubies, C. Calberg, R. Jérôme, Ph. Dubois, *Polymer* 2002 , *43*, 4017.

[30] N. Pantoustier, M. Alexandre, Ph. Degée, C. Calberg, R. Jérôme, C. Henrist, R. Cloots, A. Rulmont, Ph. Dubois, *e-Polymers* 2001, n° 009.

[31] D. Kubies, N. Pantoustier, Ph. Dubois, A. Rulmont, R. Jérôme, *Macromolecules* 2002, *35*, 3318.

[32] R. Jérôme, C. Calberg, F. Stassin, O. Halleux, Ph. Dubois, N. Pantoustier, M. Alexandre, B. Lepoittevin *Eur. Pat.* EP1247 829 A1 ; WO 02/08/541 A1.

[33] B. Lepoittevin, N. Pantoustier, M. Devalckenaere, M. Alexandre, D. Kubies, C. Calberg, R. Jérôme, Ph. Dubois, R. Jérôme, *Macromolecules* 2002, *35*, 8385.

[34] B. Lepoittevin, N. Pantoustier, M. Devalckenaere, M. Alexandre, C. Calberg, R. Jérôme, C. Henrist, A. Rulmont, Ph. Dubois, *Polymer* 2003, *44*, 2033.

Macromol. Symp. **2004**, *215*, 339-352

Functionalization of Living Polymers via Ethoxysilane Based Compounds: Synthesis and Interaction with Silica Particles

Joël Hoffstetter,[1] Ellen Giebeler,[2] Rolf Peter,[2] Pierre J. Lutz*[1]

[1] Institut C. Sadron, CNRS, UPR22 F-67083 Strasbourg Cedex, France
E-mail: lutz@ics.u-strasbg.fr
[2] Bayer AG, D-51368 Leverkusen, Germany

Summary: To optimize the reaction of ω-carbanionic styrene or butadiene/styrene polymers with ethoxysilane based compounds, the influence of several experimental parameters on the orientation of the functionalization reaction and its yield was examined. The resulting end-functionalized polymers were systematically investigated by SEC, ^{1}H NMR and elementary analysis. The orientation of the reaction was found to be directly depending on the chemical nature of the chain end and / or on the type of additive (ethers, LiCl). Best results were obtained with tetraethoxysilane, provided the functionalization is conducted around 5°C, and the active chain end of isoprenyl type. These reactions were extended to bifunctional polymers, the reaction product of butyllithium with m-diisopropenylbenzene being used as an initiator. The efficiency of this initiator for the synthesis of well-defined bifunctionalized polydienes almost quantitatively fitted with three alkoxy functions at both chain ends was demonstrated. Some preliminary results on the mechanical properties of mixtures of functionalized polymers with silica were mentioned.

Keywords: anionic polymerization; bifunctional initiator; ethoxysilanes; functionalization; rubber

Introduction

The improvement of performance of tires has been a continuous challenge in the last hundred years. In such materials, three components play an important role: The styrene-butadiene rubber polymer (SBR) eventually functionalized, the presence of polymers exhibiting other topologies such as star-shaped structures and the filling agent. In carbon black based materials, extensively used as filling agent until recently, crosslinking is achieved upon formation of sulphide bridges and the chains strongly interact with carbon black. The

DOI: 10.1002/masy.200451126

replacement of carbon black in rubbers by silica-based compounds has been shown to still improve their properties. One limitation of the system is due to the fact, that in rubbers charged with silica, phase separation may occur due to the non-compatibility between the rubber (SBR chains) and the polar groups of the silica surface. The heterogeneous dispersion of silica may also contribute to phase separation. To overcome these difficulties several solutions were introduced with more or less success: Use of selected silica compounds who do not aggregate, silica based polysulphide crosslinking agents linking the rubber matrix to silica and /or anionically synthesized SBRs[1-4]. The use of SBRs prepared by anionic polymerization and quantitatively fitted at one or better at the two chain ends with polar groups may represent an interesting alternative[5,6]. The polar group should interact with the surface of the silica particles and, therefore, limit the mobility of the chain ends.

The major part of the present work deals with a systematic investigation of the functionalization reactions of ω-carbanionic polymers with various heterofunctional micromolecular species, such as (3-chloropropyl)triethoxysilane, in order to design polymers fitted quantitatively at one chain end with alkoxysilane functions. The extension of the reaction to bifunctional polymers will also be discussed, the prerequisite being a bifunctional initiator, efficient and stable even in the present of polar additives. Some preliminary results on the behavior of ethoxysilane functionalized SBRs mixed with silica will be mentioned.

Functionalization of monofunctional polymers with ethoxysilanes

End-alkoxylated polymers have been shown to be well adapted for the surface modification of silica particles. Contrary to end-functionalized polymers having isopropyloxydimethylsilyl end-groups[7], silyl-ether functions can be hydrolyzed also with bases, whereupon grafting on the hydroxyl functions of the silica particles can be achieved and stable covalent Si-O-Si bonds created. In addition, the presence of the polymer chain on the silica particles favors their dispersion and should prevent them from aggregation. Once crosslinked, the mobility of the polymer chain is largely restricted and improved mechanical properties of the rubbery materials are expected. As mentioned above, anionic polymerization is the method of choice to achieve these functionalization reactions. In the following, we first examined the reaction of ω-carbanionic polymers with (3-chloropropyl)triethoxysilane (1), the chlorine function of the Cl-CH$_2$- group being the deactivating site. We tested several other ethoxysilane based compounds (3-bromopropyl)trimethoxy or triethoxysilane (2), 2-(3,4-cyclohexyl)ethyl-trimethoxysilane, o-/p-chlorophenyltrimethoxysilane (3), tetraethoxysilane (Figure 1). Most of these compounds exhibit moderately hydrolysable groups. Therefore, the attack of the ethoxysilane function by the living polymer during the functionalization reaction should be prevented.

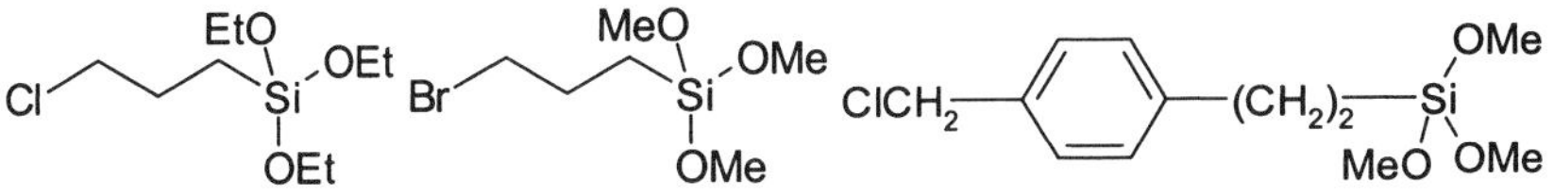

Figure 1 . Structures of some deactivating agents

Monocarbanionic polymers : polystyrene (PS), polybutadiene (PB) and styrene/butadiene copolymers (SBR) were synthesized in hydrocarbon medium either in the presence or without ether (1,2-diethoxyethane) according to well-established procedures. We examined the influence of the following parameters who may directly affect the orientation of the functionalization reaction: Nature of the chain end (styryllithium, α-methylstyryllithium, diphenylalkyllithium, butadienyllithium, isoprenyllithium), steric hindrance, functionalization temperature, polar additives (potassium alcoholates), salts (LiCl)... The reaction has been schematically represented in Figure 2.

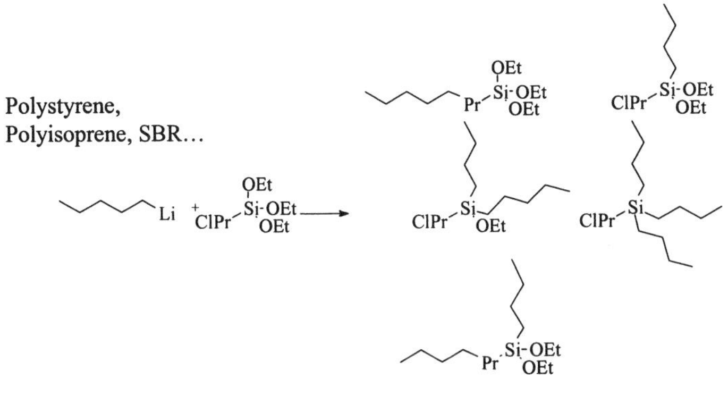

Figure 2 . Schematic representation of the deactivation reaction with component 1

Functionalization with (3-chloropropyl)triethoxysilane

We first examined the functionalization of low molar mass ω-carbanionic PSs wit (3-chloro-propyl)triethoxysilane. Chemical titration and ^{1}H NMR were extensively used to achieve that aim. In addition, Size Exclusion Chromatography (SEC) characterization was performed on all the samples to determine the amount of coupling product whenever present. The following conclusions could be drawn from these experiments :

For living polystyrenes, over a temperature range from 0 to 60°C, both the chlorine function and the ethoxy group have reacted as it is revealed from ^{1}H NMR measurements (Figure 3 b). The amount of coupling product remains low provided the reaction temperature is kept below 60°C. The chain ends of the polystyrene were also modified by an intermediate addition of isoprene before coupling with (3-chloropropyl)triethoxysilane. Under these conditions even at 60°C no coupling product was detected by SEC. Therefore, no substitution of the chlorine function was noted, whatever the temperature was. This can be explained by the lower reactivity of the chain end.

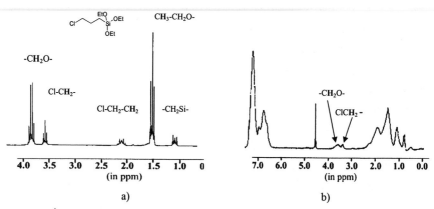

Figure 3 . ^{1}H NMR measured in CDCl$_3$ a) : component (1), b) : PS functionalized with (1)

These preliminary results prompted us to study, among the different parameters, the influence of the nature of the chain end on the efficiency and the orientation of the coupling reaction. The polystyrene chain end was modified by an intermediate addition of 1,1-diphenylethylene (DPE) to lower its reactivity. These PSs were reacted in pure benzene with

(3-chloropropyl)triethoxysilane. In that case, quantitative reaction of the chain end with the chlorine function of the deactivating agent was noted (Figure 4 and 5). This confirms earlier results obtained by Quirk[8].

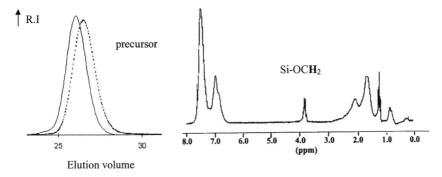

Figure 4 . SEC traces of a PS sample   Figure 5 . ^{1}H NMR of the same PS (200MHz-CDCl$_3$)

To control the composition i.e. the statistical incorporation of styrene in SBR based materials and the microstructure, the reaction may be conducted in the presence of ether or of an alkali alcoholate. On another hand, the presence of that ether may enhance the reactivity of the chain end and could affect the orientation of the functionalization reactions. To verify that point, ω-carbanionic polybutadienes or polystyrenes were reacted with (3-chloropropyl)-triethoxysilane) in the presence of ether. In fact, SEC characterization confirmed the presence of high amounts of coupling products and chain-end analysis evidenced the implication of the chlorine and the ethoxy functions in the functionalization process.

Along the same line, we modified the chain ends with α-methylstyrene end groups. This compound should not polymerize at the selected reaction temperature, but its presence should enhance the reactivity of the chain end. The addition reaction was followed by UV spectroscopy. Once quantitative addition was observed, (3-chloropropyl)triethoxysilane was introduced in the reaction medium. SEC characterization confirmed the decrease of the amount of coupling product over a temperature domain from 0 to 60°C. In fact only partial substitution of the chlorine function was noted and at least one ethoxy function was involved in the process leading to a mixture of mono to tetra substituted species. The addition of ether does not improve the results: No reaction with the chlorine function was detected, quantitative substitution of one ethoxy function was noted.

The presence of LiCl has been shown to improve the polymerization of alkyl(meth)acrylates and to limit the occurrence of side reactions. Therefore, we performed a series of functionalization reactions in the presence of LiCl : The amount of coupling product could be reduced, but no reaction at all on the chlorine function was observed.

Functionalization with other triethoxysilane based compounds

We attempted also functionalization reactions with (3-bromopropyl)triethoxysilane and with 2-(3,4-epoxycyclohexyl)ethyltrimethoxylsilane. As in the case of the chloro compound, selected polymers PS, SBRs with various chain ends were synthesized and their functionalization subsequently performed. In most cases, no quantitative functionalization could be achieved. In addition the amount of coupling product was much higher.

From these studies the following conclusions could be drawn : As expected, in most cases, the chlorine and the ethoxy functions of the micromolecular compound are involved in the process. The orientation of the reaction is directly depending on the chemical nature of the chain end and on the type of additive (ethers, LiCl): For polymers fitted with styryllithium, α-methylstyryllithium or diphenylmethyllithium chain ends, in the presence of ether the chlorine function is only partially substituted, and the ethoxy group is also involved in the process. In the case of isoprenyllithium chain ends, only one ethoxy group is substituted.

Functionalization with tetraethoxysilane

In preliminary experiments, we showed that the functionalization of PBLi or SBRLi with tetraethoxysilane (TES) yielded species exhibiting three ethoxy functions at the chain end, and also high amounts of coupling product. To try to increase the content of trifunctional product, we used higher concentrations of deactivating agent, i.e. a five molar ratio with respect to the concentration of living chain ends. The reaction is schematically represented in Figure 6. Under these conditions, provided the functionalization is conducted between 0 and 10°C and the active chain end of isoprenyl type, three ethoxy functions could be quantitatively introduced at the chain end. Some characterization data are given in Table 1.

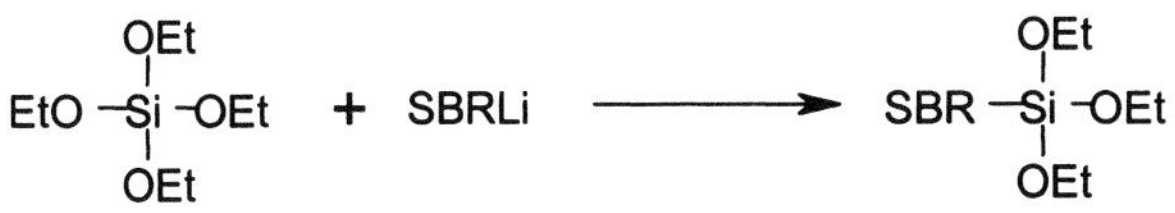

$$\underset{\overset{\displaystyle |}{OEt}}{\overset{\displaystyle OEt}{EtO-Si-OEt}} \; + \; SBRLi \quad \longrightarrow \quad \underset{\overset{\displaystyle |}{OEt}}{\overset{\displaystyle OEt}{SBR-Si-OEt}}$$

Figure 6. Schematic representation of the functionalization of SBR's with tetraethoxysilane

The reduction of the amount of coupling product with increased ratio [LE] to [TES] is confirmed unambiguously by SEC (Figure 7).

Advantage of the reaction was also taken to synthesize tetrafunctional star-shaped polymers based on TES and containing a central Si atom. It has been reported in the literature that, upon addition of living polymers to $SiCl_4$ under appropriate conditions, grafting can be realized. Strong differences in reactivity were yet observed depending upon the nature of the chain end. When PSLi is reacted with $SiCl_4$, a mixture of trifunctional (74 %) and tetrafunctional (26 %) species was detected by SEC, whatever the polarity of the solvent was. The reaction time was 48 hours in benzene at 50°C or in a mixture of benzene /THF). On the contrary when conducted in cyclohexane at 50°C and with a short polybutadiene spacer at the chain end, four arm star-shaped polymers could be obtained. We made similar observations when tetraethoxysilane was used as a coupling agent instead of $SiCl_4$ (Table 1).

Table 1 . Characterization data of SBRs functionalized with TES (5 molar excess)

Reference	SBR1	SBR1A	SBR2B	SBR3A	SBR3B
Polar Additives		Bis-Ether		Bis-Ether//Potassium Alcoholate	
[TES]/[Li]	5	-	5	-	5
$M_{n, th}$ (g.mol^{-1})	80 000	80 000		80 000	
M_w (SEC/LS)	87 000	87 000	89 000	95 000	105 000
P (PS/RI)	1,12	1,08	1,11	1,11	1,2
Styr. Content %	15 %		9 %		9 %
Microstructure (^{1}H NMR)	(1,2) : 39% (1,4) : 61%	(1,2) : 44% (1,4) :56%		(1,2) : 34% (1,4) : 66%	
Main product (SEC)	87 %	-	93 %	-	89 %
Funct. Yield	100 %	-	99 %	-	99 %

SBRA and SBRB correspond respectively to the precursor and the functionalized product

P(PS/RI) : polymolecularity based on refractive index detection (SEC)

346

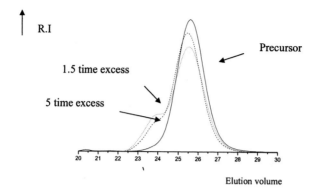

Figure 7 . SEC traces of the reaction product of a SBR chain with tetraethoxysilane

Extension to bifunctional polymers

The replacement of monofunctionalized SBRs by SBRs quantitatively functionalized at both chain ends with ethoxy groups would still improve the mechanical properties of SBR / silica materials. The prerequisite for an efficient functionalization is the existence of a bifunctional initiator allowing the controlled synthesis of the α,ω-living SBR precursors. However, until now, no bifunctional initiator, exhibiting carbon-lithium bonds and yet soluble and efficient in hydrocarbon solvents has proved really satisfactory for the synthesis in the presence of ethers of α,ω-bifunctional SBR elastomers. Many years ago, one of us[9] has used the reaction of a stoichiometric amount of sec-butyllithium (BuLi) to one molecule of m-diisopropenylbenzene (DIB) to produce a bifunctional initiator, which is soluble in non-polar solvents. In the absence of polar additives, this initiator has been shown to be quite efficient for the polymerization of isoprene and butadiene and the subsequent synthesis of well-defined elastomeric materials. More recently similar work was done along that line[10]. Provided m-DIB is reacted with BuLi at high temperature, no polymerization should occur. Therefore no oligomeric species should be formed. In the present work, the reaction of BuLi with m-DIB was studied in the presence of different types of ethers at a temperature such as to prevent DIB polymerization. The reaction is schematically represented in Figure 8.

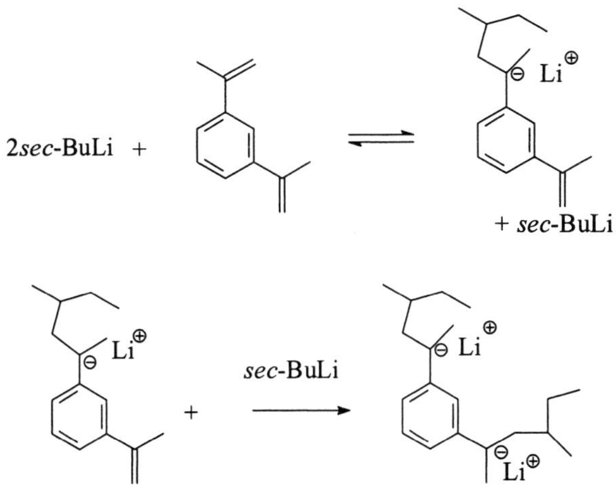

Figure 8 . Schematic representation of the reaction of m-DIB with 2 *sec*-BuLi

-We examined first the reaction in hexane (or in cyclohexane) in the presence of selected ethers (1,2-diethoxyethane ...) at 60°C. At that temperature no propagation of DIB should occur and the reaction limited to the addition of two BuLi to one DIB unit. The reaction products were examined by SEC and mass spectroscopy. The presence of the diadduct in higher quantities than under identical conditions in the absence of polar additives could be confirmed. The amount of diadduct formed was found to be directly related to the DIB concentration.

It was also of interest to investigate the kinetics of this reaction to establish whether the reactivity of the two double bonds with respect to BuLi is the same or not and, whether the bifunctional initiator once formed is stable, especially in the presence of ether. The reaction product is characterized by the presence in UV spectroscopy of a maximum of optical density around 320 nm (Figure 9). Therefore, the evolution of the optical density versus time could be followed directly and the influence of the reaction conditions on the nature of the products formed could be determined.

Optical density

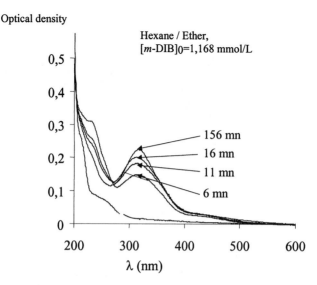

Figure 9 .Evolution of the optical density versus reaction time for the addition of *sec*-Buli onto m-DIB. (60°C, [*m*-DIB]₀=1,168 mmol/L)

Active site titration evidenced that the concentration of living sites is lower than the initial concentration of *sec*-BuLi. Therefore, that concentration corresponds to the concentration of diadduct determined by other methods. When the reaction is conducted at 40°C, high quantities of oligomers are formed during the first stage of the reaction as it can be seen by SEC measurements (Figure 10). After some time the amount of diadduct increases.

Selected reaction products were used for the polymerization of styrene. They were also tested for the (co-)polymerization of dienes with styrene. Conditions were identical to those employed for the polymerization of the same monomers with butyllithium. From the characterization of the resulting polymers, it can be concluded that :

- This initiator is efficient, since the polymer samples obtained exhibit narrow molar mass distributions and, in most cases, quite symmetric molar mass distributions, which indicates that the initiation process is rapid and quantitative.

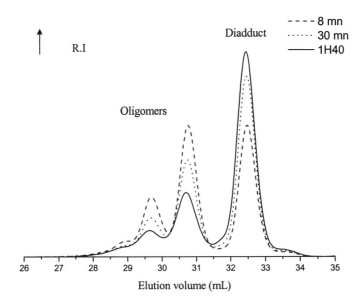

Figure 10 . SEC traces of the reaction products of 2 *sec*-BuLi /*m*-DIB at 40 °C (Hexane : 300mL ; DIB : 5,83 mmol ; sec-BuLi : 11,66 mmol ; ether: 0,58 mL)

- That initiator is bifunctional, since the molar masses calculated using the mole ratio monomer converted to initiator are close to the measured number average molar masses. Account has yet to be taken of the concentration of bifunctional initiator really present in the medium. This statement does not originate solely from comparisons between experimental and calculated molar mass.

- The "bifunctionality" of the living polymers was also verified by reacting them with a stoichiometric amount of α,α'-dibromo-*p*-xylene. The observed increase of the molar mass of the resulting chain extension product with respect to the precursor, by a factor of 10, is a strong indication in favour of the bifunctional nature of the precursor. In addition, the values of the radii of gyration are comparable to those observed for linear polymers of similar molar masses. Reaction of the living polymer with divinylbenzene resulted in network formation. Gel formation generally occurs at room temperature in less than 1 hour.

- Addition of ethylene oxide to the medium yields reversible gel formation. Once deactivation is achieved by introduction of methanol, the medium gets again soluble. The presence of the OH functions was confirmed by UV titration after modification with naphtyl-1-isocyanate. That point will be discussed in a forthcoming paper. Functionalization of these polymers with ethoxysilane-based compounds was achieved according to the strategy developed for monofunctional polymers.

Table 2 . Characteristics of functionalized polymers used in viscosimetry studies

Reference	M_n g.mol^{-1} (SEC)	M_w/M_n (SEC)	Wt.- % main product	Funct. yield	Styrene % (molar)	(1,2) % (molar)
SBR1A [a]	1630	1.09	-	-	13	84
SBR1B [a]	1820	1.2	82	93	13	84
SBR 2A [b]	2350	1.09	-		11	-
SBR 2B [b]	2500	1.11	-	100	11	-
SBR 3A [c]	2230	1.09	-		14	83
SBR 3B [c]	2540	1.22	78	97	14	83
SBR 4A [d]	2250	1.15	-	-	11	-
SBR 4B [d]	2470	1.18	90	100	11	-

[a] Functionalized with component 1
[b] SBR-Si-(OEt)$_2$PrCl
[c] SBR-Si(Oet)$_3$
[d] SBR-PI-Si(OET)$_3$

SBRA and SBRB correspond respectively to the precursor and the functionalized product

In some preliminary studies we could show that the mechanical properties of mixtures of functionalized polymers with silica are improved as compared to systems where unfunctionalized polymers were used. The characteristics of the sample are given in Table 2 and the viscosity behavior is represented in Figure 11. Further work has to be done along that line.

Viscosity Pa.s⁻¹

Viscosity Pa.s⁻¹

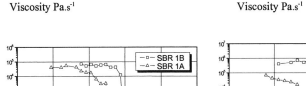

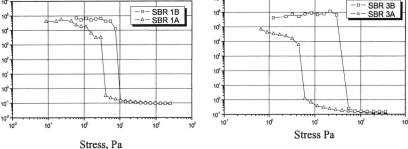

Stress, Pa

Stress Pa

Figure 11 . Evolution of the viscosity versus stress for selected SBR samples (see Table 2)

Conclusion

In the present work, the functionalization of α- or α,ω-living polystyrenes or related polymers with ethoxysilane derivates in hydrocarbon solvents has been examined. It has been shown that the nature of the chain end and /or the presence of additives in the reaction medium directly influence the orientation of the reaction. For polymers fitted with styryllithium or α-methylstyryllithium living chain ends only partial substitution of the chlorine function has been observed, whereas with isoprenyllithium end groups, only substitution of one ethoxy group occurred. On the contrary, provided appropriate experimental conditions are used the reaction of tetraethoxysilanes with monofunctional polymers lead to quantitative substitution of one ethoxy function. Once the functionalization conditions optimized for monofunctional polymers, they have been extended to bifunctional polymers. Based on the reaction of sec-butyl lithium with m-diisopropenyl benzene an efficient bifunctional initiator stable and efficient even in the presence of ether has been obtained. These polymers, characterized by well-controlled molar and by sharp molar distributions have been functionalized almost quantitatively with ethoxy groups. Some preliminary results on the solid-state behavior of these polymers in the presence of silica have been presented. More detailed information on the last point will be given in a forthcoming paper.

Acknowledgments

The authors thank Bayer Polymers and the CNRS for financial support. They thank J.Ph. Lamps for his contribution to the preparation of the samples and A. Rameau, R. Meens and M. Mottin for their help in characterization.

[1]   Rhone-Poulenc Patent, **1989**, EP 0 157 703
[2]   Y. Bomal, P. Cochet, B. Dejean, P. Fourré, D. Labarre, *L'Actualité Chimique*, **1996**, *1*, 42
[3]   Michelin Patent , **1992**, EP 0 501 277 A1
[4]   Degussa Patent, **1975**, US 3 873 489
[5]   W.E. Lindsell, K. Radha, I. Soutar, *Polym. Int.*, **1991**, *25*, 1
[6]   Brevet Bridgestone, **1996**, EP 0 447 066
[7]   M. Ohata, M. Yamamoto, Y. Isono, *J. Appl. Polym. Sci.*, **1995**, *55*, 517
[8]   R.P. Quirk, H. Yang, *Polymer Prepint,* **1996**, *37(2),* 645
[9]   G. Beinert, P. Lutz, E. Franta, P. Rempp, *Makromol. Chem.*, **1978**, *179*,551
      P. Lutz, E. Franta, P. Rempp, *Polymer*, **1982**, 23, 1953
[10] S.S. Yu, Ph. Dubois, R. Jérome, Ph. Teyssié, *Macromolecules* **1996**, *29*, 2738
[11] J. Hoffstetter, Thesis, Strasbourg, **1999**
      J. Hoffstetter, E. Giebeler, R. Peter, P.J. Lutz, to be sent for publication

Macromol. Symp. **2004**, *215*, 353-367

Reactivities of Carbocations and Carbanions

Dedicated to Professor *Henry K. Hall, Jr.*, on the occasion of his 80th birthday

Armin R. Ofial, Herbert Mayr*

Department Chemie der Ludwig-Maximilians-Universität München, Butenandtstr. 5–13 (Haus F), 81377 München, Germany

Summary: Initiation and propagation rate constants of carbocationic and carbanionic polymerizations can be predicted by the correlation equation $\log k_{20\,°C} = s(N + E)$, where E characterizes the electrophilicity of carbocations and electron-deficient alkenes, and N characterizes the nucleophilicity of carbanions and electron-rich alkenes. Since the nucleophile-specific slope parameter s is generally close to 1, it can be neglected in a first approximation, and the two-dimensional representation in Figure 3 illustrates the gradual change from carbanionic to carbocationic polymerizations with Hall's "initiation by bond-formation" as the link connecting the two ranges. The value of model studies for understanding ionic polymerizations is illustrated.

Keywords: anionic polymerization; cationic polymerization; kinetics (polym.); reactivity scales; structure-reactivity relationships

Introduction

Carbocationic and carbanionic vinyl polymerizations follow closely related mechanisms (Scheme 1).[1] Initiators, which may be carbocations or carbanions, can reversibly or irreversibly be generated from alkyl halides or CH acidic compounds, respectively. In the next step, the carbocations add to electron-rich "nucleophilic" vinyl monomers while carbanions attack electron-deficient "electrophilic" vinyl monomers. In both cases

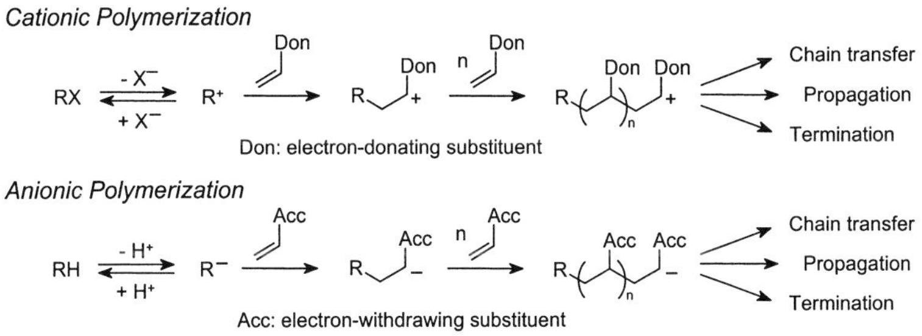

Scheme 1. Mechanisms of cationic and anionic polymerizations of vinyl monomers.

 DOI: 10.1002/masy.200451127

propagating ions are generated which may exist in equilibrium with dormant species. Chain transfer and termination processes compete with the propagation step and thus determine the molecular weight.[1]

In previous conferences of this series, we have reported about the kinetics of the reactions of carbocations with π-systems (Scheme 2).[2, 3]

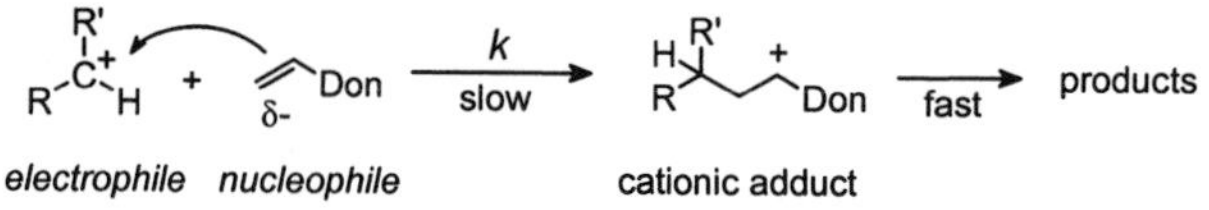

Scheme 2. Reactions of π-nucleophiles with cationic electrophiles.

We have shown that the rates of the reactions of carbocations with alkenes, arenes, and other π-nucleophiles can be calculated by the equation

$$\log k_{20°C} = s(N + E) \qquad (1)$$

where E represents the electrophilicity of the carbocations, N represents the nucleophilicity of the π-systems, and s (usually close to 1) is a nucleophile-specific slope parameter.[4–6] If very bulky systems, like tritylium ions, are excluded, equation 1 reproduces rate constants with an accuracy better than a factor of $10–10^2$; this deviation includes ordinary steric, solvent, and counterion effects.

By using benzhydrylium ions as reference electrophiles,[5] we have derived the nucleophilicity parameters of a large variety of π-systems, which have been summarized in a recent review.[6] A small collection of data is summarized in Scheme 3 which allows one to directly compare the nucleophilicities of olefinic and aromatic π-systems.

Electrophilicity parameters of carbocations were analogously determined from the kinetics of their reactions with π-systems,[6] which have been defined as reference nucleophiles in 2001.[5] Scheme 4 shows the electrophilicity parameters of some carbocations determined in this way.

Carbocationic Vinyl Polymerizations

The E parameters of eventual inititators in Scheme 4 and the N parameters of eventual monomers in Scheme 3 can be combined to predict initiation rate constants. If the E

parameters of the propagating species and the N parameters of the monomers are available, one can also predict propagation rate constants, as shown for the cationic polymerization of N-vinylcarbazole.[7]

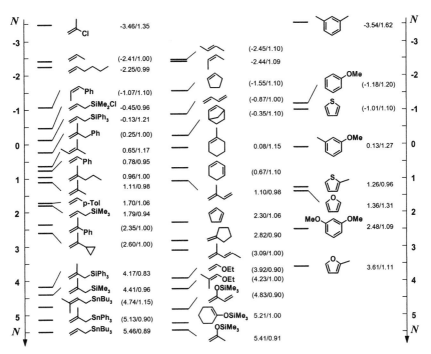

$$R^+ + \qquad \xrightarrow[\text{CH}_2\text{Cl}_2]{20\,°\text{C}} \qquad \tag{2}$$

Equation 2 shows the initiation of the polymerization of N-vinylcarbazole by the carbocation R^+. Knowledge of N and s for N-vinylcarbazole would therefore allow the prediction of the rate constants of the reactions of N-vinylcarbazole with any carbocations R^+ of known E parameters. If R^+ is the propagating cation, equation 2 corresponds to the propagation step.

Scheme 3. Nucleophilicity and slope parameters N/s for π-nucleophiles (parentheses indicate estimated values of s).[6]

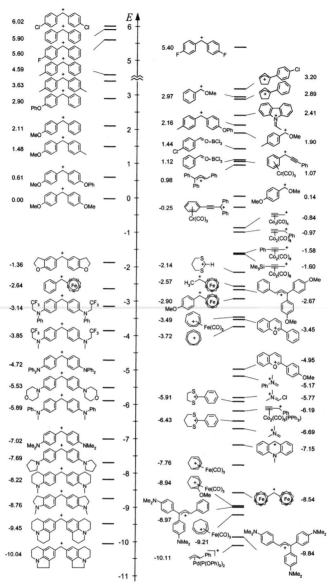

Scheme 4. Electrophilicity parameters E for carbocationic electrophiles derived from reactions with reference π-nucleophiles.[6]

Determination of the nucleophilicity parameters of N-vinylcarbazole:

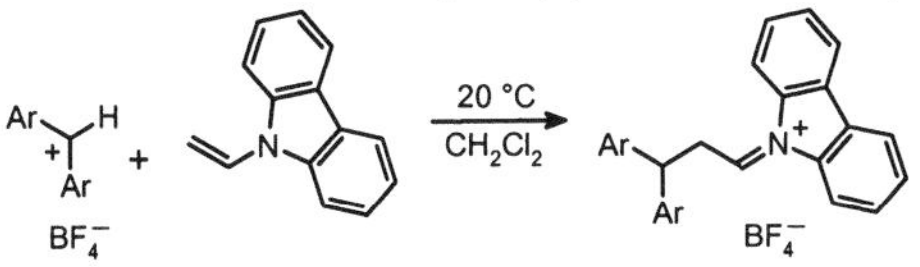

Ar	E	k / $M^{-1} s^{-1}$
p-(Ph$_2$N)-C$_6$H$_4$	-4.72	1.93
p-(Me$_2$N)-C$_6$H$_4$	-7.02	1.35×10^{-2}

Determination of the electrophilicity parameter of the propagating species:

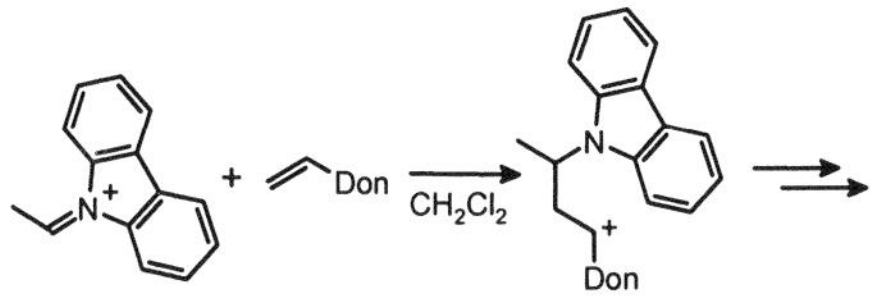

nucleophile	N	s	$k(20°C)$ / $M^{-1} s^{-1}$
SiMe$_2$Cl	-0.45	0.96	197
SiPh$_3$	-0.13	1.21	500
Ph	0.78	0.95	475

Scheme 5. Model reactions for the prediction of rate constants relevant for the carbocationic polymerization of N-vinylcarbazole.[7]

As shown in Scheme 5, the kinetics of the reactions of N-vinylcarbazole with benzhydryl cations Ar$_2$CH$^+$ have been investigated in order to determine the nucleophilicity parameters N and s of N-vinylcarbazole (N = 5.02, s = 0.94). The N-ethylidenecarbazolium ion has been used as a model for the cationically propagating N-vinylcarbazole chain (Scheme 5). From the rate constants of its reactions with three reference π-nucleophiles an electrophilicity parameter E = 2.41 has been derived.[7]

With equation 1, a rate constant of $k(20$ °C$)$ = 9.6×10^6 M^{-1} s^{-1} can be calculated for the reaction of the N-ethylidenecarbazolium ion with N-vinylcarbazole which can be considered to be close to the propagation rate constant k_p for the cationic polymerization of N-vinyl-carbazole. Comparison with the values determined calorimetrically[8–10] in polymerization experiments at 0 and 20 °C[9, 10] indicates agreement within a factor of 10^2, the error limit postulated for predictions of rate constants by equation 1. The consistency obtained for this propagation rate constant by completely different methods (Table 1) confirms the fidelity of the rate constant as well as the validity of the different methods used for its determination.

Table 1. Propagation rate constants k_p (in CH$_2$Cl$_2$) for the cationic polymerization of N-vinylcarbazole.

laboratories	methods	k_p / $M^{-1} s^{-1}$	T / °C	ref.
Rooney	calorimetry	$(1.0 ... 2.4) \times 10^4$	-40	8
Ledwith et al.	calorimetry	$(2.2 ... 4.6) \times 10^5$	0	9
Rodriguez & León	calorimetry	9.5×10^5	20	10
Mayr et al.	linear free energy relationship (equation 1)	9.6×10^6	20	7

In an earlier contribution to this conference, J.-P. Vairon compared the propagation rate constants for the carbocationic polymerization of 2,4,6-trimethylstyrene determined by different methods.[11, 12] This system polymerizes via stabilized benzyl cations and cannot undergo indanic cyclizations, and the directly measured k_p values are in perfect agreement with those determined by the competition method by Faust.[13] They deviate by less than a factor of 10^2 from the value calculated by equation $1^{[14]}$ (Scheme 6).

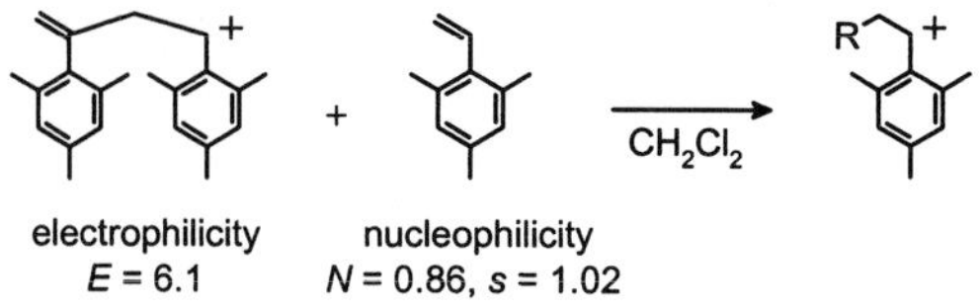

electrophilicity   nucleophilicity
$E = 6.1$ $N = 0.86$, $s = 1.02$

laboratories	methods	$k_p(20°C) / M^{-1} s^{-1}$
Vairon et al.	polymerization (ln [M_0]/[M]) vs t)	3.5×10^{5} [a]
Faust et al.	competitive capping	5.9×10^{5} [a]
Mayr et al.	linear free energy relationship (equation 1)	1.3×10^{7}

[a] Extrapolated from rate constants at –70 °C.

Scheme 6. Reactivity parameters for the cationic polymerization of 2,4,6-trimethylstyrene.

While equation 1 is an excellent tool for predicting the approximate magnitude of the rate constants of any initiation and propagation reaction as well as of the relative reactivities of vinyl monomers in copolymerization reactions within an accuracy of factor 10–100, in many cases more precise values are needed.

For that purpose we have introduced the diffusion clock method in 1995[15] which has the additional advantage that it avoids the generation of persistent carbocations. This method, which is a modification of the azide clock method by Richard and Jencks,[16] determines the degree of oligomerization of a vinyl monomer in the presence of a trapping agent, which reacts with diffusion control. Scheme 7 illustrates the 2-chloro-2,4,4-trimethylpentane/TiCl4 initiated oligomerization of isobutylene in the presence of trimethyl(2-methylallyl)silane. The transient *tert.*-alkyl cations R-CH2-C(CH3)2+ either propagate by reaction with isobutylene or are irreversibly trapped by the diffusion-controlled reaction with the organosilicon compound. Since diffusion controlled rate constants are known to be in the order of $(3 \pm 1) \times 10^9$ M^{-1} s^{-1} in typical organic solvents,[17] the ratio **A:B:C** allows one to calculate a propagation rate constant of $k_p = 6 \times 10^8$ M^{-1} s^{-1} for the cationic polymerization of isobutylene.[18]

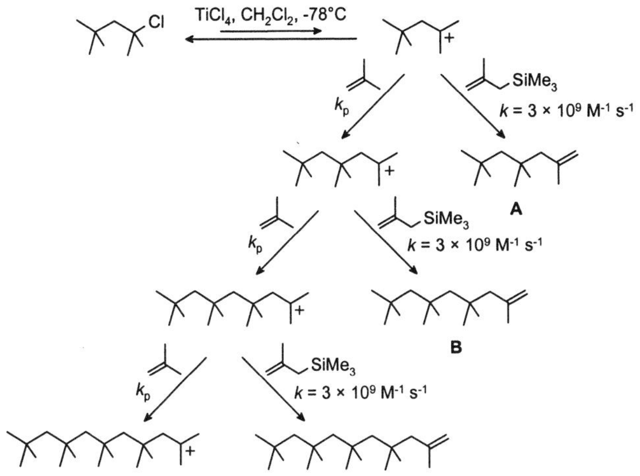

Scheme 7. Determination of the propagation rate constant k_p in the cationic polymerization of isobutylene by the diffusion clock method.[18]

Though this value is in good agreement with kinetic models by Puskas,[19] and the diffusion clock method has successfully been employed for other cationic vinyl polymerizations by Faust,[13, 20] the reliability of our approach has been questioned, because the rate constant determined by us was considerably larger than previously accepted values.[21, 22]

Previously, it had already been shown that the rates of attack of carbocations at π-nucleophiles are independent of the complex counterions,[6] and recent UV-Vis and NMR investigations allowed us to conclude that free and paired carbocations with closely similar spectroscopic properties also exhibit similar reactivities.[23, 24] Finally, different reactivities of low molecular weight and polymeric carbocations had to be excluded.

In a trinational collaboration it was recently shown that structurally analogous carbocations with $M = 195$ g mol^{-1} and $M_n \approx 2400$ g mol^{-1} exhibit the same reactivities towards weak nucleophiles (slow reactions) as well as towards strong nucleophiles (fast reactions) as illustrated in Table 2.[25] In this way, it has unequivocally been demonstrated that the conclusions derived from low-molecular weight model compounds can directly be transferred to polymerizing systems.[26]

Table 2. Rate constants (M^{-1} s^{-1}) for the reactions of benzhydrylium ions with π-systems.[25]

electrophiles	nucleophiles			
	⌁SiMe$_2$Cl	⌁SiPh$_3$	⬡–OSiMe$_3$	⬠–OSiMe$_3$
(4,4'-dimethyl benzhydrylium)	25.6[a]	385[a]	1.60×10^8 [b]	4.62×10^8 [b]
(benzhydrylium with pyrrolidine 39)	24.6[a]	391[a]	1.26×10^8 [b]	3.56×10^8 [b]

[a] Conventional UV-vis kinetic measurements in dichloromethane at −70 °C.
[b] Laser flash kinetics in acetonitrile/dichloromethane mixtures (1:3, v/v) at 20 °C.

Anionic Vinyl Polymerizations

Recently, we have shown that simple second-order kinetics can also be observed for the additions of stabilized carbanions to acceptor substituted ethylene derivatives.[27] When quinone methides, structural analogues of the previously investigated benzhydrylium ions, were used as reference electrophiles, rate determining CC-bond formation by addition of the carbanions $R^1R^2CH^-$ to the electron deficient CC-double bond could be achieved in the presence of an excess of its conjugate acid $R^1R^2CH_2$ which almost quantitatively protonates phenolates in DMSO (Scheme 8).

The high dielectric constant $\varepsilon = 46.45$ of DMSO[28] is responsible for the fact that in 10^{-4} to 10^{-3} M solutions ion-pairing is unimportant as indicated by the equal reactivities of potassium and tetrabutylammonium salts and the independence of the rate constants of the presence of crown ethers or cryptands (Scheme 8).[27]

Since we have also succeeded to directly measure the rates of the reactions of stabilized carbanions with stabilized benzhydrylium ions in DMSO,[29] it was feasible to link the two sets of data: Carbocations + neutral nucleophiles on one side and carbanions + neutral electrophiles on the other side. As shown by Figure 1, there is no noticeable break in the linear correlations as the solvent is changed from dichloromethane on the left of Figure 1 to DMSO on the right. It thus became possible to develop a single nucleophilicity scale ranging from toluene, the weakest nucleophile in the series to nitroethyl anion, the strongest

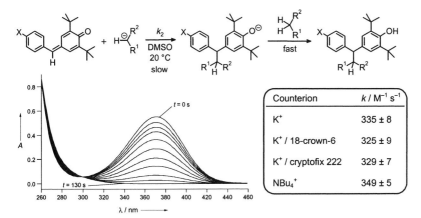

Scheme 8. Irreversible reactions of quinone methides with carbanions $R^1R^2CH^-$ in the presence of the conjugate CH acid $R^1R^2CH_2$. The UV-vis spectra and the rate constants refer to the reaction with $X = CH_3$, $R^1 = CO_2Et$, and $R^2 = CN$.[27]

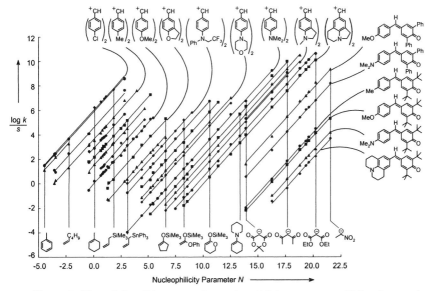

Figure 1. Plot of $(\log k)/s$ versus the nucleophilicity parameter N for the reactions of benzhydryl cations or quinone methides with π-nucleophiles or carbanions.[29]

nucleophile. The 26 orders of magnitude covered by these nucleophiles correspond to relative reactivities of 1 min to 10^{20} years.

For the parametrization, the electrophilicity parameters E according to equation 1 have been defined as solvent-independent quantities. Extensive investigations of the solvent dependence of the kinetics have shown that the rates of the reactions of carbocations with neutral π-systems and hydride donors are almost independent of the solvent.[4c, 6, 30, 31] Since the nucleophilicities of amines, alcohols, and carbanions are strongly affected by the solvents,[32–34] their N parameters have to be specified with respect to a certain solvent.

The reactivity parameters N and s derived from the reactions of the corresponding nucleophiles with benzhydrylium ions and structurally analogous quinone methides can only be considered as useful parameters if they are also applicable to reactions with other types of electrophiles. This has repeatedly been demonstrated for other types of carbocations[4–6] and recently also for neutral Michael acceptors. Figure 2 shows that the relative reactivities of carbanions which have been observed versus benzhydrylium ions and quinone methides are also observed with respect to arylidenemalodinitriles.[35]

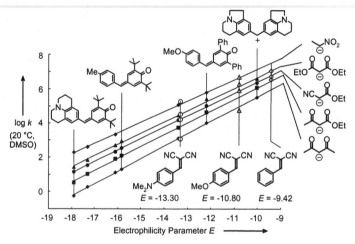

Figure 2. Determination of the electrophilicity parameters E of benzylidenemalodinitriles.[35]

While Figures 1 and 2 show that initiation and propagation rate constants of carbanionic polymerizations can also be predicted by equation 1 if ion-pairing is avoided, this approach can definitely not be employed for analyzing alkyllithium initiated polymerizations of styrenes, dienes, etc. which do not proceed via free carbanions. In these cases, the dependence of the spectral properties of the organometallics on metal, solvent, and concentration is also reflected by the different reactivities.[1]

A Framework for Describing Reactivities of Carbocations and Carbanions

Since the s parameters of π-nucleophiles, which correspond to the slopes of log k vs E correlations (e.g., in Figure 2), are generally close to 1,[6] an alternative representation of electrophile nucleophile combinations becomes possible. In Figure 3, nucleophiles are ordered according to increasing N parameters from top to bottom, and electrophiles are arranged according to increasing reactivity from left to right. The diagonal from bottom left to top right corresponds to $E + N = 0$ which implies a rate constant of 1 M^{-1} s^{-1} according to Equation 1. Starting from this diagonal, one can move upwards (towards weaker nucleophiles) or left (towards weaker electrophiles) and arrive in the sector where electrophile nucleophile combinations are predicted to be so slow that reactions cannot be expected. Moving from this diagonal downwards or to the right one gets into the diffusion controlled sector where selectivities often become low. We have discussed that most synthetically used organic reactions are found in the activation controlled corridor jacketing the drawn diagonal ($10^{-6} < k < 10^{10}$ M^{-1} s^{-1}).[29]

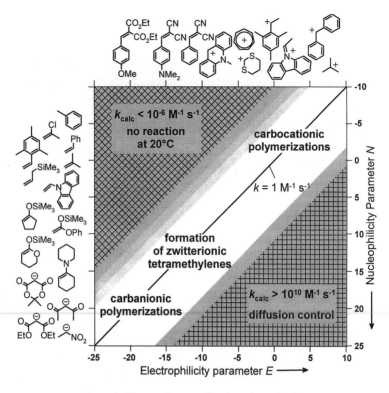

Figure 3. The continuum of ionic polymerizations.

Applying this model to ionic vinyl polymerizations, one can see that carbocationic polymerizations are found in the upper right part of this corridor while carbanionic polymerizations are found in the lower left part. In the center of this diagram, one can find strong neutral nucleophiles and strong neutral electrophiles, which are predicted by Equation 1 to give 1,4-zwitterions with reasonable rates.

Twenty years ago, H. K. Hall Jr. summarized polymerizations of alkenes initiated by bond formations (Scheme 9).[36] Using chemical experience, the reactivities of alkenes have been arranged in the form of a periodic table.[36a, 36b] Thus, Figure 4 arranges nucleophilic alkenes according to increasing donor ability from top to bottom; electrophilic alkenes are arranged from left to right with increasing acceptor ability.

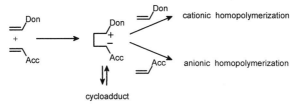

Scheme 9. Initiation of cationic and anionic homopolymerizations by bond formation according to Hall.[36]

Strong donors (on the bottom) spontaneously react with strong acceptors (on the right) to yield zwitterionic tetramethylenes which then are able to act as initiators for the homopolymerization of the corresponding olefins (Scheme 9).[36]

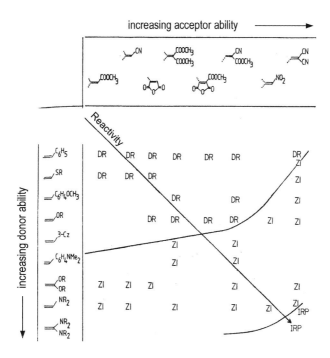

Figure 4. The sector of Hall's "periodic table of alkenes" where mechanistic changes are observed on variation of the donor and/or acceptor ability of the alkenes (DR: diradical formation, ZI: zwitterion formation, IRP: ion-radical pair formation).[36a, 36b]

Initiation by bond formation as proposed by Hall thus is the link between carbocation and carbanion initiated polymerizations,[36] and our electrophilicity and nucleophilicity scales[5, 6, 29] provide the quantitative basis for Hall's concept. In recent work we have shown that the whole range of polar organic reactions, from Friedel-Crafts alkylations (typical acid catalyzed reactions) to Michael additions (typical base catalyzed reactions) can be described by a single set of reactivity parameters.[29] We now demonstrated that the same concept can also be employed for ordering ionic polymerizations.

Acknowledgement

Financial support by the Deutsche Forschungsgemeinschaft and the Fonds der Chemischen Industrie is gratefully acknowledged.

[1]   M. Szwarc, M. Van Beylen, "*Ionic Polymerization and Living Polymers*", Chapman & Hall, New York 1993.
[2]   H. Mayr, M. Patz, *Macromol. Symp.* **1996**, *107*, 99.
[3]   H. Mayr, M. Roth, *Macromol. Symp.* **1998**, *132*, 103.
[4]   a) H. Mayr, M. Patz, *Angew. Chem.* **1994**, *106*, 990; *Angew. Chem. Int. Ed. Engl.* **1994**, *33*, 938. b) H. Mayr, in "*Cationic Polymerization: Mechanisms, Synthesis and Applications*", K. Matyjaszewski, Ed., Marcel Dekker, New York 1996, p. 51. c) H. Mayr, M. Roth, G. Lang, in "*Cationic Polymerization, Fundamentals and Applications*", R. Faust, T. D. Shaffer, Eds., ACS Symposium Series Vol. 665, Washington, DC 1997, p. 25. d) H. Mayr, O. Kuhn, M. F. Gotta, M. Patz, *J. Phys. Org. Chem.* **1998**, *11*, 642. e) H. Mayr, M. Patz, M. F. Gotta, A. R. Ofial, *Pure Appl. Chem.* **1998**, *70*, 1993.
[5]   H. Mayr, T. Bug, M. F. Gotta, N. Hering, B. Irrgang, B. Janker, B. Kempf, R. Loos, A. R. Ofial, G. Remennikov, H. Schimmel, *J. Am. Chem. Soc.* **2001**, *123*, 9500.
[6]   H. Mayr, B. Kempf, A. R. Ofial, *Acc. Chem. Res.* **2003**, *36*, 66.
[7]   H. Schimmel, A. R. Ofial, H. Mayr, *Macromolecules* **2002**, *35*, 5454.
[8]   J. M. Rooney, *Makromol. Chem.* **1978**, *179*, 165.
[9]   P. M. Bowyer, A. Ledwith, D. C. Sherrington, *Polymer* **1971**, *12*, 509.
[10] M. Rodriguez, L. M. León, *Eur. Polym. J.* **1983**, *19*, 585.
[11] P. De, R. Faust, H. Schimmel, H. Mayr, M. Moreau, B. Charleux, J.-P. Vairon, *Book of Abstracts*, Int. Symposium on Ionic Polymerization (IP '03), Boston, MA, 2003, oral presentation O2.
[12] See also: J.-P. Vairon, M. Moreau, B. Charleux, A. Cretol, R. Faust, *Macromol. Symp.* **2002**, *183*, 43.
[13] P. De, R. Faust, H. Schimmel, H. Mayr, M. Moreau, B. Charleux, J.-P. Vairon, *Polym. Prep.* **2003**, *44*, 804.
[14] H. Schimmel, Dissertation, Ludwig-Maximilians-Universität München, 2000.
[15] M. Roth, H. Mayr, *Angew. Chem.* **1995**, *107*, 2428; *Angew. Chem. Int. Ed. Engl.* **1995**, *34*, 2250.
[16] a) J. P. Richard, W. P. Jencks, *J. Am. Chem. Soc.* **1982**, *104*, 4689. b) J. P. Richard, W. P. Jencks, *J. Am. Chem. Soc.* **1982**, *104*, 4691. c) J. P. Richard, M. E. Rothenberg, W. P. Jencks, *J. Am. Chem. Soc.* **1984**, *106*, 1361.
[17] J. Bartl, S. Steenken, H. Mayr, *J. Am. Chem. Soc.* **1991**, *113*, 7710.
[18] M. Roth, H. Mayr, *Macromolecules* **1996**, *29*, 6104.
[19] a) J. E. Puskas, G. Kaszas, *Prog. Polym. Sci.* **2000**, *25*, 403. b) J. E. Puskas, H. Peng, *Polym. React. Eng.* **1999**, *7*, 553. c) J. E. Puskas, M. Lanzendörfer, *Macromolecules* **1998**, *31*, 8684.
[20] a) M. S. Kim, R. Faust, *Macromolecules* **2002**, *35*, 5320. b) H. Schlaad, Y. Kwon, L. Sipos, R. Faust, B. Charleux, *Macromolecules* **2000**, *33*, 8225.

[21] P. Sigwalt, M. Moreau, A. Polton, *Macromol. Symp.* **2002**, *183*, 35.

[22] a) P. H. Plesch, *Macromolecules* **2001**, *34*, 1143. b) P. H. Plesch, *Prog. React. Kinet.* **1993**, *18*, 1.

[23] K. Koszinowski, Diplomarbeit, Ludwig-Maximilians-Universität München, 2000.

[24] In contrast, the spectroscopic properties of iminium ions depend on the counterions: H. Mayr, A. R. Ofial, E.-U. Würthwein, N. C. Aust, *J. Am. Chem. Soc.* **1997**, *119*, 12727.

[25] H. Mayr, H. Schimmel, S. Kobayashi, M. Kotani, T. R. Prabakaran, L. Sipos, R. Faust, *Macromolecules* **2002**, *35*, 4611.

[26] M. Roth, H. Mayr, R. Faust, *Macromolecules* **1996**, *29*, 6110.

[27] R. Lucius, H. Mayr, *Angew. Chem.* **2000**, *112*, 2086; *Angew. Chem. Int. Ed.* **2000**, *39*, 1995.

[28] C. Reichardt, "*Solvents and Solvent Effects in Organic Chemistry*", 3rd ed., Wiley-VCH, Weinheim 2003.

[29] R. Lucius, R. Loos, H. Mayr, *Angew. Chem.* **2002**, *114*, 97; *Angew. Chem. Int. Ed.* **2002**, *41*, 91.

[30] H. Mayr, R. Schneider, C. Schade, J. Bartl, R. Bederke, *J. Am. Chem. Soc.* **1990**, *112*, 4446.

[31] M.-A. Funke, H. Mayr, *Chem. Eur. J.* **1997**, *3*, 1214.

[32] E. M. Arnett, K. E. Molter, *Acc. Chem. Res.* **1985**, *18*, 339.

[33] a) C. D. Ritchie, *Can. J. Chem.* **1986**, *64*, 2239. b) C. D. Ritchie, *Acc. Chem. Res.* **1972**, *5*, 348.

[34] S. Minegishi, H. Mayr, *J. Am. Chem. Soc.* **2003**, *125*, 286.

[35] T. Lemek, H. Mayr, *J. Org. Chem.* **2003**, *68*, 6880.

[36] a) H. K. Hall, Jr. *Angew. Chem.* **1983**, *95*, 448; *Angew. Chem. Int. Ed. Engl.* **1983**, *22*, 440. b) H. K. Hall, Jr., A. B. Padias, *Acc. Chem. Res.* **1990**, *23*, 3. c) H. K. Hall, Jr., A. B. Padias, *Acc. Chem. Res.* **1997**, *30*, 322. d) H. K. Hall, Jr., A. B. Padias, *J. Polym. Sci. Part A: Polym. Chem.* **2001**, *39*, 2069.

Confinement Effects on the Crystallization Kinetics and Self-Nucleation of Double Crystalline Poly(p-dioxanone)-*b*-poly(ε-caprolactone) Diblock Copolymers

Alejandro J. Müller,[1] *Julio Albuerne,*[1] *Luis M. Esteves,*[1] *Leni Marquez,*[1] *Jean-Marie Raquez,*[2] *Philippe Degée,*[2] *Philippe Dubois,*[2] *Stephen Collins,*[3] *Ian W. Hamley*[3]

[1]Grupo de Polímeros USB, Departamento de Ciencia de los Materiales, Universidad Simón Bolívar, Apartado 89000, Caracas 1080-A, Venezuela
[2]Laboratory of Polymeric and Composite Materials (LPCM), University of Mons-Hainaut, Place du Parc 20, 7000 Mons, Belgium
[3]Department of Chemistry, University of Leeds, Leeds LS2 9JT, UK

Summary: The morphology, crystallization and self nucleation behavior of double crystalline diblock copolymers of poly(p-dioxanone) (PPDX) and poly(ε-caprolactone) (PCL) with different compositions have been studied by different techniques, including optical microscopy (OM), atomic force microscopy (AFM) and differential scanning calorimetry (DSC). The two blocks crystallize in a single coincident exotherm when cooled from the melt. The self-nucleation technique is able to separate into two exotherms the crystallization of each block. We have gathered evidences indicating that the PPDX block can nucleate the PCL block within the copolymers regardless of the composition. This effect is responsible for the lack of homogeneous nucleation or fractionated crystallization of the PCL block even when it constitutes a minor phase within the copolymer (25% or less). Nevertheless, we were able to show that decreasing amounts of PCL within the diblock copolymer still produces confinement effects that retard the crystallization kinetics of the PCL component and decrease the Avrami index. On the other hand evidence for confinement was also obtained for the PPDX block, since as its content is reduced within the copolymer, a depression in its self-nucleation and annealing temperatures were observed.

Keywords: confinement; diblock copolymers; poly(ε-caprolactone); poly(p-dioxanone); self nucleation

Introduction

Poly(p-dioxanone), PPDX, is a poly(ester-ether) that is biodegradable, bioabsorbable and commercially available as suture material.[1] For some biodegradation applications, the hydrolytic degradation time of PPDX may be too fast and its copolymerization with poly(ε-

    DOI: 10.1002/masy.200451128

caprolactone), PCL, a more resistant polymer to hydrolysis, allows tuning the lifetime of the polymer for specific applications.[2]

P(CL-*b*-PDX) diblock copolymers are also interesting since they can phase separate in the melt (the order-disorder temperature is most probably above 140 °C[3]), and the two microdomains can later crystallize upon cooling. The crystallization of block copolymer microdomains is a topic that has attracted much recent attention because of intrinsic interest in the crystallization of confined materials and also because the interplay between phase separation and crystallization can be very complicated since it depends on copolymer composition, polymer-polymer interaction parameter, molecular weight and relative position of the thermal transition of each microdomain with respect to the order-disorder transition.[4-6]

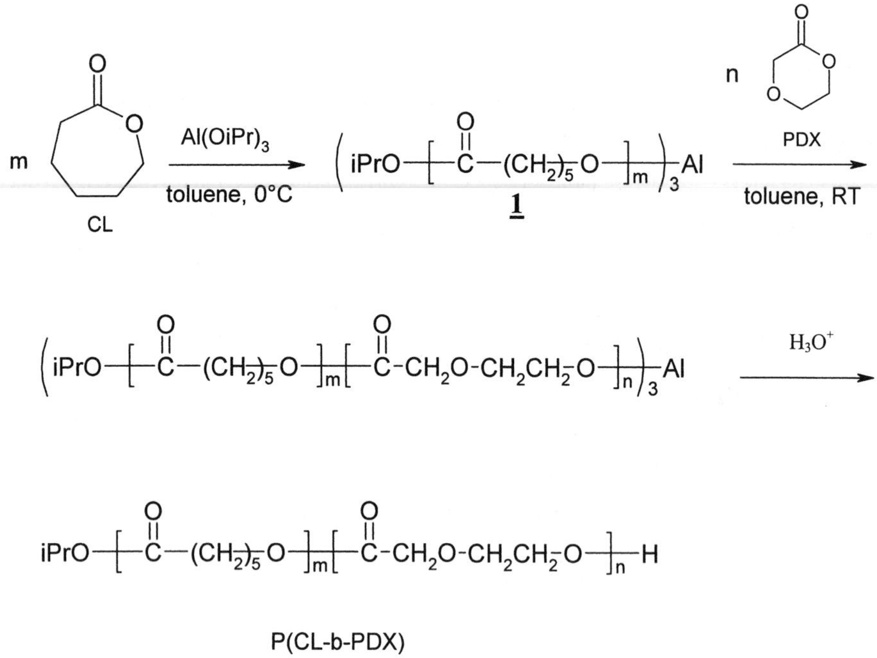

P(CL-b-PDX)

Scheme 1. Controlled ring-opening polymerization of 1,4-dioxan-2-one quantitatively initiated by ω-Al alkoxide poly(ε-caprolactone) chains (1) in toluene at room temperature.[7]

The preparation of narrowly polydispersed PCL and PPDX homopolymers employed here has been reported previously.[7] The synthesis of poly(ε-caprolactone-*b*-1,4-dioxan-2-one) block copolymers (P(CL-*b*-PDX)) is summarized in scheme 1. The polymerization of ε-caprolactone was performed in dry toluene at 0°C and was initiated by Al(OiPr)$_3$. After 1h, part of the living poly(ε-caprolactone) (PCL) solution was recovered, purified and characterized by Size Exclusion Chromatography (SEC). In a second step, purified PDX dissolved dry toluene at room temperature was added to the rest of the living PCL in toluene solution under a nitrogen atmosphere. After a given time, the polymerization was stopped by adding a few of drops of HCl (0.1 M). The poly(ε-caprolactone-b-1,4-dioxan-2-one) block copolymer (P(CL-b-PDX)) was recovered by precipitation into heptane at room temperature, filtration and drying under vacuum. Catalyst residues were extracted as previously reported.[8] The molar mass of the PPDX block was determined by ^{1}H NMR spectroscopy from the relative intensities of methylene protons of the PDX repeat units at 4.35 ppm and methylene protons of CL repeat units at 2.32 ppm knowing the molar mass of the PCL macroinitiator. The relative proportion of PDX and CL repetitive units also enabled weight fractions to be calculated.

Table 1 lists all materials employed and their molecular weights and polydispersities. The notation employed is explained as follows: the capital letters refer to PPDX and PCL blocks respectively, the superscripts are indicative of the molar mass (in thousands) and the subscripts indicate the weight fractions.

Table 1. Molecular Characteristics of the Materials Employed.

Composition[a]	M_n (PCL)	M_w/M_n	M_n (PPDX)	F_{PCL} (wt)
$D_{23}^8C_{77}^{27}$	27 200	1.06	7 500	0.77
$D_{28}^{10}C_{72}^{24}$	24 400	1.16	9 500	0.72
$D_{40}^5C_{60}^7$	7 100	1.17	4 800	0.60
$D_{55}^7C_{45}^6$	5 950	1.16	7 300	0.45
$D_{77}^{32}C_{23}^{10}$	10 280	1.07	32 280	0.23
PCL^{11}	11 400	1.07		1.00
PPDX^5			5 480	0.00

[a]D and C denote PPDX and PCL blocks, respectively; the superscripts are indicative of their molar masses (in thousands), and the subscripts refer to their weight fractions.

In a previous work[3] we have shown that P(CL-b-PDX) diblock copolymers are double crystalline materials, with two well defined melting peaks (whose temperatures are around the values for the homopolymers) and just one coincident crystallization exotherm (upon cooling from the melt) that occurs at lower temperatures than that of PPDX homopolymer crystallization and is independent of copolymer composition. In this coincident crystallization process, the PPDX starts to crystallize first, and then the PCL segment follows, apparently nucleated by the PPDX crystallization. Evidence for this behaviour was obtained from real time WAXS experiments during isothermal crystallization and DSC measurements[3]. In this work, we present results on how the confinement of the PCL block affects its crystallization kinetics even when the nucleation effect of PPDX prevents homogeneous nucleation and fractionated crystallization of the PCL block in copolymers where PCL is present as a minor phase. Another type of confinement effect can be identified for the PPDX block, when it is a minor component, by applying the self-nucleation technique.

Morphology

The susceptibility to degradation of PPDX makes the study of the morphology challenging. The P(CL-*b*-PDX) diblock copolymers cannot be annealed for long times at high temperatures to produce highly ordered morphologies because the PPDX block degrades, and it is similarly susceptible to staining agents.[3] Therefore, the samples are used as synthesized, without any prior thermal treatment. Previous attempts to obtain SAXS patterns in the melt on these samples failed possibly because they are not treated (and the morphology may lack the degree of order required for SAXS to produce a reflection).[3]

Previous Polarized Optical Microscopy (POM) experiments revealed that most samples crystallized with spherulitic superstructures at intermediate to high undercoolings and with granular aggregate superstructures at low undercoolings. The samples for POM studies were prepared by melting (at around 130°C) between two microscope cover slips for 10 min, and then they were quickly cooled to the crystallization temperature, Tc. At this temperature micrographs were taken with a digital camera at different time intervals.

In this work, we have also performed Atomic Force Microscopy (AFM) with a Veeco Instruments Nanoscope III in tapping mode. The samples were spin cast over silica or mica substrates from a 2% weight solution in 1,1,2,2-tetrachloroethane, at 2500 rpm for 1 min. The thermal treatment of the AFM samples was performed using two vacuum ovens, where they were first held in the melt for 3 min at 130°C, and then transferred to the other oven at a constant crystallization temperature, Tc; the samples remained at this temperature for a fixed amount of time before they were cooled down to room temperature to observe them in the AFM.

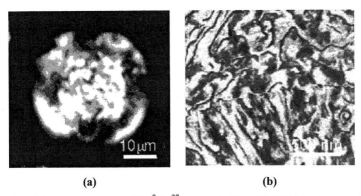

(a)   **(b)**

Figure 1. Spherulitic superstructure of $D_{23}{}^{8}C_{77}{}^{27}$ as revealed by: (a) POM during growth at 44°C, and (b) AFM phase contrast image of a sample that was crystallized at 42°C before quenching to room temperature. Note that an inter-spherulitic border runs almost horizontally through the middle of the image.

Figure 1 (a) presents an example (for $D_{23}{}^{8}C_{77}{}^{27}$) of the type of spherulitic superstructure that is seen with POM at undercoolings that are high enough to allow the crystallization of both PCL and PPDX microdomains. Figure 1(b) shows an AFM image that reveals that the spherulites are composed of radial lamellae of PCL (since PCL constitutes 77% of the material) that grow until they impinge with one another. The PPDX phase cannot be clearly identified since it is a minor phase that must be located within the interlamellar region of the dominant PCL lamellae where the amorphous PCL phase, together with all of the PPDX, must be accommodated.

Using POM we have identified spherulites in the crystallization region where both blocks crystallize (high undercoolings). Conversely, in the region where just the PPDX segment may crystallize (at Tc above 48°-50°C), POM revealed a transformation from spherulites to granular aggregates (for sample $D_{40}{}^{5}C_{60}{}^{7}$) or from rounded spherulites to ellipsoidal ones (for sample $D_{77}{}^{32}C_{23}{}^{10}$), or the superstructure cannot be seen (e.g., in sample $D_{23}{}^{8}C_{77}{}^{27}$ if the PCL domain is molten, the PPDX cannot form spherulites since it only amounts to 23% by weight and does not form aggregates that are large enough to be seen under POM).

POM allows the growing process of the superstructures to be monitored at different crystallization temperatures, and thus their growing rate (G) can be calculated (Figure 2). At lower temperatures (below 49°C) both blocks can crystallize into mixed spherulites similar to those of Figure 1. It can be appreciated that the growth rates of the diblock copolymers are lower than that exhibited by PCL homopolymer by a factor that is less than ten in this temperature range. However, in the region in which the PCL block is molten, we observe a depression of the growing rate of the copolymers by a factor of 10 as compared to PPDX homopolymer. This is probably due to the presence of the molten PCL in the system, which slows down the PPDX segments as they form the superstructural array. It is interesting to observe that a minimum in growth rate is observed around the transition temperature of 50°C at which both blocks start to crystallize together; it is at this temperature where the difference in undercooling experienced by each block goes through a maximum (i.e., it represents the highest temperature at which the PCL block can crystallize but a rather low crystallization temperature for PPDX).

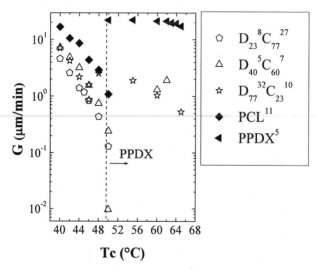

Figure 2. Superstructural growth rate (G) as a function of isothermal crystallization temperature (Tc) for the indicated samples. The dashed line indicates the approximate temperature limit beyond which only the PPDX microdomain can crystallize.

Overall Isothermal Crystallization Kinetics

The overall isothermal crystallization kinetics was studied in a Perkin-Elmer DSC-7. We had previously demonstrated[3] that if both blocks are simultaneously crystallized by quenching the diblock copolymer from the melt directly to Tc, a slower kinetics is observed than when the PPDX block is allowed to retain its crystalline morphology and then the PCL block is crystallized. These experiments demonstrated that a nucleation effect of the PPDX on the PCL was operating. In the first case, we started from the molten state (around 120°C), and then quickly cooled down to the crystallization temperature, in a similar way to that previously described for POM and AFM experiments. The second type of measurement was performed starting from the molten state at 120°C, then cooling down to –25°C (at 10°C/min), only to subsequently heat the sample above the PCL melting point (62°C) for a fixed period of time (70 min), and finally quench it to the crystallization temperature. In the present case, only the second type of procedure was employed where we monitored the PCL block isothermal crystallization after the PPDX had been allowed to fully crystallize. Figure 3 shows the overall crystallization rate, expressed as the inverse of the crystallization half-time, of PCL[11] and of the PCL block for selected diblock copolymers. Figure 3 shows that when the PCL content in the copolymer is high enough to form the matrix, for $D_{23}^8C_{77}^{27}$ and $D_{40}^5C_{60}^7$, the overall crystallization rate of the PCL block is higher than that of PCL[11]. This result confirms the previous finding that the crystallized PPDX nucleates the PCL; otherwise the acceleration of the overall kinetics (that includes nucleation and growth) could not have been observed since superstructural growth of the PCL block is most likely slower than that of PCL[11] in view of the evidence provided in Figure 2.

The nucleation effect of the PPDX block seems to be overtaken by topological confinement when the PCL content within the copolymer is lower than 50%. Figure 3 shows how the PCL block of the diblock copolymer with 47% PCL ($D_{53}^{21}C_{47}^{21}$) and that with 23% PCL ($D_{77}^{32}C_{23}^{10}$) crystallize at a slower overall rate than neat PCL. It is very interesting that a diblock copolymer like $D_{77}^{32}C_{23}^{10}$ with only 23% PCL exhibits crystallization of the PCL block in the temperature range shown in Figure 3, which is very similar to that of the heterogeneously nucleated PCL homopolymer. This means that even though the PCL is a minor component it does not nucleate homogeneously nor even display fractionated crystallization.[4-5] We have interpreted this result by considering that the previously crystallized PPDX heterogeneously nucleates the PCL, therefore preventing the nucleation problems of this block component.[3] Another possibility could be that the PCL phase is percolated or interconnecting; however all the evidences suggesting a nucleation effect by the PPDX lead us to the above quoted conclusion. In any case, both effects may be present.

The crystallization temperature range at which the PCL block within $D_{77}^{32}C_{23}^{10}$ was able to crystallize is comparable to that of PCL homopolymer, indicating that a heterogeneous nucleation mechanism is present. Another corroboration of the heterogeneous nucleation mechanism can be made by calculating the slope of the crystallization half time versus temperature (in a semi-log scale) and comparing its value with that of the PCL homopolymer[9].

In our case, PCL[11] exhibited a slope of 0.14 min/°C. The crystallization of the PCL blocks within the diblock copolymers exhibited slopes range between 0.13 to 0.24 min/°C and in the case of $D_{77}{}^{32}C_{23}{}^{10}$ the value was 0.20 corroborating a heterogeneous nucleation mechanism for the minor PCL component.

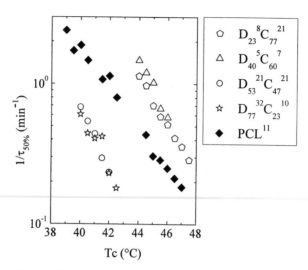

Figure 3. Reciprocal crystallization half-time at different crystallization temperatures, for the PCL homopolymers and the PCL block in the copolymers (the PPDX was previously crystallized according to the procedure stated in the text).

The most important contribution of the present work is that in spite of the nucleation effect of PPDX on PCL, that causes a heterogeneous nucleation mechanism for all the diblock copolymers employed regardless of composition, we can still find evidence for "confinement" for the PCL component. We will also show below that in the composition range where PCL is the major component, PPDX can also exhibit signs of confinement.

In the case of the PCL block, one of the first indicators of confinement is the reduction in the overall crystallization kinetics presented in Figure 3 and discussed above. The second is shown in Figure 4 where the Avrami index, obtained after fitting the Avrami equation to the overall crystallization data[3] obtained by DSC, is plotted as a function of the PCL content in the diblock copolymer. The spreading of the data corresponds to the different crystallization temperatures employed and to differences in molecular weight (we have found higher Avrami indexes in PCL homopolymers of higher molecular weights; this may explain the values displayed by the PCL block within $D_{23}{}^8C_{77}{}^{27}$). Nevertheless a clear trend can still be observed; the Avrami index decreases as the PCL content in the copolymer decreases. Such a decrease in Avrami index can be interpreted as a decrease in the dimensionality of growth from spherulites to two-dimensional structures or even to one-dimensional structures (for an sporadic nucleation case). This decrease in growth dimensionality is the result of confinement since the size of the PCL block decreases as its content in the copolymer decreases.

The classical Avrami interpretation takes into account the nucleation and dimensionality of growth during the crystallization process. It is now well known that when a block copolymer is confined into a great number of isolated microdomains (e.g, spheres or cylinders) it crystallizes from homogeneous nuclei that form at the largest possible supercooling.[5] Loo et al.[9] have shown that when homogeneous nucleation takes place, the Avrami index is usually 1. They have interpreted this result by considering that the growth rate is so fast at large supercoolings that only nucleation controls the rate of overall transformation.

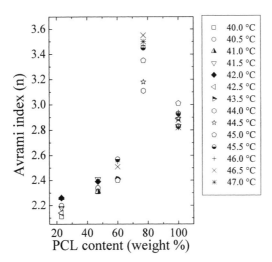

Figure 4. Avrami index as a function of PCL composition, for the isothermal crystallization of the PCL block at different temperatures (the PPDX was previously crystallized according to the procedure stated in the text). The 100% values correspond to PCL[11].

Another evidence that in our case the PCL block is not homogeneously nucleated is the fact that even for $D_{77}{}^{32}C_{23}{}^{10}$ the Avrami index was not found to be 1. Nevertheless, for this copolymer, Figure 4 shows that the Avrami index is of the order of 2, and this could correspond to one-dimensional crystals grown from sporadic nuclei, or two-dimensional growth from instantaneous nuclei. As the PCL content in the copolymer increases the Avrami index increases to values that are similar to those of PCL homopolymer, i.e., around 3, that indicate spherulitic growth from instantaneous nuclei, since the spherulites in such cases are easily observed in POM.

Self Nucleation

The self-nucleation technique was employed in order to ascertain the self-nucleation temperature Domains for the homopolymers and copolymers. Details of the experimental procedure can be found in previous works.[5]

Samples are first heated to a temperature high enough to completely melt the polymer in order to erase thermal history; then they are cooled at 10 °C/min down to –25 °C to provide them with a standard thermal history. Then they are heated to a temperature denoted Ts (or self-nucleation temperature) and isothermally kept there for 3 min. After treatment at Ts, the sample is cooled down to –25 °C (examples of such cooling scans can be seen in Figure 5) and subsequently heated at 10 °C/min until full melting occurs. Depending on Ts, the sample can be in one of three general Domains: In Domain I the sample is completely molten; in Domain II the sample is self-nucleated since Ts is high enough to melt almost all crystals but low enough to leave small fragments that can act as self-nuclei (or to induce memory effects) upon subsequent cooling and therefore the nucleation density can be enormously increased; in Domain III the sample is partially molten so that self-nucleation and annealing of unmelted crystals will take place at Ts.

Figure 5 (a) shows DSC cooling scans after 3 min at Ts for $D_{55}{}^{7}C_{45}{}^{6}$. At 125°C the Ts is too high and the sample is said to be in Domain I or complete melting Domain. Notice how after complete melting, the DSC cooling scan exhibits a single crystallization exotherm where both PPDX and PCL crystallize in a coincident fashion[3]. The origin of such coincident crystallization is related to the decrease in growth rate experience by the PPDX block when it tries to crystallize at low undercoolings (in the region where PCL has not started to crystallize).[3] Figure 5(b) shows the subsequent heating scans after the cooling runs shown in Figure 5(a). The two melting endotherms are due to the melting of the PCL component (at lower temperatures) and the PPDX component (at higher temperatures).

A novel result is shown in Figure 5(a) thanks to the self-nucleation technique. When a Ts of 112°C is used, only the PPDX is self-nucleated (since PCL is molten at that temperature, see Figure 5(b)) and upon cooling from 112°C a new high temperature exotherm, labeled 1 in Figure 5(a), starts to develop while the PCL component still crystallizes at lower temperatures (exotherm 2). The PPDX block of $D_{55}{}^7C_{45}{}^6$ at 112°C is in Domain II or self-nucleation Domain. Lowering Ts leads to a clear separation of the crystallization of the PPDX and the PCL block in temperature.

The location of Domain III is easily detected by observing the subsequent heating scans of Figure 5 (b).[5, 10] At a Ts temperature of 106°C a very small high temperature melting peak signaled with an arrow can be seen that is the trademark of the first signs of annealing at Ts. The PPDX block of $D_{55}{}^7C_{45}{}^6$ at 106°C is in Domain III or the self-nucleation and annealing Domain. For the PPDX block of $D_{55}{}^7C_{45}{}^6$, Figure 5 demonstrates that all three self-nucleation Domains can be observed.

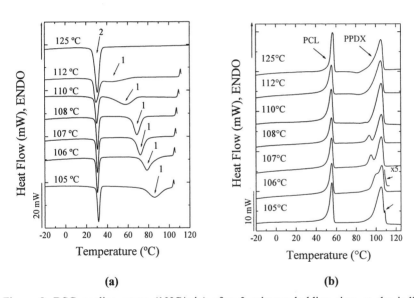

(a) (b)

Figure 5. DSC cooling scans (10°C/min) after 3 minutes holding time at the indicated Ts temperatures for $D_{55}{}^7C_{45}{}^6$. (a) Ts Temperature range where only the PPDX block can be self-nucleated. (b) Subsequent melting.

Müller et al. have previously shown[5, 10-11] that Domain II disappears for systems where the crystallizable block (PE, PCL or PEO) is strongly confined into small isolated microdomains. For self-nucleation to occur, a higher density of self-nuclei is then necessary and self-nucleation can only take place when Ts is lowered well into Domain III, where already annealing of remaining crystallites takes place. This is a direct result of the extremely high number density of microdomains that need to be self-nucleated when the crystallizable block is confined within small isolated microdomains.

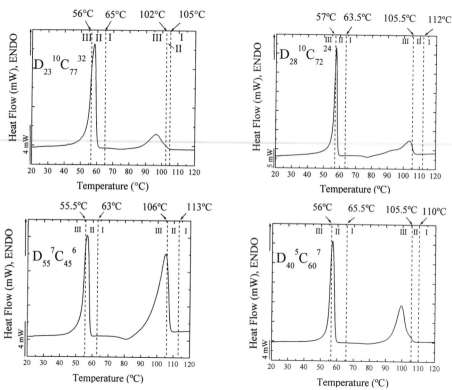

Figure 6. Self nucleation domains for PPDX-*b*-PCL diblock copolymers of different compositions. The temperature boundaries between the domains as well as the compositions are denoted.

Examples of the location of the self-nucleation Domains are shown in Figure 6 for a PPDX homopolymer and 3 diblock copolymers of varying compositions. The examples of Figure 6 illustrate an important finding since the three self-nucleation Domains were observed for both

PPDX and PCL blocks within all diblock copolymers examined regardless of composition. This is yet another evidence that homogeneous nucleation is not taking place for either the PPDX or the PCL blocks because Domain II was always observed.[5] Once more, even if Domain II were present for all copolymers examined, we found another evidence of confinement since the temperatures for the transition between Domain I and II, and between Domain II and III, for the PPDX block were found to increase as the PCL content in the copolymer decreases. This indicates that the processes of self-nucleation and annealing are increasingly more difficult as the PPDX is confined in smaller quantities in a matrix of PCL. A detailed study on the self-nucleation behaviour of these copolymers will soon be reported.

Conclusions

P(CL-*b*-PDX) diblock copolymers of a wide composition range have been studied by DSC, POM and AFM. The two blocks crystallize in a single coincident exotherm when cooled from the melt. The self-nucleation technique was able to separate the crystallization of each block into two exotherms depending on the self-nucleation temperature employed. We have gathered evidence indicating that the PPDX block can nucleate the PCL block within the copolymers regardless of composition. This effect is responsible for the lack of homogeneous nucleation or fractionated crystallization of the PCL block even when it constitutes a minor phase within the copolymer (25% or less). It also accounts for the presence of the three self-nucleation Domains in both the PCL and the PPDX block regardless of composition. Nevertheless, we were able to show that decreasing the amount of PCL within the diblock copolymer still produces confinement effects that retard the crystallization kinetics of the PCL component and decrease its Avrami index. For copolymers with a majority of PCL, decreasing amounts of PPDX within the copolymer make the self-nucleation and annealing of this block component more difficult, thereby reflecting confinement restrictions.

Acknowledgements

The USB team acknowledges financial support from Fonacit through grant No. S1-20001000742. LPCM members thank the *Service Fédéraux des Affaires Scientifiques, Techniques et Culturelles* in the frame of the PAI-5/03, as welle as the *Région Wallonne and Fonds social Européen* in the frame of the Materia Nova program, for general support. J-M. Raquez is grateful to F.R.I.A. for his PhD grant.

382

[1]   M.A. Sabino, J.L. Feijoo, A.J. Müller, *Macromol. Chem. Phys.*, **2000**, *201*, 2687.
[2]   M. A. Sabino, L. Sabater, G. Ronca, A.J. Müller, *Polym. Bull.*, **2002**, *48*, 291.
[3]   J. Albuerne, L. Márquez, A.J. Müller, J.M. Raquez, Ph. Degée, Ph. Dubois, V. Castelletto, I. Hamley, *Macromolecules*, **2003**, *36*, 1633.
[4]   Y.-L. Loo, R.A. Register, A.J. Ryan, *Macromolecules*, **2002**, *35*, 2365 (and references therein).
[5]   A.J. Müller, V. Balsamo, M.L. Arnal, T. Jakob, H. Schmalz, V. Abetz, *Macromolecules*, **2002**, *35*, 3048 (and references therein).
[6]   L. Zhu, S.Z.D. Cheng, B.H. Calhoun, Q. Ge, R.P. Quirk, E.L. Thomas, B.S. Hsiao, F. Yeh, B. Lotz, *Polymer*, **2001**, *42*, 9121.
[7]   J.-M. Raquez, P. Degée, R. Narayan, P. Dubois, *Macromol. Rapid. Commun.*, **2000**, *21*, 1063.
[8]   N. Ropson, Ph. Dubois, R. Jerome, Ph. Teyssie, *J. Polym. Sci. Polym. Chem. Ed.*, **1997**, *35*, 183.
[9]   Y.-L. Loo, R. Register, A.J. Ryan, *Phys. Rev. Lett.*, **2000**, *84*, 4120.
[10] V. Balsamo, Y. Paolini, G. Ronca, A.J. Müller, *Macromol. Chem. Phys.*, **2000**, *201*, 2711.
[11] H. Schmalz, A.J. Müller, V. Abetz, *Macromol. Chem. Phys.*, **2003**, *204*, 111.

Macromol. Symp. **2004**, *215*, 383-393

Advances in the Synthesis and Characterization of Polypeptide-Based Hybrid Block Copolymers

Ivaylo Dimitrov, Hildegard Kukula, Helmut Cölfen, Helmut Schlaad*

Max Planck Institute of Colloids and Interfaces, Colloid Department, Research Campus Golm, Am Mühlenberg 1, 14476 Golm, Germany

Summary: Linear polystyrene-*block*-poly(Z-L-lysine) copolymers with a very narrow molecular weight distribution (polydispersity index < 1.03) could be obtained *via* the ring-opening polymerization of Z-L-lysine-*N*-carboxyanhydride using ω-(primary amino hydrochloride)-polystyrenes as macroinitiators in *N,N*-dimethylformamide as the solvent at 40-80 °C. The block copolymer samples were analyzed by means of NMR, size exclusion chromatography, and analytical ultracentrifugation.

Keywords: analytical ultracentrifugation; block copolymers; liquid chromatography (SEC/GPC); polypeptides; ring-opening polymerization

Introduction

Amphiphilic block copolymers are known for their ability to form thermodynamically stable hybrid materials with a diverse ordering on the nanometer length scale.[1,2] Among the large number of block copolymers being investigated, the ones comprising a polypeptide and a synthetic segment (so-called "hybrids" or "molecular chimeras")[3] are of special interest as they may combine the advantageous properties of polypeptides, i.e. the structure formation, mutual recognition, high mechanical performance, and biodegradability, with the solubility, melt processability, and rubber elasticity of synthetic polymers. As a result, new materials with superior properties and complex hierarchical superstructures can be obtained.[3-5] Copolymers with polypeptide segments have as well a great potential in the field of biomedical applications and are used as model systems to study biophysical processes.[6,7]

Basically, there are two different ways to synthesize linear polypeptide-based block copolymers. The first one involves the solid-phase synthesis of a polypeptide (Merrifield synthesis),[8] followed by a coupling of the peptide chain with a carboxyl-endfunctional polymer. The second route is the primary amine-initiated ring-opening polymerization of amino acid-*N*-carboxyanhydrides (NCA, Leuchs Anhydride), which is the most frequently applied technique to produce polypeptides on a larger scale.[9-18] Here, chain growth should preferably proceed

DOI: 10.1002/masy.200451129

via the nucleophilic ring-opening of the NCA, the so-called "amine" mechanism, as depicted in Scheme 1.[19]

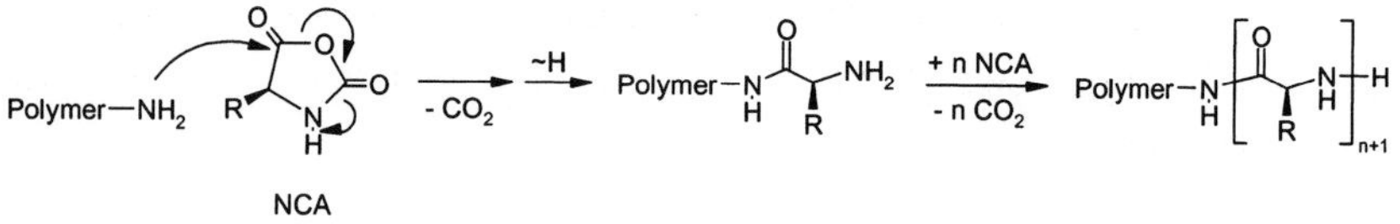

Scheme 1. Synthesis of polypeptide-based block copolymers by ring-opening polymerization of NCAs, initiated by primary amino-endfunctional polymers.

The so obtained polypeptides usually exhibit a very broad molecular weight distribution (polydispersity index, PDI > 2)[20] because NCA polymerization is not only proceeding *via* the "amine" mechanism but is suffering from side reactions. The most likely one is the "activated monomer" process, initiated by the deprotonation of an NCA, which then becomes the nucleophile and initiates by itself polymerization. Condensation of the produced *N*-aminoacyl NCA intermediates can finally give high-molecular weight polypeptides when monomer conversion approaches 100%. Since primary amines can act as both nucleophile and base, polymerization can switch back and forth between the "amine" and "activated monomer" mechanism.[8,19] However, side reactions can be avoided when metal-amine complex catalysts like bipyNi(COD) are used instead of primary amine initiators. Such a coordination polymerization of NCAs yields well-defined block copolypeptides with PDI < 1.2.[21]

For the characterization of polypeptides, methods like spectroscopy, viscometry, osmometry, light scattering, mass spectrometry, electrophoresis, and analytical ultracentrifugation (AUC) are commonly applied. Note, all these methods are well established to characterize natural proteins that are monodisperse. In order to gain information about molecular weight distributions, samples were submitted to time-consuming fractionation procedures prior to further analyses.[13,20,22,23] The characterization of amphiphilic polypeptide block copolymers demands even greater efforts and is complicated by the chemical dispersity of the samples and the presence of aggregates in most common organic solvents. To the best of our knowledge, the only applied methods to determine molecular weights of this kind of block copolymers were spectroscopy (NMR, IR, UV), elemental analysis, standard size exclusion chromatography (SEC), and AUC.[9-18] However, no adequate procedures have so far been reported that allow determination of absolute molecular weight distributions.

In the present contribution, we wish to describe our experiences in the polymerization of Z-L-lysine (ZLLys)-NCA initiated by ω-primary amino-functional polystyrenes (PS-NH₂) or corresponding amine hydrochlorides (PS-NH₂·HCl). Molecular weights and molecular weight distributions of the prepared PS-PZLLys block copolymer samples were determined using NMR, SEC, and AUC.

Experimental Part

Chemicals. All reagents were purchased from Aldrich or Fluka and were used as received; solvents were purified following standard procedures as described elsewhere in the literature. PS-NH₂(·HCl) macroinitiators were prepared by quenching polystyryl lithium with 1-(chloro-dimethylsilyl)-3-[*N*,*N*-bis(trimethylsilyl)amino]-propane and subsequent hydrolysis of the trimethylsilyl protecting groups.[24] ZLLys-NCA was prepared from ZLLys and triphosgene as described by Poché and coworkers.[25,26] *Polymerizations.* Mixtures of PS-NH₂(·HCl) and ZLLys-NCA in *N*,*N*-dimethylformamide (DMF) were stirred for 3 days at 40-80 °C under a dry argon atmosphere. PS-PZLLys products were precipitated in petrolether or methanol/water 70:30 (v/v), extracted with cyclohexane, and finally dried in vacuum at 40 °C.

NMR. ¹H NMR spectra of PS-PZLLys copolymer samples were recorded with a Bruker DPX-400 spectrometer in DMF-d₇ (99.5% d, Euriso-top) at 25-100 °C. *Size Exclusion Chromatography.* SEC was performed in *N*,*N*-dimethylacetamide (DMA + 0.5 wt % LiBr; flow rate: 1.0 mL/min) at 70 °C on four 300 × 8 mm PSS GRAM 10-μm columns (30, 30, 100, 3000 Å). Detectors employed were TSP UV1000 (λ = 270 nm), Shodex RI-71, and Viscotek model H502B on-line viscometer. The comonomer-specific UV and RI detector response factors were determined by analyzing PS and PZLLys homopolymer samples. The universal calibration curve (log [η]M *vs.* elution volume) was recorded with commercial PS standards (PSS, Mainz, Germany). *Analytical Ultracentrifugation.* AUC measurements were performed on an Optima XL-I ultracentrifuge (Beckman-Coulter, Palo Alto, CA) with Rayleigh interference and UV/visible absorption optics. Sedimentation velocity experiments were done with 0.15 wt % polymer solutions in DMF at 40 °C and 60K rpm. Time-dependent concentration profiles were evaluated with correction for diffusion broadening using the SEDFIT 5 software (Peter Schuck).[27] Partial specific volumes of the copolymers in DMF solution were obtained with a density meter DMA5000 (Anton Paar, Graz, Austria).

Results and Discussion

In a first series of experiments, four different PS-PZLLys copolymer samples (**1-4**) were prepared by ring-opening polymerization of ZLLys-NCA in DMF at 40 °C. Polymerizations were initiated with PS_x-NH_2 (x = 52 (→ **1, 2**) or 217 (→ **3, 4**); PDI ~ 1.03), and the initial concentration of the NCA was ~8 wt %. The NCA was prepared from ZLLys and triphosgene in THF, re-crystallized three times from THF/petrolether 1:2 (v/v), and dried in high vacuum (method A).[25] As indicated by SEC and AUC analyses (see below), the isolated copolymer samples did not contain any homopolymer impurities.

^{1}H NMR was applied to confirm the chemical structure of the PS-PZLLys samples. The average composition of copolymers was determined from the peak intensities of meta phenyl protons of PS (*8*) and benzylic methylene protons of PZLLys (*19*, see Figure 1). From the mole fraction of ZLLys in the copolymer and the molar mass of the PS block segment (SEC), the absolute number-average molecular weight (M_n) was calculated—the results for samples **1-4** are listed in Table 1. Note that measurements were performed in DMF-d_7, and it was confirmed by temperature-dependent NMR experiments (25-100 °C) and by AUC that in DMF copolymers were dissolved on a molecular level and not forming aggregates.

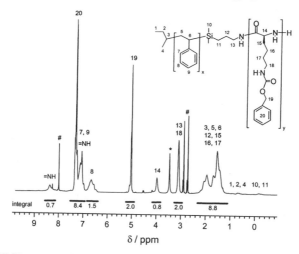

Figure 1. ^{1}H-NMR spectrum (400 MHz) of the PS-PZLLys sample **1** recorded in DMF-d_7 at 25 °C (#DMF, *water).

The samples were further characterized by SEC. For an adsorption-free fractionation of the samples, measurements were performed at 70 °C using DMA (+ 0.5 wt % LiBr) as the eluent and a polyester gel (PSS GRAM) stationary phase. The chromatograms obtained for **1-4** are shown in Figure 2. Also included in these plots are the results of compositional analysis of SEC fractions; the mole fractions of ZLLys were determined from the signal intensities of UV and RI detectors as described elsewhere.[28,29] Having this information and knowing the molar mass of the PS precursor, it was possible to determine the molar mass of every SEC fraction and thus the molecular weight distribution of the complete PS-PZLLys copolymer sample (SEC-UV/RI).[28] Molecular weight distributions were also determined by SEC with on-line tracing of the differential viscosity and application of the concept of universal calibration (SEC-DV/UC).[30,31] The M_n and PDI values of the PS-PZLLys samples **1-4**, as obtained by these two SEC methods, are summarized in Table 1. Evidently, the M_n values obtained by SEC agree well with the ones determined by NMR. PDI values were found to be in the range of 1.2-1.6 (SEC-UV/RI) or 1.4-1.8 (SEC-DV/UC), thus confirming that the polypeptide segments are rather polydisperse with respect to molar mass (cf. Introduction).

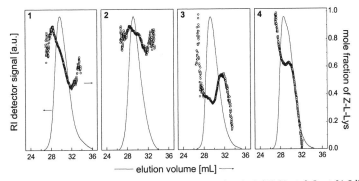

Figure 2. SEC chromatograms of the PS-PZLLys samples **1-4** (DMA + 0.5 wt % LiBr, 70 °C, PSS-GRAM); lines: RI detector signal, circles: mole fraction of ZLLys.

Table 1. Number-average molecular weights (M_n) and polydispersity indexes (PDI) of the PS-PZLLys samples **1-4** as determined by NMR and SEC.

sample	M_n [kg/mol]			PDI	
	NMR	SEC-UV/RI	SEC-DV/UC	SEC-UV/RI	SEC-DV/UC
1	23.7	23.6	21.7	1.6	1.7
2	34.7	38.4	35.8	1.3	1.6
3	47.2	54.6	51.0	1.3	1.4
4	100	101	76.4	1.5	1.8

Aiming to confirm SEC data, the PS-PZLLys samples were analyzed by AUC sedimentation-velocity. The sedimentation coefficient distributions obtained for **1-4** in DMF at 40 °C are shown in Figure 3. Note that the appearance of a single sedimenting species excludes the presence of homopolymer contaminants in the block copolymer samples. Also, sedimentation coefficients were found to be very low, which indicates that under the given experimental conditions copolymer chains were sedimenting as unimers.

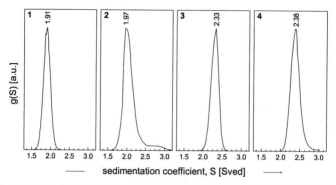

Figure 3. Sedimentation coefficient distributions, g(S), of the PS-PZLLys samples **1–4** (sedimentation-velocity; DMF, 40 °C, 60K rpm, UV absorption optics).

Assuming a constant partial specific volume and frictional ratio, evaluation of AUC data should provide absolute molar mass distributions and thus absolute values of the weight-average molecular weight (M_w).[27] However, the latter were about 10-60 % smaller than the corresponding M_n values obtained by NMR (or SEC). The reason for this nonsensical result is not known yet. It should be mentioned that the PZLLys segment might not only be poly-disperse with respect to molar mass but as well as with respect to conformation or shape. This should not affect NMR analyses but could have considerable impact on translational diffusion coefficients and partial specific volumes and thus on the modeling of the time-dependent concentration profiles to the Lamm equation.[27] Whether or not this is the case is subject of current investigations. In addition, samples shall be characterized by static light scattering (SLS). Note that the determination of the absolute molecular weight (M_w) of a polydisperse copolymer system requires the performance of SLS measurements in at least three different solvents (contrast variation method) to compensate the difference in specific refractive in-crements of the two block segments.[32]

Nevertheless, there seems to be good evidence that the applied SEC methods provide reliable information about the true molecular weight distribution of PS-PZLLys copolymers. This information is most important when aiming to optimize reaction conditions for a controlled polymerization of NCAs. The products, which were prepared throughout these studies, were characterized by means of NMR and SEC-UV/RI. The same polymerization conditions were applied as for the preparation of sample **1** (see above and Table 2). Mixtures of the PS_{52}-NH_2 macroinitiator and ZLLys-NCA in DMF were stirred for 3 days at 40-80 °C. In any case, conversion of the NCA went to completion as indicated by SEC analysis of the crude reaction solutions.

It is well known that acidic or other contaminants in the NCA sample can have severe impact on the polymerization results.[19] Recently, Poché et al.[26] reported an advanced procedure that makes it possible to obtain NCAs in a very high purity. Following this method B, the NCA was prepared from ZLLys and triphosgene in ethyl acetate as the solvent. The crude product was successively washed with cold water and aqueous $NaHCO_3$, precipitated from petrol-ether, and dried in high vacuum. The polymerization of the so obtained ZLLys-NCA using PS_{52}-NH_2 as the macroinitiator in DMF at 40 °C yielded PS-PZLLys sample **5** with $M_n = 21.4$ kg/mol (NMR) and PDI = 1.24 (SEC-UV/RI; cf. Figure 4, Table 2). This PDI value is considerably lower than that of reference sample **1** (PDI = 1.6), which was made using a ZLLys-NCA prepared by method A (see above). Performing the polymerization reaction at 80 °C produced a PS-PZLLys with even narrower molecular weight distribution (PDI = 1.12, sample **6**). It should further be noted that the initiator efficiency, as calculated from the ratio of the targeted molecular weight over the experimental molecular weight, was virtually the same for entries **1** (0.63), **5** (0.64), and **6** (0.66).

Hence, the NCA should be prepared according to method B in order to achieve a better control of the polymerization reaction. However, as mentioned in the Introduction, there is still the possibility that the polymerization proceeds not only *via* the desired "amine" mechanism but is competing with the "activated monomer" process. The key species of the latter process is an NCA anion (NCA⁻), resulting from the deprotonation of the NCA by the basic amino moiety of the macroinitiator or growing peptide chain end. It is obvious that the suppression of this side reaction would require a chemical modification of the growing species, e.g. transformation into a transition metal-amine complex (Deming et al.).[8] We were thinking of adding protons (HCl) to the reaction solution in order to shift the equilibrium back to the protonated NCA. Basically, this idea goes back to work of Knobler et al. published in the 1960s.[33,34] These authors employed the stoichiometric reaction between NCAs and the hydrochlorides of primary amines to synthesize α-aminoacyl compounds. It was emphasized that the nucleophilic substitution reaction proceeded smoothly without producing any polymeric by-products (presumably *via* an "activated monomer" process).

Since HCl could not be added as an aqueous solution (H_2O can promote polymerization of the NCA *via* the "activated monomer" mechanism),[19] we prepared a water-free hydrochloride of PS_{52}-NH_2. The polymerization of ZLLys-NCA (prepared by method B) with PS_{52}-$NH_2 \cdot HCl$ in DMF at 40 °C afforded PS-PZLLys sample **7** with M_n = 17.9 kg/mol and a near monodisperse molecular weight distribution (PDI < 1.03; cf. Figure 4). Like in the previous runs, NCA conversion went to completion (SEC) after 3 days of stirring. According to SEC and AUC (not shown), the sample was free of PZLLys homopolymer, which might have eventually been formed by a chloride-initiated polymerization of the NCA.[19]

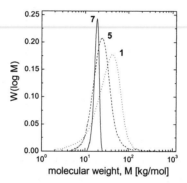

Figure 4. Mass distributions (SEC-UV/RI) of the PS-PZLLys samples **1** (PS_{52}-NH_2, ZLLys-NCA / method A, DMF, 40 °C), **5** (PS_{52}-NH_2, ZLLys-NCA / method B, DMF, 40 °C), and **7** (PS_{52}-$NH_2 \cdot HCl$, ZLLys-NCA / method B, DMF, 40 °C).

The tentative mechanism of the reaction is depicted in Scheme 2 (cf. Knobler et al.). The amine hydrochloride chain ends are considered as a dormant species, dissociating into the growing free primary amine and $H^+(Cl^-)$. The released protons should re-protonate any NCA^- present in solution. Note that this process should be faster than the nucleophilic attack of another NCA molecule to avoid chain growth *via* the "activated monomer" mechanism. However, kinetic studies have not yet been performed to confirm this mechanism. It is expected that the rate of polymerization will vastly depend on the position of the hydrochloride/amine equilibrium and thus on the polarity of the reaction medium and on the temperature.

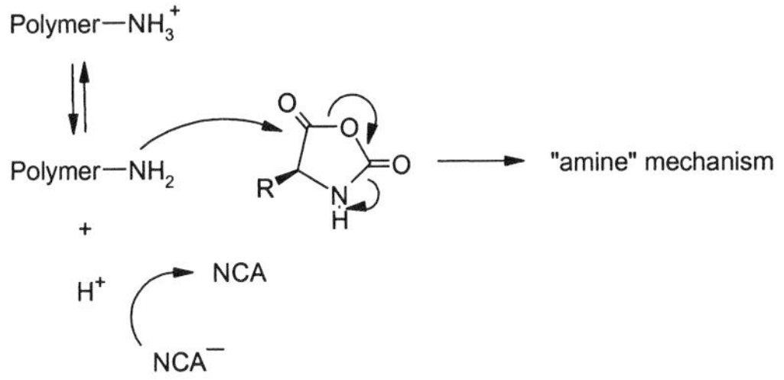

Scheme 2. Tentative mechanism of the ring-opening polymerization of NCAs using primary amine hydrochlorides (chloride ions omitted).

Table 2. Experimental results of the polymerization of ZLLys-NCA (prepared by method B) with either PS$_{52}$-NH$_2$ (**5, 6**) or PS$_{52}$-NH$_2$·HCl (**7-9**) ([NCA]$_0$/[-NH$_2$] = 31, [NCA]$_0$ ~ 8 wt %) in DMF at 40-80 °C.

sample	T [°C]	M$_n$ [kg/mol] NMR	PDI SEC-UV/RI	initiator efficiency
5	40	21.4	1.24	0.64
6	80	20.8	1.12	0.66
7	40	17.9	< 1.03	0.78
8	60	18.0	< 1.03	0.77
9	80	16.3	< 1.03	0.84

Polymerization of ZLLys-NCA with PS$_{52}$-NH$_2$·HCl in DMF proceeded in a controlled manner in the temperature range of 40-80 °C. All the prepared PS-PZLLys copolymers exhibit a very narrow molecular weight distribution, close to a Poisson distribution (PDI < 1.03; see Table 2). The initiator efficiency was found to be about 0.8, which is somewhat higher as for the free amine initiating system (< 0.7).

Conclusion

The synthesis of PS-PZLLys block copolymers by the NCA polymerization technique and their characterization with NMR, SEC, and AUC has been described. Advanced SEC analysis (SEC-UV/RI and SEC-DV/UC) was applied to determine the molecular weight distribution of the copolymers. SEC results (M$_n$) were found to be in good agreement with the ones obtained from NMR. However, AUC sedimentation-velocity runs provided M$_w$ values lower than M$_n$, which is presumably due to the fact that the applied standard algorithm to evaluate AUC data was not applicable to conformationally disperse copolymer systems.

Regarding the NCA polymerization reaction, it was found that whether or not polymerization was proceeding in a controlled manner depends very much on the quality of the NCA. It was further advantageous to employ PS-NH$_2$·HCl instead of the free amine macroinitiator. Dissociation of the hydrochloride produces the propagating primary amine and H$^+$(Cl$^-$). The proton being released is supposed to re-protonate eventually formed NCA$^-$, thus eliminating a non-desired chain growth *via* the "activated monomer" process. The PS-PZLLys copolymers, which were so obtained in DMF solution at 40-80 °C, exhibited a very narrow molecular weight distribution, close to a Poisson distribution (PDI < 1.03; SEC-UV/RI).

Acknowledgment

The authors would like to thank Markus Antonietti, Ines Below, Jana Falkenhagen, Marlies Gräwert, Magdalena Łosik, Olaf Niemeyer, Rémi Soula, Antje Völkel, and Erich C. for their valuable contributions to this work. Financial support was given by the *Max-Planck-Gesellschaft* and the *Deutsche Forschungsgemeinschaft (Sfb 448, A5)*.

[1] Förster, S.; Antonietti, M. *Adv. Mater.* **1998**, *10*, 195.
[2] Bates, F. S.; Fredrickson, G. H. *Physics Today* **1999**, *52*, 32.
[3] Schlaad, H.; Antonietti, M. *Eur. Phys. J. E* **2003**, *10*, 17.
[4] Gallot, B. *Prog. Polym. Sci.* **1996**, *21*, 1035.
[5] Klok, H.-A.; Lecommandoux, S. *Adv. Mater.* **2001**, *13*, 1217.
[6] Kataoka, K.; Harada, A.; Nagasaki, Y. *Adv. Drug Deliv. Rev.* **2001**, *47*, 113.
[7] Hayashi, T. In *Developments in block copolymers*; Goodman, I., Ed.; Elsevier Applied Science Publishers: London, 1985; pp 109.
[8] Deming, T. J. *Adv. Mater.* **1997**, *9*, 299.
[9] Billot, J.-P.; Douy, A.; Gallot, B. *Makromol. Chem.* **1977**, *178*, 1641.
[10] Douy, A.; Gallot, B. *Polymer* **1982**, *23*, 1039.
[11] Ovsiannikova, L. A.; Rudkovskaya, G. D.; Vlasov, G. P. *Makromol. Chem.* **1986**, *187*, 2351.
[12] Janssen, K.; van Beylen, M.; Samyn, C.; Scherrenberg, R.; Reynaers, H. *Makromol. Chem.* **1990**, *191*, 2777.
[13] Yoda, R.; Hiorkawa, Y.; Hayashi, T. *Eur. Polym. J.* **1994**, *30*, 1397.
[14] Harada, A.; Kataoka, K. *Macromolecules* **1995**, *28*, 5294.
[15] Klok, H.-A.; Langenwalter, J. F.; Lecommandoux, S. *Macromolecules* **2000**, *33*, 7819.
[16] Klok, H.-A.; Hernández, J. R.; Becker, S.; Müllen, K. *J. Polym. Sci, Part A: Polym. Chem.* **2001**, *39*, 1572.
[17] Kukula, H.; Schlaad, H.; Antonietti, M.; Förster, S. *J. Am. Chem. Soc.* **2002**, *124*, 1658.
[18] Schlaad, H.; Kukula, H.; Smarsly, B.; Antonietti, M.; Pakula, T. *Polymer* **2002**, *43*, 5321.
[19] Kricheldorf, H. R. *α-Aminoacid-N-Carboxyanhydrides and Related Heterocycles*; Springer: Berlin, 1987.
[20] Cosani, A.; Peggion, E.; Scoffone, E.; Verdini, A. S. *Makromol. Chem.* **1966**, *97*, 113.
[21] Deming, T. J. *Nature* **1997**, *390*, 386.
[22] Rinaudo, M.; Domard, A. *Biopolymers* **1976**, *15*, 2185.
[23] Jeannot, M. A.; Zeng, J.; Li, L. *J. Am. Soc. Mass Spectrom.* **1999**, *10*, 512.

[24] Kukula, H.; Schlaad, H.; Falkenhagen, J.; Krüger, R.-P. *Macromolecules* **2002**, *35*, 7157.
[25] Daly, W. H.; Poché, D. S. *Tetrahedron Lett.* **1988**, *29*, 5859.
[26] Poché, D. S.; Moore, M. J.; Bowles, J. L. *Synth. Commun.* **1999**, *29*, 843.
[27] Schuck, P. *Biophys. J.* **2000**, *78*, 1606.
[28] Schlaad, H.; Kilz, P. *Anal. Chem.* **2003**, *75*, 1548.
[29] Pasch, H.; Trathnigg, B. *HPLC of Polymers*; Springer: Heidelberg, 1997.
[30] Mori, S.; Barth, H. G. *Size exclusion chromatography*; Springer: Berlin Heidelberg, 1999.
[31] Yau, W. W. *Chemtracts, Macromol. Chem.* **1990**, *1*, 1.
[32] Bushuk, W.; Benoit, H. *Can. J. Chem.* **1958**, *36*, 1616.
[33] Knobler, Y.; Bittner, S.; Virov, D.; Frankel, M. *J. Chem. Soc.* **1969**, 1821.
[34] Knobler, Y.; Bittner, S.; Frankel, M. *J. Chem. Soc.* **1964**, 3941.